Progress in Optical Science and Photonics

Volume 40

Indexed by Scopus

The purpose of the series Progress in Optical Science and Photonics is to provide a forum to disseminate the latest research findings in various areas of Optics and its applications. The intended audience are physicists, electrical and electronic engineers, applied mathematicians, biomedical engineers, and advanced graduate students.

Mithun Kuniyil Ajith Singh

Photoacoustic Imaging Using Affordable Resources

Bridging the Gap to Clinical Translation

Springer

Mithun Kuniyil Ajith Singh
CYBERDYNE INC.
Rotterdam, The Netherlands

ISSN 2363-5096 ISSN 2363-510X (electronic)
Progress in Optical Science and Photonics
ISBN 978-981-95-7174-1 ISBN 978-981-95-7175-8 (eBook)
https://doi.org/10.1007/978-981-95-7175-8

This Springer imprint is published by the registered company Springer Nature Singapore Pte Ltd.
The registered company address is: 152 Beach Road, #21-01/04 Gateway East, Singapore 189721, Singapore

In loving memory of my father, whose selflessness and life lessons shaped who I am and continue to guide me every day.

Foreword

Medical imaging continues to evolve toward higher precision, safety, and accessibility. Each innovation in this field carries the shared goal of enabling earlier diagnosis, improving patient outcomes, and reducing healthcare disparities worldwide. Among the many imaging modalities, photoacoustic imaging has distinguished itself by combining the molecular sensitivity of optics with the spatial resolution and penetration depth of ultrasound. This unique hybrid nature allows visualization of tissue function and molecular composition at depths unattainable by purely optical techniques.

During the past two decades, research in photoacoustics has advanced from fundamental physics to sophisticated imaging platforms, spanning microscopy to tomography. The translation of this technology, however, has long been constrained by the high cost and complexity of conventional laser systems and proprietary instrumentation. The current global movement toward *affordable and open photoacoustic imaging* represents a critical turning point, one where innovation meets inclusivity. Low-cost solid-state sources such as LEDs and laser diodes, coupled with compact detectors, open hardware, and open-source reconstruction tools, are reshaping what is technically and economically feasible.

This book, *Photoacoustic Imaging Using Affordable Resources—Bridging the Gap to Clinical Translation*, arrives at precisely this transformative moment. It has been written by Dr. Mithun Kuniyil Ajith Singh, whose experience spans the full spectrum of photoacoustic imaging from academic research and system development to commercialization and clinical validation. His deep involvement across these stages brings a uniquely comprehensive perspective that bridges science, engineering, and real-world application. The chapters gathered here provide a practical and forward-looking resource, addressing illumination, detection, image reconstruction, open-source frameworks, and pathways for clinical adoption. Together, they document how affordability, interoperability, and collaboration can accelerate the journey from bench to bedside.

The community is witnessing the democratization of photoacoustic imaging, a transition from specialized laboratories to educational settings, clinical departments, and even global-health applications. I am confident that the knowledge shared in this book will empower the next generation of scientists and engineers to develop innovative, cost-effective systems that broaden access to this remarkable technology.

Photoacoustics continues to exemplify how curiosity-driven science can lead to life-saving innovation. Let us continue to build on this momentum, combining technical excellence with the commitment to make advanced imaging available to all who need it.

Lihong V. Wang, Ph.D.
Bren Professor of Medical Engineering
and Electrical Engineering
California Institute of Technology
Pasadena, USA

Preface

My journey in photoacoustic imaging began in 2012, when I started developing preamplifiers and data acquisition systems for detector arrays in Twente Photoacoustic Mammoscope. What began as an introduction to an emerging imaging modality soon became a lifelong pursuit to make this technology more accessible, affordable, and globally relevant. During my doctoral work within the European project FULLPHASE, I focused on improving the image quality of handheld laser-diode based photoacoustic imaging systems, with the goal of translating this promising technique from the laboratory to practical, point-of-care use.

After completing my PhD, my career continued at CYBERDYNE INC., where I had the privilege of contributing to the democratization of LED-based photoacoustic imaging. Over the years, I have been involved in nearly every aspect of this field—research, development, commercialization, marketing, and clinical translation. Each phase reinforced my belief that meaningful technological progress is as much about people and shared vision as it is about science or engineering.

I have been extraordinarily fortunate to work alongside some of the pioneers and most generous minds in photoacoustics, individuals who not only shaped the science but also embodied a spirit of openness, mentorship, and collaboration. Their example has left a lasting influence on my own approach to research and community building.

Coming from a small town in India, I have always felt that advanced medical technologies should not remain the privilege of a few but must serve the many. Affordability has never been just a research theme for me; it has been a guiding principle. The idea that an imaging system could be powerful, precise, and yet within reach for clinics across the world continues to motivate my work.

Earlier in my career, I had the opportunity to edit a book dedicated to LED-based photoacoustic imaging, where I explored how simple light sources could open new possibilities for accessibility and adoption. Since then, the field has moved forward in exciting ways, with affordable laser diodes, VCSELs, and other low-cost components expanding what is technically and clinically achievable. This new monograph builds on the foundations of that earlier work but takes a broader view, looking across illumination, transducers, acquisition, and system design to reflect how the landscape of affordability has grown and how much closer we are to practical, scalable solutions.

This book is a culmination of my experiences, spanning more than a decade of learning, collaboration, and discovery across academic, industrial, and clinical environments. It brings together practical knowledge, open resources, and collective insights that, I hope, will help others develop, adapt, and expand photoacoustic imaging using affordable and accessible approaches. To complement this perspective, the book also includes brief reflections from experts representing research, clinical, industrial, and community viewpoints, offering readers a broader understanding of how collaborative effort continues to shape the field.

I am grateful to all my mentors, collaborators, and students who have been part of this journey and to the broader photoacoustic community for embracing openness and cooperation as its foundation.

It is my hope that this book serves not only as a technical reference but also as a reflection of what can be achieved when curiosity, integrity, and collaboration come together, proving that true innovation begins with the intent to make knowledge and technology accessible to all.

Rotterdam, The Netherlands — Mithun Kuniyil Ajith Singh, Ph.D.

Acknowledgments This book represents the culmination of my experiences and learning across more than a decade in photoacoustic imaging. If I have achieved anything in this journey, it is only because I have been fortunate to work alongside many generous and brilliant individuals who have shaped my scientific and personal growth.

I begin by expressing my deepest gratitude to Prof. Wiendelt Steenbergen, my Ph.D. supervisor, for his guidance and mentorship; Prof. Srirang Manohar, who first introduced me to the world of photoacoustics; and Dr. Peter Brands, from whom I learned how to bridge academia, industry, and clinical practice with ease. I am equally indebted to Dr. Michael Jager for his constant inspiration, and to Dr. Wenfeng Xia and Dr. Altaf Hussain, both trusted colleagues and friends whose support has been invaluable since my doctoral days.

For this book, I am sincerely thankful to Prof. Lihong v. Wang for kindly contributing an inspiring foreword that sets the tone for the volume. My gratitude extends to Prof. Wiendelt Steenbergen, Prof. Srirang Manohar, Prof. Muyinatu Bell, Prof. Sarah Bohndiek, Prof. Ferdinand Knieling, Dr. Patrick Leisching, and Mr. Bryan Clingman for providing expert perspectives that bring real-world context to the discussion. Their generosity and prompt support despite busy schedules remind me how strong and collaborative this community truly is. I also wish to thank Dr. Francis Kalloor Joseph, a close friend and collaborator, for his support and constructive feedback; Dr. Min Wu for thoughtful input on educational applications and for contributing the CVENT probe image; and Prof. Manojit Pramanik, Dr. Kamran Avanaki, Dr. Anjali Thomas, Dr. Vijitha Periyasamy, and Ms. Celine Canal for valuable suggestions and permission to reuse illustrative figures.

I am deeply grateful to Naoto Sato, my colleague and partner in developing LED-based photoacoustic imaging (PAI) technology over the past decade, for his unwavering support through every success and setback. My sincere thanks also go to Prof. Yoshiyuki Sankai, CEO of CYBERDYNE INC., for providing a unique environment where research and innovation are encouraged and freely shared, and to Dr. Toshitaka Agano for his remarkable contributions to the advancement of LED-PAI.

I would like to acknowledge the Springer Nature editorial team, especially Loyola D'Silva, whose encouragement began with my first book, and Vinothini Elango, for their patience and professionalism throughout this process. Limited use of AI-assisted language tools was made for editorial refinement during manuscript preparation.

There are far more people who have supported, encouraged, and inspired me throughout this journey than I could ever name here; to all of them, I offer my sincere gratitude and warmest appreciation.

Finally, I am deeply thankful to my mother-in-law, Mrs. Latha Kesavan, for her quiet support at home, which gave me the time and space needed to complete this book. My heartfelt thanks also go to my family—Srinidhi, Vidyut, and Vedant—for their love, patience, and understanding, and to my parents, whose values continue to guide me. Above all, I am grateful to the Almighty for granting me the strength and curiosity to work with light and sound in pursuit of better diagnostic imaging.

Competing Interests The author is employed by CYBERDYNE, INC. in the field of photoacoustic imaging. The views expressed in this book are those of the author and do not necessarily reflect those of the employer.

Contents

About the Author

Mithun Kuniyil Ajith Singh is Engineering Scientist with extensive experience in preclinical and clinical photoacoustic and ultrasound imaging. He is currently Research and Business Development Manager at CYBERDYNE INC., The Netherlands, where he initiates and coordinates international collaborations with leading research institutions worldwide. His primary focus is on bridging the gap between laboratory research and clinical translation of LED-based photoacoustic imaging technology. He currently serves on the editorial board of *SPIE's Flagship Journal*, the *Journal of Biomedical Optics*, as Associate Editor. Dr. Mithun obtained his Ph.D. from the University of Twente, The Netherlands, in 2016 under the supervision of Prof. Wiendelt Steenbergen, specializing in laser-diode-based photoacoustic system design and image optimization. He has authored and co-authored over 100 scientific publications, including journal papers, book chapters, and conference proceedings, with a primary focus on advancing affordable and accessible photoacoustic imaging technologies. He has also guest-edited special journal issues dedicated to affordable photoacoustic imaging and served as Editor of the first comprehensive book on *LED-Based Photoacoustic Imaging*, published by Springer Nature in 2020. Through his research, publications, and collaborative initiatives, Dr. Mithun continues to play a leading role in shaping the global landscape of cost-effective, translational photoacoustic imaging.

Chapter 1
Introduction to Photoacoustic Imaging

Learning Objectives

After reading this chapter, the reader will be able to:

- **Explain the photoacoustic effect** and the physical conditions required for efficient conversion of absorbed optical energy into broadband ultrasonic waves.
- **Describe the principal components of a photoacoustic imaging system** and understand their respective functional roles.
- **Compare the major strengths and limitations of photoacoustic imaging** relative to other optical and acoustic imaging modalities.
- **Recognize why affordability and accessibility** are central to the next phase of photoacoustic imaging research and clinical translation.

1.1 Overview of Photoacoustic Imaging

Photoacoustic imaging (PAI), also referred to as optoacoustic imaging, is a hybrid biomedical imaging technique that leverages the interaction between light and sound [1, 2]. Unlike conventional optical imaging, which is limited by strong photon scattering in biological tissues, PAI detects the acoustic waves generated following optical absorption. Because acoustic scattering in tissue is orders of magnitude weaker than optical scattering, PAI achieves deeper penetration with preserved spatial resolution [3]. This unique combination provides optical contrast at ultrasonic resolution.

The basic principle is straightforward (Fig. 1.1): a pulsed or rapidly modulated light source illuminates biological tissue [4]. Endogenous chromophores such as hemoglobin, melanin, lipids, or water absorb part of the incident energy. The rapid deposition of energy leads to a small but measurable thermoelastic expansion, which in turn generates broadband ultrasound waves. These acoustic waves propagate through tissue and can be detected at the surface by ultrasound transducers. By

M. Kuniyil Ajith Singh, *Photoacoustic Imaging Using Affordable Resources*, Progress in Optical Science and Photonics 40, https://doi.org/10.1007/978-981-95-7175-8_1

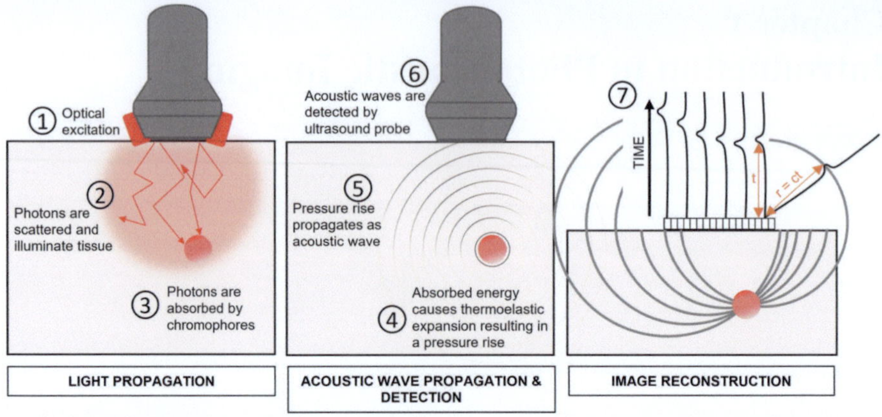

Fig. 1.1 Principle of photoacoustic imaging (PAI). (1) Light emitted from an external source illuminates the target tissue, which is typically turbid. (2) As photons propagate, scattering redistributes and attenuates them. (3) Chromophores within the tissue absorb part of the light. (4) The absorbed optical energy is converted into heat, producing a thermoelastic expansion and a local pressure rise. (5) This transient pressure launches broadband acoustic waves that propagate through the tissue. (6) The waves are detected at the surface by an ultrasound probe or array. (7) Using dedicated reconstruction algorithms that account for the time delays of signals across multiple elements, an image of the initial optical absorption distribution is reconstructed. Reproduced with permission from [5]

analyzing the time of arrival and frequency content of the signals, cross-sectional or volumetric images of tissue absorption can be reconstructed [5].

To place PAI in context, it is instructive to compare its performance envelope with other established biomedical imaging modalities (Fig. 1.2). Optical microscopy offers sub-micron resolution but is restricted to very superficial depths (<1 mm) because of strong scattering [6]. Optical coherence tomography (OCT) and confocal microscopy extend penetration slightly deeper, to 1–2 mm, but resolution remains on the order of a few microns [7, 8]. At the other extreme, ultrasound penetrates several centimeters to tens of centimeters into tissue, yet with spatial resolution typically limited to 100 μm–1 mm depending on frequency [2, 9]. Whole-body modalities such as MRI and CT provide unlimited penetration but sacrifice fine resolution and do not offer intrinsic molecular sensitivity [3, 10]. Photoacoustic imaging occupies a unique middle ground, combining optical absorption contrast with ultrasonic resolution at depths of up to several centimeters [1–3]. This balance of depth, resolution, and molecular specificity makes PAI complementary to both optical and conventional radiological techniques.

Key Advantages

Photoacoustic imaging combines several advantages that distinguish it from purely optical or acoustic methods [1–10]:

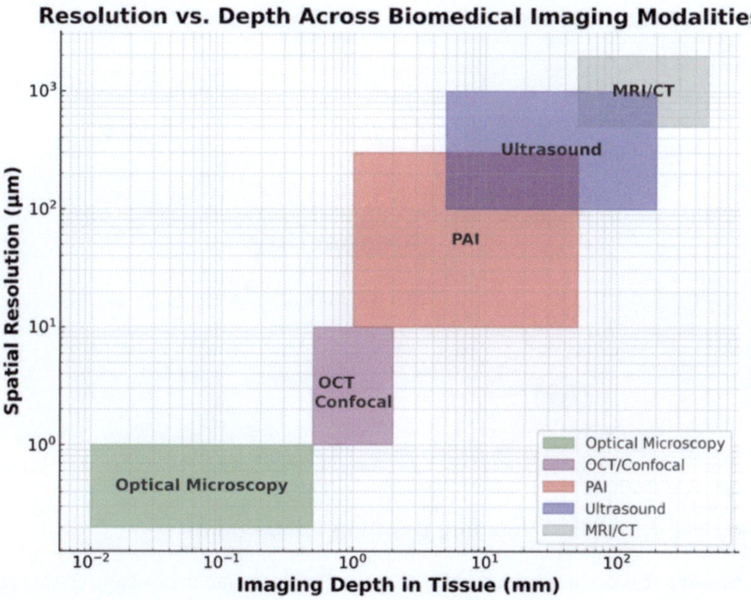

Fig. 1.2 Resolution versus depth across biomedical imaging modalities. Optical microscopy provides the highest spatial resolution but is limited to very shallow depths. OCT and confocal microscopy extend penetration slightly deeper with moderate resolution. Photoacoustic imaging occupies an intermediate regime, combining optical contrast with ultrasonic resolution at depths of up to several centimeters. Ultrasound achieves greater penetration with reduced resolution, while MRI and CT enable whole-body imaging at millimeter-scale resolution without intrinsic molecular contrast

- **Molecular sensitivity**: Because absorption depends on molecular composition, PAI can detect chromophores without exogenous contrast agents.
- **Resolution scalability**: Resolution depends on the acoustic detector frequency, not optical scattering, allowing imaging from micrometer to centimeter scales.
- **Depth penetration**: Acoustic propagation in tissue is far less attenuated than light, permitting imaging depths of several centimeters with maintained resolution.
- **Compatibility**: PAI can be seamlessly integrated with conventional ultrasound imaging, providing structural and functional information in a single handheld probe.

1.1.1 Imaging Regimes in Photoacoustic Imaging

PAI can be configured across different spatial scales depending on how the optical excitation and acoustic detection are implemented. Three main regimes are commonly recognized (Fig. 1.3):

Photoacoustic Microscopy (PAM) [11]:

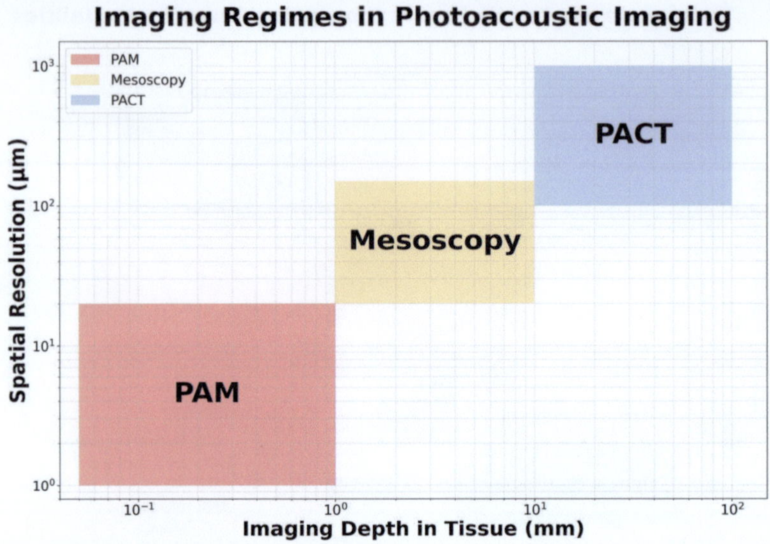

Fig. 1.3 Imaging regimes in photoacoustic imaging. Photoacoustic microscopy (PAM) achieves micrometer-scale resolution at shallow depths. Mesoscopy bridges the gap between microscopy and tomography, enabling tens to hundreds of micrometers resolution at depths up to ~ 1 cm. Photoacoustic computed tomography (PACT), including linear-array–based imaging, extends penetration to several centimeters with resolutions on the order of 100 μm–1 mm

- *Excitation*: tightly focused optical beam.
- *Detection*: single-element or confocal acoustic detection.
- *Depth*: superficial (hundreds of micrometers to ~1 mm for optical-resolution PAM, a few mm for acoustic-resolution PAM).
- *Resolution*: micrometer-scale.
- *Applications*: capillary-level angiography, microvascular studies.

Photoacoustic Mesoscopy [12, 13]:

- *Excitation*: wide-field or weakly focused light.
- *Detection*: high-frequency, wideband transducers, often mechanically scanned.
- Depth: 1 mm to ~1 cm.
- *Resolution*: tens to hundreds of micrometers.
- *Applications*: small animal skin and ear imaging, zebrafish studies, developmental biology.
- *Note*: Mesoscopy bridges the gap between microscopy and tomography but is less standardized in clinical use.

Photoacoustic Computed Tomography (PACT) [14–16]:

- *Excitation*: diffuse or broad-field optical illumination.
- *Detection*: tomographic detector geometries such as circular, spherical, or linear arrays.

- *Depth:* up to several centimeters (depending on wavelength and tissue type).
- *Resolution*: ~100 μm–1 mm, depending on array aperture and frequency.
- *Applications*: preclinical whole-body imaging, small animal brain imaging, breast cancer detection, human extremity imaging, and vascular mapping.
- *Note*: Linear-array–based imaging, widely used in handheld clinical probes, is considered a form of PACT because images are reconstructed tomographically from multi-element detection.

1.1.2 Applications in Biomedicine

The versatility of PAI has enabled applications in a wide range of biomedical domains. In oncology, it provides label-free visualization of tumor angiogenesis [10, 17, 18]. In vascular biology, it enables quantification of blood oxygenation and flow dynamics [1, 2]. In dermatology, it helps assess lesions and pigmentation [19]. More recently, PAI has been explored for monitoring inflammation, assessing metabolic status, and guiding interventional procedures [9, 10].

1.1.3 Why Affordable Systems Matter

Traditional PAI systems, while powerful, rely heavily on expensive pulsed lasers, multi-channel data acquisition units, and custom reconstruction software. This has limited widespread adoption to well-funded laboratories. Affordable alternatives using LEDs, LDs, repurposed ultrasound hardware, and open-source reconstruction algorithms have opened new avenues for resource-limited settings, teaching laboratories, and point-of-care diagnostics [20–22]. By lowering barriers to entry, affordable PAI systems can democratize access and accelerate clinical translation.

1.2 Historical Background and Evolution

The concept of photoacoustics traces back more than a century to the pioneering work of Alexander Graham Bell in 1880 [4]. Bell observed that when a solid material was illuminated with a modulated beam of sunlight, it produced audible sounds detectable with a simple microphone. This observation of the "photophone effect" laid the foundation for what later became known as the photoacoustic effect. While Bell's invention was ahead of its time, practical applications were limited by the absence of powerful and controllable light sources.

For much of the twentieth century, photoacoustics was explored primarily in physics and chemistry laboratories, where it was used for spectroscopic measurements of gases, liquids, and solids [23]. With the development of continuous-wave and pulsed lasers in the 1960s, photoacoustic spectroscopy emerged as a sensitive analytical tool, capable of detecting trace gases at parts-per-billion concentrations [24, 25]. This period established the photoacoustic effect as a robust scientific principle, but applications remained mostly in material sciences and environmental monitoring rather than in medicine [2].

The transition toward biomedical applications began in the late twentieth century, when researchers realized that photoacoustics could exploit the strong optical absorption of biological chromophores, particularly hemoglobin and melanin [1, 2]. In the 1990s, with the increasing availability of Q-switched Nd:YAG lasers and tunable optical parametric oscillators (OPOs), the first demonstrations of photoacoustic imaging of tissues and small animals were reported [14]. These early studies established the feasibility of visualizing vascular networks with high contrast, a capability unmatched by conventional ultrasound [2, 3].

The early 2000s marked a turning point as PACT and PAM emerged as distinct imaging modalities [11, 15]. Tomographic systems, often using arrays of ultrasound detectors arranged around the target, enabled deep-tissue imaging of small animals and human breast tissue [16]. In parallel, PAM systems demonstrated microvascular imaging with capillary-level resolution, opening a new window into functional and structural biology [11]. This period firmly positioned PAI within the broader field of biomedical imaging, attracting increasing attention from both academia and industry [12, 13].

In the past decade, PAI has evolved rapidly toward clinical translation. Handheld hybrid photoacoustic–ultrasound probes have been developed for applications ranging from dermatology to breast cancer detection and vascular imaging [26]. Several prototypes have advanced into early-phase human studies, demonstrating safety and feasibility. At the same time, the field has diversified beyond bulky laser systems, with the emergence of LED- and laser diode (LD)–based excitation sources and affordable ultrasound front ends [20–22, 27–42]. These developments have dramatically reduced system cost and complexity, paving the way for PAI to move from research laboratories into routine use [20–22].

The historical trajectory of PAI highlights three recurring themes: the importance of light source innovation, the parallel advances in ultrasound detection technologies, and the continuous push from spectroscopy toward imaging and eventually clinical translation. Today, PAI stands as a hybrid modality uniquely poised to complement both optical and radiological techniques. Its evolution underscores how the marriage of fundamental physics with engineering advances can transform a century-old physical phenomenon into a modern biomedical imaging tool. Figure 1.4 highlights key milestones in the evolution of PAI, including the establishment of PAM and PACT in the 2000s and the recent push toward affordable excitation sources.

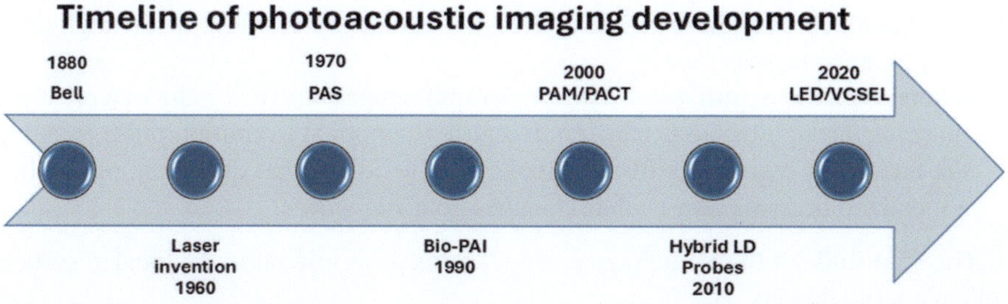

Fig. 1.4 Timeline of photoacoustic imaging development. The first observation of the photoacoustic effect was reported by Alexander Graham Bell in 1880. The invention of the laser in 1960 provided the enabling light source for modern photoacoustics. During the 1970s, photoacoustic spectroscopy (PAS) matured as a tool for gas detection and materials analysis. In the 1990s, pulsed-laser–based approaches demonstrated the first biomedical applications in tissue and vascular imaging. Around the 2000s, distinct imaging modalities emerged: photoacoustic microscopy (PAM) and photoacoustic computed tomography (PACT), enabling both microscopic and deep-tissue imaging. The 2010s introduced hybrid LD-based photoacoustic–ultrasound probes, leading to preclinical and early clinical studies. Most recently, in the 2020s, low-cost excitation sources such as LEDs and VCSELs have enabled portable, affordable PAI systems

1.3 Basic Principles of the Photoacoustic Effect

The photoacoustic effect arises when short optical pulses are absorbed in tissue and converted into measurable acoustic waves [1–3]. It is fundamentally a process of energy conversion: electromagnetic energy from photons is deposited locally, thermalized into heat, and subsequently converted into mechanical pressure through thermoelastic expansion. The resulting transient pressure waves propagate as ultrasound and can be detected at the tissue surface.

This section outlines the basic physics of the photoacoustic effect, highlighting the underlying equations and conditions that govern efficient generation and detection.

1.3.1 Energy Conversion Pathway

The chain of events underlying photoacoustic signal generation can be summarized in three steps:

1. **Optical absorption**—An incident optical pulse penetrates tissue. A fraction of the photons is absorbed by endogenous chromophores such as hemoglobin, melanin, lipid, or water, or by exogenous contrast agents if administered. The local absorption coefficient μa (cm^{-1}) quantifies the probability of photon absorption per unit path length.
2. **Thermalization**—The absorbed optical energy is converted into heat within picoseconds to nanoseconds via non-radiative relaxation processes. For most

biological chromophores, this conversion efficiency is high, with negligible fluorescence losses.

3. **Thermoelastic expansion**—The local, rapid temperature rise induces a pressure increase through tissue expansion. Because the optical excitation pulse is short, the heating is essentially instantaneous relative to thermal diffusion, producing a localized pressure source that launches acoustic waves.

The amplitude of the initial pressure rise, p_0, is proportional to the local absorbed optical energy density, H:

$$p_0 = \Gamma H = \Gamma \mu_a \Phi$$

where Φ is the optical fluence (J/cm^2) and Γ is the dimensionless **Grüneisen parameter**, which encapsulates tissue-specific thermomechanical properties (thermal expansion coefficient, heat capacity, and speed of sound).

1.3.2 Stress and Thermal Confinement

Efficient photoacoustic signal generation requires two important confinement conditions:

- **Stress confinement**: The optical pulse duration τ must be shorter than the time it takes for the pressure wave to propagate out of the heated region, i.e., $\tau < d/c_s$, where d is the characteristic absorber size and c_s is the speed of sound. If this condition is not met, pressure partially relaxes during heating, reducing signal amplitude.
- **Thermal confinement**: The pulse duration must also be shorter than the characteristic thermal diffusion time, $\tau < d^2/\alpha$, where α is the thermal diffusivity. This ensures that absorbed heat remains localized during excitation rather than dissipating.

In practice, nanosecond-scale optical pulses satisfy both conditions for tissue structures ranging from submicron chromophores to millimeter-scale vessels.

1.3.3 Acoustic Wave Propagation

Once generated, the pressure rise evolves according to the acoustic wave equation:

$$\nabla^2 p(\mathbf{r}, t) - \frac{1}{c_s^2} \frac{\partial^2 p(\mathbf{r}, t)}{\partial t^2} = -\frac{\beta}{C_p} \frac{\partial H(\mathbf{r}, t)}{\partial t}$$

where p is the acoustic pressure, β is the thermal expansion coefficient, and C_p is the heat capacity at constant pressure. The source term on the right-hand side corresponds to the time derivative of the deposited heating function H.

The acoustic waves generated are broadband, with frequency content determined by the size of the absorber:

- Small structures produce higher-frequency signals (tens of MHz).
- Larger structures emit lower-frequency components (hundreds of kHz to a few MHz).

Thus, PAI intrinsically encodes multiscale information, but system design (detector bandwidth, sampling rate) determines how much of this spectrum is captured.

1.3.4 Key Determinants of Signal Strength

Several factors govern the amplitude and detectability of photoacoustic signals:

- **Optical properties**—Absorption coefficient μ_a and local fluence distribution strongly influence signal magnitude.
- **Tissue thermomechanical properties**—The Grüneisen parameter varies with temperature, tissue type, and oxygenation state, leading to measurable contrast.
- **Illumination geometry**—Homogeneous versus focused illumination affects fluence distribution and signal uniformity.
- **Acoustic attenuation**—As ultrasound propagates, higher-frequency components attenuate more strongly, limiting resolution at greater depths.

1.3.5 Implications for Imaging

The physics of the photoacoustic effect confers several important implications for imaging:

- **Optical absorption contrast** enables visualization of blood oxygenation, melanin distribution, lipid content, and other functional or molecular targets.
- **Ultrasonic resolution scaling** ensures that lateral and axial resolution depend on acoustic detection rather than photon scattering.
- **Broadband nature of signals** provides opportunities for spectroscopic and quantitative analysis, but also imposes challenges for detection bandwidth and reconstruction.

Summary of Key Points

- The photoacoustic effect is a three-step process: optical absorption → thermalization → thermoelastic expansion, leading to acoustic wave generation.

- Efficient signal generation requires both stress confinement (pulse shorter than stress relaxation time) and thermal confinement (pulse shorter than heat diffusion time).
- The initial pressure rise is proportional to the local absorption coefficient and optical fluence, scaled by the Grüneisen parameter.
- The frequency content of the generated ultrasound depends on the size of the absorber: smaller structures produce higher frequencies.
- Signal detectability is determined by a combination of optical fluence, tissue properties, and acoustic detection bandwidth.

1.4 Photoacoustic Imaging System Architectures

PAI systems can be implemented in multiple ways, depending on the choice of illumination geometry and acoustic detection strategy. While the underlying physical principle, the conversion of absorbed optical energy into propagating ultrasound waves remains the same, the system architecture strongly influences spatial resolution, imaging speed, penetration depth, and application domain. This section outlines three canonical architectures that form the foundation of most PAI systems: single-element scanning systems, linear-array systems, and tomographic systems [2, 15, 16].

1.4.1 Single-Element Scanning Systems

The earliest demonstrations of photoacoustic imaging employed a single ultrasound transducer mechanically scanned around the target. In this configuration, a pulsed laser illuminates the tissue, and a point-like acoustic detector records the photoacoustic signals at multiple positions. By combining these signals with knowledge of the detector geometry, cross-sectional or three-dimensional images can be reconstructed.

- **Advantages**:
  - High sensitivity, as large-aperture focused transducers can be used.
  - Flexibility in adjusting detection position and focal depth.
  - Straightforward system design, suitable for proof-of-concept experiments.
- **Limitations**:
  - Slow acquisition speed due to mechanical scanning.
  - Motion artifacts in living tissue.
  - Limited field of view if the scanning aperture is small.

Single-element systems remain popular for preclinical imaging, ex vivo studies, and fundamental method development.

1.4.2 Linear-Array–Based Systems

Modern clinical and preclinical PAI systems often employ linear arrays of ultrasound transducers, similar to those used in conventional ultrasonography. In this configuration, the optical source illuminates tissue broadly or in a strip overlapping the array aperture, while all array elements detect photoacoustic signals simultaneously.

- **Advantages**:
  - High frame rate: images can be reconstructed in real time.
  - Clinical familiarity: arrays and front-ends are borrowed directly from ultrasound systems.
  - Integration: hybrid photoacoustic–ultrasound probes allow anatomical and functional imaging in a single device.
- **Limitations**:
  - Restricted field of view due to linear geometry.
  - Reconstruction artifacts for sources outside the array acceptance angle.
  - Dependence on illumination geometry; shadowing may occur in deeper structures.

Linear-array systems are particularly well suited for applications such as dermatology, breast imaging, vascular imaging, and point-of-care diagnostics.

1.4.3 Tomographic Systems

PACT employs large detection apertures—circular, spherical, or hemispherical arrays surrounding the target. The tissue is illuminated broadly, and detectors collect signals from multiple angles simultaneously. Reconstruction algorithms such as back-projection or model-based inversion yield tomographic images analogous to those in X-ray CT.

- **Advantages**:
  - Isotropic resolution in cross-section due to full angular coverage.
  - Deep penetration with wide detection apertures.
  - Well suited for small-animal whole-body imaging and for breast or brain imaging in humans.
- **Limitations**:
  - Complexity and cost of building large-aperture detector arrays.
  - Bulkiness, limiting portability.
  - Data acquisition and reconstruction demand increase with aperture size.

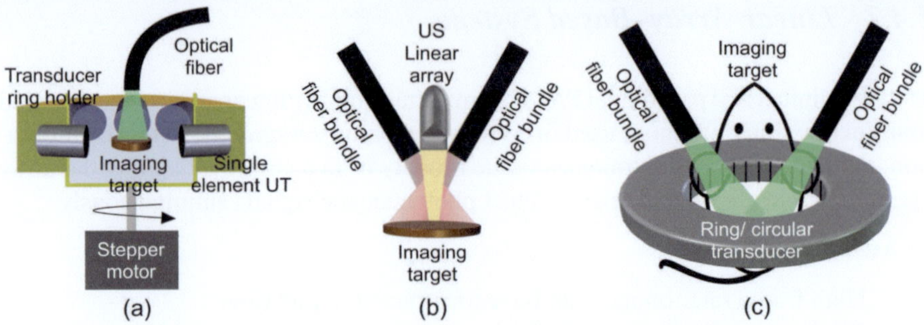

Fig. 1.5 Representative system architectures for photoacoustic imaging: **a** single-element scanning, **b** linear-array imaging with potential for handheld dual-mode PAI and ultrasound, and **c** tomographic circular detection. Reproduced with permission from [43]

PACT represents the most powerful form of PAI in terms of coverage and image quality, but its clinical adoption has been slower due to hardware complexity (Fig. 1.5).

1.4.4 Hybrid and Emerging Configurations

Beyond these three canonical architectures, hybrid configurations are increasingly common. Examples include endoscopic photoacoustic probes, and handheld systems with curved or two-dimensional arrays. Recent advances in affordable light sources and compact detectors are accelerating the development of portable systems, which blur the boundaries between traditional categories [20–22].

1.5 Image Contrast and Spectroscopy

The unique strength of PAI lies in its ability to combine the molecular sensitivity of optical absorption with the spatial resolution of ultrasound detection. Unlike purely optical methods that often rely on scattering contrast, or purely ultrasonic methods that are sensitive to acoustic impedance, PAI derives its contrast primarily from the distribution of optical absorption in tissue. This makes it highly versatile for mapping endogenous chromophores and exogenous contrast agents [2, 3].

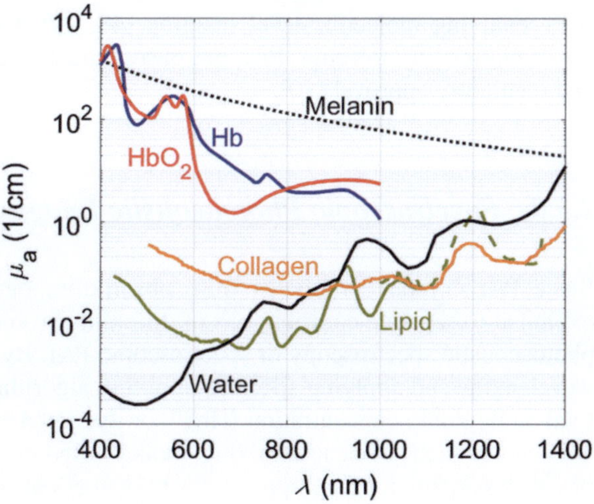

Fig. 1.6 Absorption spectra of major endogenous chromophores relevant to photoacoustic imaging, including hemoglobin (oxy- and deoxy-), water, lipids, collagen, and melanin. The spectral region between ~650 and 950 nm represents the optical "transparency window," where tissue absorption is reduced, allowing deeper light penetration. Reproduced with permission from [5]

1.5.1 Endogenous Contrast Sources

Several intrinsic tissue components serve as strong absorbers in the visible and near-infrared (NIR) spectrum (Fig. 1.6). Hemoglobin (Hb) is one of the most important, with oxyhemoglobin (HbO_2) and deoxyhemoglobin (Hb) having distinct absorption spectra, particularly in the 500–600 nm range and around 900 nm [2]. This spectral dependence enables functional imaging of blood oxygen saturation (sO_2) and vascular morphology. Melanin is another broadband absorber responsible for pigmentation in skin and hair follicles, providing strong contrast in dermatological and ophthalmic applications. Lipids show absorption bands in the second NIR window (~1200, ~700 nm), allowing visualization of lipid-rich tissues such as atherosclerotic plaques. Water exhibits weak absorption in the NIR-I window but strong absorption beyond 1450 nm, important for deep-tissue mapping at longer wavelengths. Endogenous contrast makes PAI inherently label-free, which is advantageous for non-invasive and translational biomedical applications.

1.5.2 Exogenous Contrast Agents

To expand imaging capabilities beyond natural absorbers, a wide range of synthetic or engineered agents are employed. Organic dyes such as indocyanine green and methylene blue are clinically approved and widely used for lymph node mapping and perfusion imaging [44, 45]. Nanoparticles such as gold nanorods and carbon nanotubes provide tunable absorption peaks, high molar extinction coefficients, and can be functionalized with targeting ligands [46, 47]. Activatable probes are smart agents that change absorption in response to enzymatic activity, pH, or other microenvironmental cues, enabling molecular-level imaging [48, 49]. Exogenous contrast introduces opportunities for molecular PAI, but raises concerns about toxicity, biodistribution, and regulatory approval [10].

1.5.3 Spectroscopic Photoacoustic Imaging

Since PAI contrast originates from absorption, multispectral excitation can be exploited to separate contributions of different chromophores [2]. This is known as photoacoustic spectroscopy or spectroscopic PAI. By acquiring images at multiple wavelengths and applying spectral unmixing algorithms, quantitative maps of sO_2, total hemoglobin concentration (HbT), or dye uptake can be reconstructed [8, 50]. Techniques such as linear unmixing, model-based inversion, and machine learning–based classification have improved the robustness of spectroscopic analysis [51, 52]. Challenges remain, particularly due to fluence variation in tissue (light attenuation with depth), which can bias spectral quantification [53]. Correction methods are an active area of research.

1.5.4 Factors Influencing Contrast

The magnitude and reliability of PAI contrast are influenced by several factors. The optical wavelength and tissue window strongly determine penetration depth and absorption specificity. Optical fluence distribution is another important factor: depth-dependent attenuation leads to signal decay, and higher fluence near the surface can overshadow deeper signals [50–53]. Acoustic detection bandwidth also plays a role, since smaller absorbers generate higher-frequency components; if not detected, spatial resolution and contrast degrade [2]. Background noise and artifacts such as acoustic reverberation, motion, and reconstruction errors can further reduce the contrast-to-noise ratio (CNR) [20]. Optimizing these factors is crucial for both preclinical studies and clinical translation.

Summary of key points

- PAI contrast originates from optical absorption, unlike conventional ultrasound or purely optical imaging.
- Endogenous absorbers (hemoglobin, melanin, lipids, water) provide strong, label-free contrast for functional and structural imaging.
- Exogenous contrast agents (dyes, nanoparticles, activatable probes) extend imaging capabilities but raise safety and regulatory considerations.
- Spectroscopic PAI enables separation of chromophore contributions and quantitative mapping of parameters such as oxygen saturation and dye uptake.
- Image contrast depends not only on the absorber but also on illumination fluence, detection bandwidth, and noise/artifacts, which can alter contrast-to-noise ratio.

1.6 Biomedical Importance and Use Cases

PAI is more than a technical innovation, it directly addresses unmet needs in biomedical research and clinical practice by combining the rich molecular contrast of optics with the depth penetration and resolution of ultrasound. Its hybrid nature positions PAI as a bridge between microscopy and radiology, offering new opportunities across multiple scales and specialties [1, 2].

1.6.1 Vascular and Hemodynamic Imaging

Because hemoglobin is a dominant absorber in the visible and near-infrared spectrum, PAI is uniquely suited for visualizing blood vessels without exogenous contrast. Microvascular networks can be mapped label-free, providing insights into angiogenesis, ischemia, and wound healing [1, 3]. Functional extensions allow quantification of oxygen saturation (sO_2) and total hemoglobin concentration (HbT), enabling hemodynamic monitoring in real time [50, 53]. Initial clinical investigations further demonstrate the value of PAI for vascular health assessment [9].

1.6.2 Oncology and Cancer Research

Tumor growth and progression are accompanied by changes in angiogenesis, hypoxia, and metabolic activity, all of which can be detected photoacoustically. PAI enables high-resolution mapping of tumor vasculature, assessment of oxygenation heterogeneity, and monitoring of therapy response [11, 16]. Clinical studies have shown promise in breast cancer imaging [13, 17] and melanoma detection [19], establishing PAI as a valuable tool in oncology.

1.6.3 Neurological Applications

Brain imaging is traditionally challenged by skull scattering and absorption. PAI provides a complementary approach, particularly in small-animal neuroscience, where cortical hemodynamics, functional activation, and pathological processes such as stroke can be monitored with high temporal resolution [54]. Emerging human applications, including handheld transcranial angiography, suggest feasibility for bedside functional monitoring [3, 55].

1.6.4 Dermatology and Ophthalmology

Skin and eye are highly accessible organs for PAI. In dermatology, applications include melanoma detection and assessment of vascular malformations [19]. Recent handheld LED-based systems have demonstrated accurate visualization of superficial vasculature and interventional tools [29]. In ophthalmology, photoacoustic ophthalmoscopy provides novel contrast for imaging retinal and choroidal vasculature, complementing OCT and fluorescein angiography [56].

1.6.5 Cardiovascular Imaging

Atherosclerotic plaque composition, neovascularization, and lipid accumulation can be detected photoacoustically using wavelength-sensitive contrast. Intravascular photoacoustic catheters, often combined with intravascular ultrasound (IVUS), enable compositional analysis of arterial walls, which is critical for identifying vulnerable plaques [57, 58].

1.6.6 Gastrointestinal and Endoscopic Applications

Miniaturized photoacoustic probes have been developed for integration with endoscopes, enabling imaging of esophageal and gastrointestinal lesions. The combination of molecular sensitivity and minimally invasive delivery expands the scope of clinical use cases [59].

1.6.7 Preclinical Small-Animal Imaging

PAI bridges optical microscopy and whole-body modalities for small animals. It allows label-free functional imaging of tumors, cardiovascular physiology, and neurological processes [60]. Advances in multispectral PAI have further accelerated preclinical studies [12].

1.6.8 Multimodal Imaging and Therapy Guidance

PAI integrates seamlessly with ultrasound, fluorescence, and magnetic resonance imaging, providing complementary anatomical and functional information [16]. In interventional contexts, PAI can guide needle placement, monitor photothermal therapy, and track nanoparticle delivery, enhancing precision medicine approaches [61].

Summary of Key Points

- PAI provides label-free vascular and hemodynamic imaging through hemoglobin absorption.
- Oncology benefits from PAI's ability to map tumor vasculature, hypoxia, and therapy response.
- Neurological, dermatological, and ophthalmic applications exploit accessible targets for functional and structural imaging.
- Cardiovascular and intravascular PAI reveal lipid-rich plaques and vessel wall composition.
- Endoscopic and minimally invasive probes extend PAI into gastrointestinal and surgical settings.
- Preclinical small-animal imaging has become one of the most widely adopted uses of PAI.
- Integration with other imaging modalities enhances both diagnostic and therapeutic workflows.

1.7 Challenges in Traditional Photoacoustic Imaging Systems

Despite impressive progress in laboratory demonstrations and preclinical imaging, conventional PAI systems face several practical barriers that limit their widespread clinical translation [20–22, 43]. These challenges arise largely from the reliance on bulky, high-power pulsed lasers, sophisticated ultrasound detection arrays, and complex integration workflows. Addressing these issues is critical for moving the field beyond academic laboratories and into accessible, affordable healthcare solutions.

1.7.1 Cost of Light Sources and Detectors

Traditional PAI platforms rely on nanosecond-pulsed solid-state lasers, such as Nd:YAG or optical parametric oscillators (OPOs). These systems are highly tunable and provide the pulse energies necessary for deep tissue imaging but typically cost tens to hundreds of thousands of dollars [20–22, 43]. High-end ultrasound arrays, multichannel amplifiers, and fast data acquisition hardware add further expense. The cumulative cost often exceeds that of clinical ultrasound or MRI scanners, making PAI impractical for routine use in most hospitals. Representative commercially available laser-based PAI systems and their approximate costs are summarized in Table 1.1.

Table 1.1 Typical price list of commercially available laser-based PAI systems (Adapted from [43] with permission)

PAI system	Company (location)	Approx. cost (USD)	Application areas
Vevo LAZR X	Fujifilm VisualSonics Inc., Ontario, Canada	~950,000	Oncology, cardiology, molecular and neurobiology
Vevo LAZR	Fujifilm VisualSonics Inc., Ontario, Canada	~750,000	Same as Vevo LAZR X
PAFT	PST Inc., Texas, USA	~315,000	Small animal imaging
MSOT inVision 128	iThera Medical GmbH, Germany	~470,000	Real-time whole-body imaging
Nexus 128+	Endra Life Sciences, Michigan, USA	~375,000	Molecular imaging, tumor hypoxia
LOUISA 3D	TomoWave Inc., Texas, USA	~215,000	Breast cancer research

1.7.2 Size, Bulk, and Infrastructure Requirements

Conventional pulsed lasers used in PAI are large, heavy, and demand extensive infrastructure. Systems often require optical tables, vibration isolation, and active water cooling, making them incompatible with portable or bedside use. Beam delivery typically relies on free-space optics or bulky fiber bundles, which limit flexibility and increase alignment demands [15, 43].

In clinical environments, space is at a premium, and compact devices are favored. The large footprint of conventional PAI systems prevents deployment in outpatient clinics or operating rooms, confining use to specialized imaging facilities. Even when smaller fiber-coupled lasers are employed, the supporting hardware: power supplies, cooling systems, and high-voltage electronics adds significant weight and volume [6, 13]. This physical bulk remains one of the greatest obstacles to translating PAI into resource-limited settings, where infrastructure is minimal.

1.7.3 Safety Concerns with High-Energy Lasers

The nanosecond pulsed lasers commonly used in PAI are typically classified as Class IV devices, which pose substantial risks of ocular or dermal injury [20]. Strict protective measures are required, including laser safety goggles, beam interlocks, and operation within restricted-access rooms. While manageable in a controlled laboratory environment, these protocols are cumbersome in a clinical setting, where patient flow, nurse access, and physician safety must be prioritized [27].

Moreover, many clinicians are reluctant to adopt imaging systems that require specialized laser safety training. The added administrative and regulatory burden of maintaining compliance with laser safety standards has further deterred adoption in hospitals. These concerns highlight the pressing need for low-power, eye-safe light sources that can be integrated into routine workflows without compromising safety.

1.7.4 Limited Repetition Rates and Frame Rates

Most conventional Q-switched Nd:YAG lasers operate at 10–50 Hz, which restricts frame rates and limits the ability to capture rapid physiological events such as cardiac motion or oxygenation changes [20–22]. Although kilohertz-rate fiber and diode-pumped lasers exist, they are significantly more expensive, bulky, and often lack the broad tunability needed for spectroscopic PAI [43].

In preclinical imaging, this limitation forces compromises between temporal resolution and image quality. Real-time functional imaging, especially in applications like interventional guidance or monitoring hemodynamics, is often not feasible with traditional lasers [9]. By contrast, affordable LED or LD-based systems operating at higher repetition rates have demonstrated the ability to produce video-rate imaging in superficial tissues, suggesting a path forward for clinical adoption [21, 29].

1.7.5 Complexity of System Integration

Traditional PAI systems are composed of multiple stand-alone components: lasers, beam-shaping optics, fiber bundles, ultrasound probes, amplifiers, and high-speed data acquisition systems. Aligning and synchronizing these components requires expertise in optics, acoustics, and electronics, as well as regular calibration [60].

This complexity undermines the goal of developing plug-and-play clinical imaging devices. In practice, systems often require an operator trained in laser physics and biomedical imaging, which makes them unsuitable for busy hospitals. Maintenance is also costly: misaligned optics or failing pump lasers can result in long downtime, further discouraging clinical uptake [6, 27].

1.7.6 Regulatory and Maintenance Barriers

High-energy laser systems face stringent regulatory oversight. Obtaining medical device clearance involves extensive safety testing, clinical validation, and documentation, which significantly increases development costs and delays translation [27, 65]. Even after approval, ongoing regulatory compliance requires regular inspection and reporting, adding to institutional overhead.

Maintenance adds another layer of difficulty. Traditional lasers require frequent calibration, periodic replacement of flashlamps or diodes, and careful alignment, which are resource-intensive and impractical in clinical settings [20, 43]. For hospitals with limited technical staff, these requirements are prohibitive. Affordable and robust alternatives such as solid-state laser diodes or LEDs reduce these burdens dramatically, providing a more sustainable path for widespread adoption [21, 62].

Summary of Key Points

- **Cost**: Lasers and detectors in conventional systems are prohibitively expensive.
- **Bulk**: Large, non-portable hardware limits bedside and point-of-care use.
- **Safety**: Class IV lasers raise significant regulatory and workflow concerns.
- **Frame rate**: Low repetition rates restrict real-time and functional imaging.
- **Complexity**: Multi-component integration requires technical expertise.
- **Regulation**: Stringent requirements for approval delay clinical translation.

1.8 The Need for Affordable and Accessible Photoacoustic Imaging

PAI has matured significantly in the laboratory and preclinical domains, yet translation into routine clinical use remains limited. The barriers outlined in Sect. 1.7: cost, bulk, safety concerns, and regulatory hurdles underscore the urgent need for alternative strategies that lower complexity and increase accessibility without compromising image quality [20–22].

1.8.1 Bridging the Gap Between Innovation and Translation

While PAI provides unique contrast mechanisms and has demonstrated high diagnostic potential, its dependence on expensive, high-power lasers and sophisticated hardware makes it inaccessible to most hospitals and research groups. Affordable alternatives are essential to democratize the technology and ensure that its benefits extend beyond a few specialized centers [20–22].

1.8.2 Expanding Clinical Reach

Many medical applications of PAI such as breast cancer screening, vascular health monitoring, dermatological evaluation, and intraoperative guidance occur in settings where portability, ease of use, and low cost are critical [9, 10]. Affordable PAI systems could be deployed at the point of care, integrated into handheld ultrasound devices, or adapted for outpatient clinics, making them viable in both developed and resource-limited healthcare environments [62, 63].

1.8.3 Leveraging Emerging Light Sources

LEDs, VCSELs and LDs have emerged as promising, low-cost excitation sources. While their pulse energy is lower than solid-state lasers, advances in packaging, synchronization, and multi-diode arrays are steadily improving their performance. These sources are inherently safer, smaller, and cheaper, enabling designs that are compact and user-friendly [20–22].

Recent studies have demonstrated that overdriving LED arrays with optimized pulse widths can achieve fluences sufficient for vascular imaging, while maintaining safe operating conditions. Similarly, multi-diode assemblies can be spectrally combined to achieve multi-wavelength illumination, enabling functional photoacoustic spectroscopy. In addition, the use of high-repetition-rate laser diodes provides video-rate imaging capabilities, addressing one of the major limitations of traditional nanosecond lasers. Together, these innovations are making diode-based systems increasingly competitive with conventional sources while retaining advantages in cost and portability.

1.8.4 Affordable Detection and Data Acquisition

Low-cost piezoelectric transducers, capacitive micromachined ultrasonic transducers (CMUTs), and optical microphones offer opportunities for inexpensive detection modules [1]. Paired with commodity data acquisition hardware and open-source software, these solutions can significantly reduce overall system costs, lowering the barrier to entry for new laboratories and clinics.

Several groups have shown that USB-based and low-channel-count DAQs (<$1000) can be adapted for PAI without significant loss in image quality for superficial imaging applications. Coupled with open-source reconstruction libraries written in MATLAB, Python, and C++, researchers can bypass expensive proprietary systems and accelerate innovation. These software ecosystems also promote transparency and reproducibility, making affordable PAI not just a cost-driven strategy but a movement toward democratized science.

1.8.5 Importance for Global Health

In many regions of the world, particularly in low- and middle-income countries, advanced diagnostic imaging modalities are prohibitively expensive and therefore unavailable. Affordable PAI systems could provide noninvasive, portable diagnostic tools that address unmet clinical needs, such as early cancer detection, maternal health monitoring, and infectious disease diagnosis [9, 10]. By focusing on scalability and cost reduction, PAI has the potential to contribute to equitable healthcare worldwide [64].

1.8.6 Catalyzing Interdisciplinary Collaboration

Developing accessible PAI solutions encourages collaboration across engineering, physics, medicine, and industry. Efforts to standardize low-cost components, open-source software, and shared databases will not only accelerate technical innovation but also create a broader community of practice [65]. These collaborations are vital to moving from feasibility studies to real-world implementation.

Summary of Key Points

- Conventional PAI systems remain prohibitively expensive and complex for routine clinical use.
- Affordable systems are needed to expand PAI into outpatient and point-of-care settings.
- LEDs and laser diodes provide compact, safe, and low-cost light sources.
- Cost-effective detectors and open-source data acquisition reduce system complexity.
- Global health applications stand to benefit most from affordable solutions.
- Interdisciplinary collaboration will accelerate translation and standardization.

Academic Insight: Upholding Rigor, Diversity, and Accessibility in Biomedical Optics Interdisciplinary innovation holds great potential to transform healthcare, with photoacoustic imaging being on exemplary use case, merging optical and acoustic technologies. As the field advances toward widespread clinical adoption, the scientific community must advance commitments to rigor, transparency, and inclusion in research and publishing. Open access tools, equitable designs, and technology developed through international collaborations are three linchpins that will help to ensure global photoacoustic benefits, empowering researchers, patients, and healthcare infrastructures through shared discovery.

—Muyinatu A. Lediju Bell, PhD, Editor-in-Chief of the Journal of Biomedical Optics, John C. Malone Professor at Johns Hopkins University, and Director of the Photoacoustic & Ultrasonic Systems Engineering (PULSE) Lab

1.9 Scope of This Book

PAI is a promising biomedical modality, but its translation into clinical and routine research use has been slowed by reliance on expensive, bulky, and complex hardware. The scope of this book is deliberately focused on strategies, technologies, and applications that lower the barrier to entry and make PAI more accessible to a broader community of users, from engineers and physicists to clinicians and industry developers.

1.9.1 Emphasis on Affordability

The central theme of this book is affordability: not simply in terms of reducing component cost, but in designing systems that are simpler, safer, and more deployable. Affordability is treated as a multidimensional concept covering acquisition, maintenance, portability, and usability, because each of these factors influences whether a technology can realistically be adopted outside specialized laboratories. By concentrating on LEDs, LDs, and cost-effective detection schemes, the book highlights approaches that can be implemented in outpatient clinics, bedside monitoring devices, and resource-limited healthcare settings. This perspective recognizes that the success of PAI in global health and point-of-care medicine depends as much on system practicality as on performance metrics.

1.9.2 Balance Between Principles and Practice

The book is structured to provide both theoretical background and practical guidance. This Chapter introduces the foundational principles of the photoacoustic effect, imaging regimes, and system architectures. This is followed by dedicated chapters that explore affordable light sources (LEDs and LDs), detection techniques, and image enhancement strategies. Each technical section emphasizes hands-on considerations such as component availability, assembly methods, calibration, and cost–benefit trade-offs. Wherever possible, practical design examples and case studies are provided to help researchers translate concepts into working systems. The goal is to ensure that readers not only understand the underlying physics and engineering but also acquire the tools and insights needed to build, adapt, or deploy affordable PAI systems in their own laboratories or clinics.

1.9.3 Integration with Biomedical Applications

Although the physics and engineering aspects of PAI form the technical foundation of the book, the scope extends deliberately into biomedical use cases. Applications in oncology, vascular imaging, dermatology, ophthalmology, and endoscopy are discussed to illustrate how affordable solutions can address real-world clinical challenges. For example, low-cost PAI systems have been tested for imaging tumor vasculature, detecting melanoma, or monitoring blood oxygenation in peripheral vessels. Preclinical small-animal studies and early pilot trials are highlighted as steppingstones toward clinical translation, showing how even modest, affordable systems can provide meaningful biological insights. This application-oriented perspective ensures that readers appreciate not just the hardware innovations, but also their potential impact on patient care and biomedical research.

1.9.4 Clinical Translation and Global Impact

Finally, the book moves beyond technology to consider the pathways of clinical translation. Later chapters address regulatory considerations, validation studies, and common barriers to adoption in healthcare systems. Special emphasis is placed on global health applications, where affordable PAI systems may have the greatest impact. By reducing cost and complexity, PAI can be positioned as a diagnostic tool for underserved communities, enabling early detection of cancers, vascular diseases, or infectious conditions that would otherwise go unmonitored. The integration of PAI into handheld ultrasound devices and other portable platforms also makes it highly suitable for rural clinics and mobile health units. In this way, affordability is not only a matter of economic efficiency but also a means of ensuring equitable access to advanced biomedical imaging worldwide.

1.9.5 Audience

The intended readership includes:

- Researchers and engineers seeking to design and evaluate affordable PAI systems.
- Clinicians interested in emerging imaging technologies for diagnosis and therapy guidance.
- Graduate students entering the field of biomedical optics and ultrasound.
- Industry and translational scientists exploring commercialization pathways for accessible PAI devices.

1.10 Summary

This chapter has introduced the fundamental principles, biomedical importance, and practical challenges of PAI. Beginning with the basic physics of the photoacoustic effect, we outlined how optical absorption and thermoelastic expansion produce ultrasound signals that can be reconstructed into high-resolution images. The three main imaging regimes—microscopy, mesoscopy, and tomography—were described to situate PAI across multiple spatial scales, from microvascular mapping to whole-organ imaging.

We then explored the biomedical relevance of PAI, highlighting its ability to visualize vascular networks, quantify oxygen saturation, characterize tumors, and support dermatological, ophthalmological, cardiovascular, and gastrointestinal applications. Preclinical small-animal studies and multimodal integration further illustrate the breadth of opportunities and reinforce the modality's role as a bridge between microscopy and radiology.

Despite these advantages, conventional PAI systems remain hindered by high cost, bulk, safety concerns, and integration complexity. These barriers slow translation and restrict access to a relatively small number of well-funded laboratories. Commercial laser-based PAI platforms, while powerful, are priced well above what is feasible for routine research or clinical practice, underscoring the urgent need for affordable alternatives.

The scope of this book is therefore to examine how affordable and accessible PAI approaches—centered on light-emitting diodes, laser diodes, cost-effective detection methods, and open-source processing strategies—can lower the barrier to entry. By doing so, the field may expand beyond a niche research tool to a widely adopted biomedical modality.

The chapters that follow build upon this foundation:

- Chapter 2 introduces LEDs as practical excitation sources.
- Chapter 3 examines the role of laser diodes in achieving deeper penetration and higher resolution.
- Chapter 4 addresses affordable detection strategies, including piezoelectric transducers, CMUTs, and emerging alternatives.
- Chapter 5 discusses image reconstruction and enhancement, with emphasis on accessible computational methods.
- Chapter 6 outlines the pathway to clinical translation, including regulatory, validation, and implementation challenges.
- Chapter 7 looks forward to future trends and innovations, particularly integration with AI and applications in resource-limited settings.
- Chapter 8 concludes with practical resources, tools, and collaborative opportunities for researchers entering the field.

Together, these sections provide a roadmap for researchers, clinicians, and industry partners interested in harnessing the full potential of PAI through affordable, scalable, and clinically relevant solutions.

Key Takeaways

- **PAI** uniquely combines optical absorption contrast with ultrasonic spatial resolution, bridging the scale from microscopy to tomography.
- **Clinical and research applications** span oncology, vascular and neurological imaging, dermatology, ophthalmology, and preclinical studies.
- **Conventional laser-based systems**, while powerful, remain costly, bulky, and technically complex, limiting accessibility and widespread clinical use.
- **Affordable illumination and detection technologies** including LEDs, laser diodes, and low-cost acoustic sensors are enabling a new generation of compact, safe, and cost-effective PAI platforms.
- **This book explores these advances comprehensively**, outlining how open, affordable strategies are driving the transition of photoacoustics from the research laboratory to real-world clinical and global-health applications.

References

1. M. Xu, L.V. Wang, Photoacoustic imaging in biomedicine. Rev. Sci. Instrum. **77**, 041101 (2006)
2. P. Beard, Biomedical photoacoustic imaging. Interface Focus **1**, 602–631 (2011)
3. L.V. Wang, S. Hu, Photoacoustic tomography: in vivo imaging from organelles to organs. Science **335**, 1458–1462 (2012)
4. A.G. Bell, Upon the production of sound by radiant energy. Phil. Mag. **11**, 510–528 (1880)
5. J.J.M. Riksen, A.V. Nikolaev, G. van Soest et al., Photoacoustic imaging on its way toward clinical utility: a tutorial review focusing on practical application in medicine. J. Biomed. Opt. **28**(12), 121205 (2023)
6. V. Ntziachristos et al., Going deeper than microscopy: the optical imaging frontier in biology. Nat. Methods **7**(8), 603–614 (2010)
7. S.L. Jacques, Optical properties of biological tissues: a review. Phys. Med. Biol. **58**, R37–R61 (2013)
8. L.V. Wang, J. Yao, A practical guide to photoacoustic tomography in the life sciences. Nat. Methods **13**, 627–638 (2016)
9. I. Steinberg, D.M. Huland, O. Vermesh et al., Photoacoustic clinical imaging. Photoacoustics **14**, 77–98 (2019)
10. A.B.E. Attia, G. Balasundaram, M. Moothanchery et al., A review of clinical photoacoustic imaging: current and future trends. Photoacoustics **16**, 100144 (2019)
11. J. Yao, L.V. Wang, Photoacoustic tomography: fundamentals, advances and prospects, Laser Photonics Rev. **7**, 758–778 (2013)
12. A. Taruttis, V. Ntziachristos, Advances in real-time multispectral optoacoustic imaging. Nat. Photonics **9**, 219–227 (2015)
13. S. Manohar, M. Dantuma, Current and future trends in photoacoustic breast imaging. Photoacoustics **16**, 100134 (2019)
14. R.A. Kruger, P. Liu, Y.R. Fang, C.R. Appledorn, Photoacoustic ultrasound (PAUS) reconstruction tomography. Med. Phys. **22**, 1605–1609 (1995)
15. J. Xia, J. Yao, L.V. Wang, Photoacoustic tomography: principles and advances. Progress Electromagn. Res. **147**, 1–22 (2014)
16. V. Ntziachristos, D. Razansky, Molecular imaging by means of multispectral optoacoustic tomography. Nat. Rev. Mol. Cell Biol. **11**, 455–462 (2010)
17. M. Heijblom, D. Piras, S. Manohar, Clinical photoacoustic breast imaging: the Twente experience. IEEE Pulse **6**(3), 42–46 (2015)

18. M. Xavierselvan, M.K.A. Singh, S. Mallidi, In vivo tumor vascular imaging with light emitting diode-based photoacoustic imaging system. Sensors **20**(16), 4503 (2020)
19. I. Stoffels et al., Melanoma diagnosis by multispectral optoacoustic tomography. J. Clin. Oncol. **33**(5), 543–551 (2015)
20. T.J. Allen, High-power light emitting diodes; an alternative excitation source for photoacoustic tomography, in *LED-Based Photoacoustic Imaging. Progress in Optical Science and Photonics*, vol. 7, ed. by M. Kuniyil Ajith Singh (Springer, Singapore, 2020)
21. Y. Zhu, T. Feng, Q. Cheng, X. Wang, S. Du, N. Sato, J. Yuan, M.K.A. Singh, Towards clinical translation of LED-based photoacoustic imaging: a review. Sensors **20**(9), 2484 (2020)
22. S.K. Kalva, M. Pramanik, *Photoacoustic Tomography with High-Energy Pulsed Laser Diodes*, vol. SL56 (SPIE Press, 2020)
23. A. Rosencwaig, A. Gersho, Theory of the photoacoustic effect in solids. J. Appl. Phys. **47**(1), 64–69 (1976)
24. A. Rosencwaig, *Photoacoustics and Photoacoustic Spectroscopy* (Wiley, New York, 1980)
25. C.K.N. Patel, A.C. Tam, Pulsed optoacoustic spectroscopy of condensed matter. Rev. Mod. Phys. **53**(3), 517–550 (1981)
26. M.K.A. Singh, W. Steenbergen, S. Manohar, Handheld probe-based dual mode ultrasound/photoacoustics for biomedical imaging, in *Frontiers in Biophotonics for Translational Medicine. Progress in Optical Science and Photonics*, vol. 3, ed. by M. Olivo, U. Dinish (Springer, Singapore, 2016), pp. 209–247
27. M.K.A. Singh, N. Sato, F. Ichihashi, Y. Sankai, Clinical translation of photoacoustic imaging—opportunities and challenges from an industry perspective, in *LED-Based Photoacoustic Imaging* (Springer, Singapore, 2020), pp. 379–393
28. M.K.A. Singh, W. Xia, Portable and affordable light source–based photoacoustic tomography. Sensors **20**, 6173 (2020)
29. W. Xia, M.K.A. Singh, E. Maneas, N. Sato, Y. Shigeta, T. Agano, S. Ourselin, S.J. West, A.E. Desjardins, Handheld real-time LED-based photoacoustic and ultrasound imaging system for accurate visualization of clinical metal needles and superficial vasculature to guide minimally invasive procedures. Sensors **18**, 1394 (2018)
30. M.K.A. Singh, T. Agano, N. Sato, Y. Shigeta, T. Uemura, Real-time in vivo imaging of human lymphatic system using an LED-based photoacoustic/ultrasound imaging system, in *Proceedings SPIE Photons Plus Ultrasound: Imaging and Sensing 2018*, vol. 10494 (2018), p. 1049404
31. M.K.A. Singh, Y. Shigeta, T. Hanaoka, T. Agano, N. Sato, High-speed photoacoustic imaging using an LED-based photoacoustic imaging system, in *Proceedings of SPIE Photons Plus Ultrasound: Imaging and Sensing 2018*, vol. 10494 (2018), p. 104943N
32. M.K.A. Singh, N. Sato, Y. Sankai, F. Ichihashi, In vivo demonstration of real-time oxygen saturation imaging using a portable and affordable LED-based multispectral photoacoustic and ultrasound imaging system, in *Proceedings of SPIE Photons Plus Ultrasound: Imaging and Sensing 2019*, vol. 10878, (2019), p. 108785N
33. K.J. Francis, Y.E. Boink, M. Dantuma, M.K.A. Singh, S. Manohar, W. Steenbergen, Tomographic imaging with an ultrasound and LED-based photoacoustic system. Biomed. Opt. Express **11**, 2152–2165 (2020)
34. S. Agrawal, M.K.A. Singh, K. Johnstonbaugh, D.C. Han, C.R. Pameijer, S.-R. Kothapalli, Photoacoustic imaging of human vasculature using LED versus laser illumination: a comparison study on tissue phantoms and in vivo humans. Sensors **21**, 424 (2021)
35. K.J. Francis, M. Xavierselvan, M.K.A. Singh, S. Mallidi, C. Van Der Laken, F. Van De Loo, W. Steenbergen, LED-based photoacoustic imaging for early detection of joint inflammation in rodents: towards achieving 3Rs in rheumatoid arthritis research, in *Proceedings of SPIE Photons Plus Ultrasound: Imaging and Sensing 2020* (2020), vol. 11240, p. 112400M
36. R. Bulsink, M.K.A. Singh, M. Xavierselvan, S. Mallidi, W. Steenbergen, K.J. Francis, Oxygen saturation imaging using LED-based photoacoustic system. Sensors **21**, 283 (2021)
37. W. Xia, E. Maneas, N.T. Huynh, M.K.A. Singh, N.M. Brown, S. Ourselin, E. Gilbert-Kawai, S.J. West, A.E. Desjardins, Imaging of human peripheral blood vessels during cuff occlusion

with a compact LED-based photoacoustic and ultrasound system, in *Proceedings of SPIE Photons Plus Ultrasound: Imaging and Sensing 2019*, vol. 10878 (2019), p. 1087804
38. E. Maneas, R. Aughwane, N. Huynh, W. Xia, R. Ansari, M.K.A. Singh, J.C. Hutchinson, N.J. Sebire, O.J. Arthurs, J. Deprest et al., Photoacoustic imaging of the human placental vasculature. J. Biophotonics **13**, e201900167 (2020)
39. S. Van Heumen, J.J.M. Riksen, M.K.A. Singh, G. Van Soest, D. Vasilic, LED-based photoacoustic imaging for preoperative visualization of lymphatic vessels in patients with secondary limb lymphedema. Photoacoustics **29**, 100446 (2022)
40. T. Agano, M.K.A. Singh, R. Nagaoka, K. Awazu, Effect of light pulse width on frequency characteristics of photoacoustic signal—an experimental study using a pulse-width tunable LED-based photoacoustic imaging system. Int. J. Eng. Technol. **7**(4), 4300–4303 (2018)
41. M.K.A. Singh, N. Sato, Y. Sankai, F. Ichihashi, Real-time improvement of LED-based photoacoustic image quality using intermittent pulse echo acquisitions, in *Proceedings of SPIE Photons Plus Ultrasound: Imaging and Sensing 2020*, vol. 11240 (2020), p. 1124051
42. M.K.A. Singh, N. Sato, F. Ichihashi, Y. Sankai, Optimizing acoustic detection for deep-tissue LED-based photoacoustic imaging, in *Proceedings of SPIE Photons Plus Ultrasound: Imaging and Sensing 2021*, vol. 11642 (2021), p. 116423V
43. A. Fatima, K. Kratkiewicz, R. Manwar, M. Zafar, R. Zhang, B. Huang, N. Dadashzadeh, J. Xia, K.M. Avanaki, Review of cost reduction methods in photoacoustic computed tomography. Photoacoustics **15**, 100137 (2019)
44. J.T. Alander, I. Kaartinen, A. Laakso, T. Pätilä, T. Spillmann, V.V. Tuchin, M. Venermo, P. Valisuo, A review of indocyanine green fluorescent imaging in surgery. Int. J. Biomed. Imaging **2012**, 940585 (2012)
45. M.V. Marshall, J.C. Rasmussen, I.C. Tan, M.B. Aldrich, K.E. Adams, X. Wang, C.E. Fife, E.A. Maus, L.A. Smith, E.M. Sevick-Muraca, Near-infrared fluorescence imaging in humans with indocyanine green: a review and update. Open Surg. Oncol. J. **2**, 12–25 (2010)
46. X. Huang, P.K. Jain, I.H. El-Sayed, M.A. El-Sayed, Plasmonic photothermal therapy (PPTT) using gold nanoparticles. Lasers Med. Sci. **23**, 217–228 (2008)
47. A. de la Zerda, C. Zavaleta, S. Keren, S. Vaithilingam, S. Bodapati, Z. Liu, J. Levi, B.R. Smith, T.-J. Ma, O. Oralkan, Z. Cheng, X. Chen, H. Dai, B.T. Khuri-Yakub, S.S. Gambhir, Carbon nanotubes as photoacoustic molecular imaging agents in living mice. Nat. Nanotechnol. **3**, 557–562 (2008)
48. K. Pu, A.J. Shuhendler, J.V. Jokerst, J. Mei, S.S. Gambhir, Z. Bao, J. Rao, Semiconducting polymer nanoparticles as photoacoustic molecular imaging probes in living mice. Nat. Nanotechnol. **9**, 233–239 (2014)
49. J. Levi, S.-R. Kothapalli, T.-J. Ma, K. Hartman, B.T. Khuri-Yakub, S.S. Gambhir, Design, synthesis, and imaging of an activatable photoacoustic probe. J. Am. Chem. Soc. **132**, 11264–11269 (2010)
50. B.T. Cox, J.G. Laufer, S.R. Arridge, P.C. Beard, Quantitative spectroscopic photoacoustic imaging: a review. J. Biomed. Opt. **17**(6), 061202 (2012)
51. J. Gröhl, T. Kirchner, T.J. Adler et al., Deep learning for biomedical photoacoustic imaging: a review. Photoacoustics **22**, 100241 (2021)
52. J. Laufer, A. Jathoul, M. Pule, P. Beard, In vitro characterization of genetically expressed absorbing proteins using photoacoustic spectroscopy. Biomed. Opt. Express **4**(11), 2477–2490 (2013)
53. S. Tzoumas, A. Nunes et al., Eigenspectra optoacoustic tomography achieves quantitative blood oxygenation imaging deep in tissues. Nat. Commun. **7**, 12121 (2016)
54. X. Wang, Y. Pang, G. Ku, X. Xie, G. Stoica, L.V. Wang, Noninvasive laser-induced photoacoustic tomography for structural and functional in vivo imaging of the brain. Nat. Biotechnol. **21**, 803–806 (2003)
55. X.L. Dean-Ben, D. Razansky, Functional optoacoustic human angiography with handheld probe. Photoacoustics **1**, 68–73 (2013)
56. S. Jiao, M. Jiang, J. Hu et al., Photoacoustic ophthalmoscopy for in vivo retinal imaging. Opt. Express **18**(4), 3967–3972 (2010)

57. B. Wang, A. Karpiouk, D. Yeager et al., Intravascular photoacoustic imaging of lipid in atherosclerotic plaques in vivo. Opt. Lett. **37**, 1244–1246 (2012)
58. Y. Li, S. Yang, D. Xing, Intravascular photoacoustic imaging at 35 MHz. Opt. Express **20**, 767–776 (2012)
59. J.M. Yang, R. Chen, C. Favazza et al., A 2.5-mm diameter probe for photoacoustic and ultrasonic endoscopy. Opt. Express **20**(21), 23944–23953 (2012)
60. L. Li, L. Zhu, C. Ma et al., Small-animal whole-body photoacoustic tomography: a review. IEEE Trans. Biomed. Eng. **64**(4), 953–960 (2017)
61. X. Wang et al., Photoacoustic imaging for interventional guidance: needle and catheter visualization. J. Biomed. Opt. **17**(6), 066016 (2012)
62. H. Zhong, T. Duan, H. Lan, M. Zhou, F. Gao, Review of low-cost photoacoustic sensing and imaging based on laser diode and light-emitting diode. Sensors **18**(7), 2264 (2018)
63. X. Liu, S.K. Kalva, B. Lafci, D. Nozdriukhin, X.L. Deán-Ben, D. Razansky, Full-view LED-based optoacoustic tomography. Photoacoustics **31**, 100521 (2023)
64. WHO, *WHO Global Observatory on Health Technologies: Medical Devices* (World Health Organization, Geneva, 2016)
65. H. Assi, R. Cao, M. Castelino, B. Cox, F.J. Gilbert, J. Gröhl, K. Gurusamy, L. Hacker, A.M. Ivory, J. Joseph, F. Knieling, M.J. Leahy, L. Lilaj, S. Manohar, I. Meglinski, C. Moran, A. Murray, A.A. Oraevsky, M.D. Pagel, M. Pramanik, J. Raymond, M.K.A. Singh, W.C. Vogt, L. Wang, S. Yang, Members of IPASC, S.E. Bohndiek, A review of a strategic roadmapping exercise to advance clinical translation of photoacoustic imaging: From current barriers to future adoption. Photoacoustics **32**, 100539 (2023)

Chapter 2
LED-Based Photoacoustic Imaging

Learning Objectives

After reading this chapter, the reader will be able to:

- **Explain the physical principles of LED operation** and their relevance to photoacoustic excitation mechanisms.
- **Summarize the historical evolution of LED-based photoacoustic imaging** and recognize key technological milestones in its development.
- **Describe system architectures** that integrate LED illumination with ultrasound detection for hybrid photoacoustic and ultrasound imaging.
- **Identify major preclinical and clinical applications** made possible through LED-based excitation, including vascular, rheumatologic, and perfusion imaging.
- **Discuss the advantages, limitations, and future prospects** of LED-based photoacoustic imaging in advancing safe, compact, and affordable imaging solutions.

2.1 Introduction

Photoacoustic imaging (PAI) has emerged as one of the most versatile hybrid imaging modalities in biomedicine, combining the molecular sensitivity of optical absorption with the penetration depth and resolution of ultrasound [1, 2]. For more than two decades, however, the majority of PAI systems have relied on high-energy nanosecond pulsed lasers such as Nd:YAG or optical parametric oscillators (OPOs). While these sources provide excellent performance in terms of fluence and wavelength flexibility, they also impose significant drawbacks: cost, bulk, safety classification, and infrastructure requirements [3–6]. As a result, despite hundreds of successful preclinical studies, the clinical translation of PAI has been slower than anticipated [7].

M. Kuniyil Ajith Singh, *Photoacoustic Imaging Using Affordable Resources*, Progress in Optical Science and Photonics 40, https://doi.org/10.1007/978-981-95-7175-8_2

Light-emitting diodes (LEDs) have recently emerged as an attractive alternative to traditional laser sources. Historically considered unsuitable for photoacoustic excitation because of their relatively low optical output, LEDs have undergone rapid advances in semiconductor fabrication, packaging, and driver electronics [7]. Modern high-power pulsed LED arrays are capable of delivering nanosecond pulses with sufficient fluence to generate measurable photoacoustic signals, particularly when coupled with sensitive ultrasound detection and signal averaging strategies [8, 9]. These developments have opened the door to a new class of PAI systems that are safe, compact, and cost-effective [10, 11].

The appeal of LED-based PAI extends beyond affordability. Unlike Class IV lasers, LEDs are classified as inherently eye-safe, eliminating the need for restrictive safety protocols in clinical environments [12]. They are electrically efficient, exhibit long lifetimes, and can be driven at high repetition rates, enabling real-time imaging [13]. Furthermore, LEDs are available across a broad spectral range from visible to near-infrared wavelengths, allowing access to the absorption peaks of key endogenous chromophores such as hemoglobin, melanin, and lipids [9, 14, 15].

Clinically, LED-based systems are particularly well suited for superficial and mesoscopic applications where high fluence is not mandatory but safety, portability, and workflow integration are critical. Examples include vascular imaging, dermatology, musculoskeletal evaluation, and interventional guidance [16–19]. In these domains, LEDs offer sufficient optical power to achieve diagnostic-quality images, while the absence of bulky optical tables and cooling systems facilitates handheld and portable implementations [9]. Importantly, these systems also align with international safety regulations for photobiological exposure, as defined by the IEC 62,471 standard, which classifies LEDs as low-risk sources and facilitates clinical integration without the stringent controls required for Class IV laser systems [20].

The introduction of LED-based excitation represents a paradigm shift for PAI: moving from specialized laboratories toward bedside and point-of-care imaging. This section sets the stage for the chapter by outlining the rationale for LEDs, their unique advantages, and the motivation for exploring them as a central theme of affordable photoacoustic imaging.

2.2 Fundamentals of LED Illumination in Photoacoustic Imaging

LEDs have emerged as a compelling alternative to conventional pulsed lasers for photoacoustic excitation. Unlike bulky solid-state sources, LEDs are compact, inexpensive, and inherently eye-safe, making them suitable for portable and clinical systems. Although their pulse energies are typically orders of magnitude lower than those of nanosecond Q-switched lasers, improvements in efficiency, overdriving techniques, and spectral engineering have brought their performance within a useful range for biomedical photoacoustics. This section summarizes the key optical, electrical, and safety considerations that define the use of LEDs in PAI.

2.2.1 *Spectral and Temporal Characteristics*

LEDs are semiconductor devices that emit light when electrons recombine with holes across a p–n junction. The emission wavelength λ is primarily governed by the semiconductor bandgap $E = hc/\lambda$, where h is Planck's constant and c the speed of light. By altering the alloy composition—such as GaN and InGaN for blue to green emission, AlGaInP for red, and AlGaAs or InGaAs for near-infrared, LEDs can cover the 400–950 nm spectral range relevant for biomedical imaging. LEDs spanning the visible to near-infrared range (450–940 nm) are widely employed for photoacoustic excitation. Their broad spectral bandwidth (≈20–40 nm FWHM) enables efficient illumination across multiple absorption peaks of hemoglobin and melanin. Figure 2.1 shows representative emission spectra from multiple high-power LEDs spanning 400–950 nm, covering the visible to near-infrared range [9].

In photoacoustics, temporal behavior is as important as spectrum. LEDs can be operated in continuous-wave (CW) or pulsed mode. CW operation provides steady low-power output, suitable for spectroscopy or slow scanning. For imaging, nanosecond-to-microsecond pulses are preferred to generate sufficient pressure rise without significant thermal diffusion. The pulse duration τ should satisfy the stress and thermal confinement conditions ($\tau < L2/\alpha$ and $\tau < L/c_S$, respectively, where L is optical absorption depth, α thermal diffusivity, and c_s sound speed).

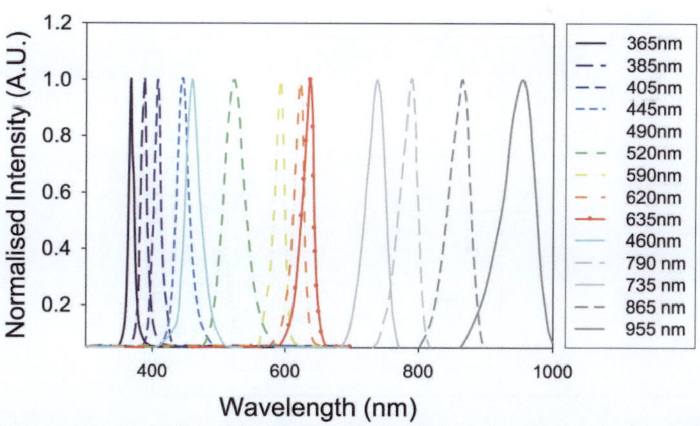

Fig. 2.1 Representative emission spectra of high-power light-emitting diodes (LEDs) used in photoacoustic imaging. The spectra span the visible to near-infrared region (≈400–950 nm), which correspond to the most frequently employed wavelength bands for biological imaging. The broad full width at half maximum (FWHM) of 20–40 nm enables efficient excitation of hemoglobin and melanin absorption bands. Reproduced with permission from [8]

2.2.2 *Pulsed and Continuous Operation*

PAI relies on rapid heating within an optical absorption volume, so the LED must be driven with high peak currents to achieve short, intense pulses. Commercial high-power packages can safely deliver several amperes in CW mode and also can be overdriven to tens or hundreds of amperes if duty cycle remains low (typically <1%). The emitted optical energy per pulse can be estimated as $E = P_p \times \tau$, where P_p is instantaneous optical power. The average optical power is then P_avg = E × f_rep, with f_rep being the pulse repetition rate. Overdriving enhances instantaneous power while keeping thermal load manageable. Typical PAI systems use pulse widths between 30 and 150 ns and repetition rates of 1–5 kHz, providing sufficient acoustic bandwidth for imaging depths up to several millimeters [7, 22].

2.2.3 *Driver Electronics and Synchronization*

Generating these nanosecond pulses requires specialized driver electronics. A common design uses a capacitor-discharge circuit controlled by a MOSFET or GaN transistor (Fig. 2.2). The capacitor is charged to a preset voltage Vcc and then rapidly discharged through the LED when triggered, producing a current pulse I_p ≈ Vcc/R, where R is the equivalent resistance of the circuit. The rise time of the current pulse should be shorter than the desired optical pulse width to avoid temporal broadening.

Precise synchronization between LED excitation and ultrasound data acquisition is crucial. A digital delay generator or microcontroller typically provides trigger

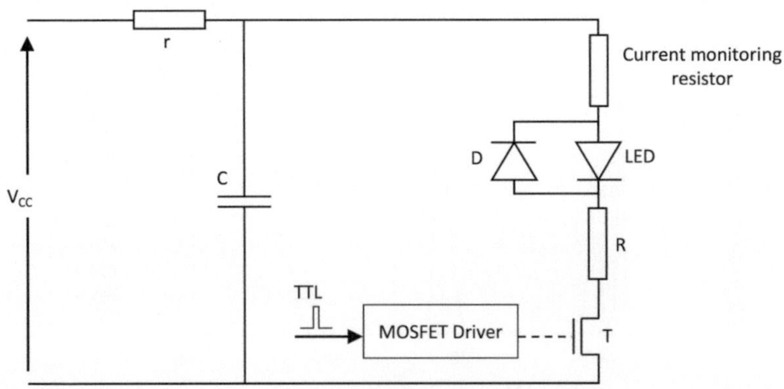

Fig. 2.2 Simplified schematic of a high-current pulsed LED driver circuit used for photoacoustic imaging. A capacitor (C) stores charge and rapidly discharges through the LED via a MOSFET (T) switch when triggered by a TTL signal. The current is monitored using a series resistor, while a flyback diode (D) protects the LED from reverse voltage. This configuration enables nanosecond-scale current pulses (tens of amperes) synchronized with ultrasound acquisition. Reproduced with permission from [8]

outputs both to the LED driver and to the acquisition card. Jitter must be less than one sampling period (≈10 ns for MHz-range transducers) to maintain image quality. Synchronization errors lead to blurring and amplitude errors in reconstructed images [7, 8, 22].

2.2.4 Optical Power Density and Fluence

The efficiency of photoacoustic signal generation depends on local optical fluence F = E/A, where E is optical energy incident on area A. The initial pressure rise p_0 is given by $p_0 = \Gamma\ \mu_a\ F$, with Γ the Grüneisen parameter and μ_a the optical absorption coefficient. To achieve similar fluence to low-energy laser systems, LEDs are often arranged in arrays that deliver tens of mJ/cm^2 at the tissue surface. Unlike laser or laser-diode–based PAI systems, which typically require carefully designed optical components such as lenses, fiber bundles, or diffusers to homogenize the light field, LED arrays inherently produce diffuse and spatially uniform illumination at the skin surface. In many practical implementations, the LED emitters are positioned directly adjacent to the imaging region without intervening optics, allowing broad, low-divergence light to penetrate the tissue naturally through scattering. This simplifies system design, minimizes optical alignment requirements, and improves safety by reducing the risk of localized hotspots. Moreover, the wide angular emission of LEDs ensures more homogeneous fluence distribution across curved or irregular anatomical surfaces, which is advantageous for clinical imaging applications such as dermatology, vascular mapping, and lymphatic visualization [9]. This intrinsic diffusivity of LED illumination contrasts sharply with the collimated beams of lasers, where optical coupling losses and non-uniform fluence often demand precise alignment and costly beam-shaping optics.

2.2.5 Light Delivery and Thermal Management

When operated in pulsed overdrive mode, junction temperature management becomes a key determinant of stability and lifetime. The temperature rise (ΔT) can be estimated as

$$\Delta T = P_d \times R_{th}$$

where P_d is the dissipated power and R_{th} the thermal resistance (°C/W). For high-power packages, R_{th} typically ranges from 5–15 °C/W. As junction temperature increases, luminous efficiency decreases and the emission peak shifts (≈0.2 nm per °C). Excessive heating therefore limits duty cycle and repetition rate. Adequate heat sinking using copper substrates or thermoelectric elements is required to maintain

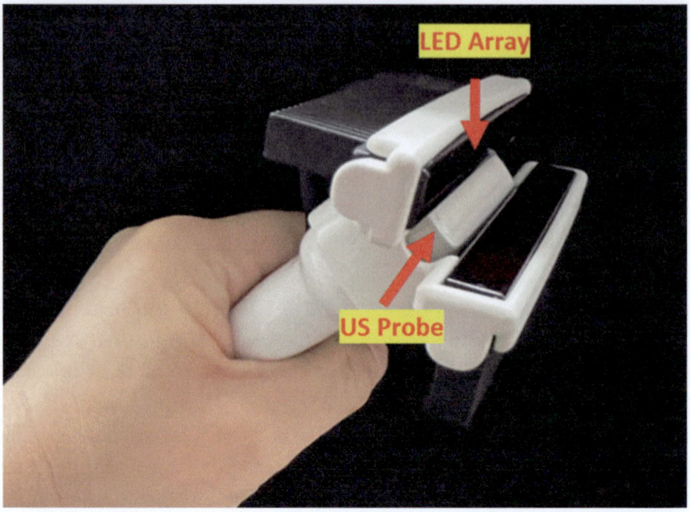

Fig. 2.3 AcousticX dual-mode photoacoustic and ultrasound imaging probe with integrated 850 nm LED arrays on either side of a linear ultrasound transducer, enabling handheld operation without additional beam-shaping optics

stability. LED lifetimes generally follow Arrhenius-type behavior, halving for each 10 °C rise in junction temperature.

One of the unique advantages of LED-based illumination in PAI is its naturally diffuse emission profile. Unlike laser or laser-diode systems that require precise optics or fiber coupling for uniform beam delivery, LED arrays can be placed directly in contact with the skin or coupled through a thin acoustic window. The wide angular spread of each LED ensures homogeneous surface illumination without the need for beam-shaping lenses or diffusers. When multiple LED packages are arranged around the ultrasound probe in circular or planar configurations, they collectively provide smooth and even fluence across the imaging area. This simplicity in light delivery significantly reduces system complexity, cost, and alignment requirements while maintaining consistent photoacoustic signal generation [7–9, 22]. Figure 2.3 shows the AcousticX dual-mode photoacoustic and ultrasound imaging probe, in which two arrays of 850 nm LEDs are integrated on either side of a linear ultrasound transducer. The design eliminates the need for complex or expensive beam-shaping optics, enabling compact, handheld PAI with uniform surface illumination.

2.2.6 *Safety Classification*

Ensuring safe light delivery is a central design consideration in PAI. Conventional laser-based systems typically operate close to the maximum permissible exposure (MPE) defined by international safety standards such as IEC 60825-1 and ANSI

Z136.1. For nanosecond pulsed lasers within the visible and near-infrared spectrum (400–1100 nm), the MPE for skin exposure generally ranges between 20 and 100 mJ/cm^2, depending on wavelength, pulse duration, and repetition rate. In contrast, LED-based excitation sources emit optical fluence that is one to two orders of magnitude lower, typically between 0.1 and 2 mJ/cm^2 per pulse, well below MPE thresholds [7–9]. Consequently, LED-driven systems are inherently safer, posing minimal risk of ocular or dermal injury, and generally do not require laser-rated safety goggles, controlled environments, or interlock mechanisms.

LED illumination is also diffuse rather than collimated, resulting in a more uniform fluence distribution across the tissue surface. This spatial homogeneity reduces the possibility of localized hot spots, further enhancing operational safety. While the lower fluence output compared with lasers can limit imaging depth, it can be mitigated through multi-LED arrays, higher repetition rates, and signal averaging strategies [8, 9]. Thermal management remains important during extended operation. Junction temperature affects both emission wavelength and output stability. Through thermoelectric cooling, optimized duty cycling, and real-time feedback control, pulse-to-pulse variations of less than 2% can be maintained, ensuring stable performance [7].

From a regulatory perspective, LED-based PAI systems fall under the category of non-laser optical radiation sources, simplifying certification under IEC and FDA frameworks. This distinction removes many of the barriers faced by Class IV laser systems and facilitates integration with handheld ultrasound probes and other clinical instrumentation. Overall, LED-based excitation offers a combination of safety, simplicity, and regulatory flexibility that makes it well suited for clinical, preclinical, and educational environments. These advantages are central to the broader goal of democratizing PAI through affordable and accessible system design [7–9, 20, 22].

2.2.7 *Multispectral Operation and Spectral Tuning*

The broadband nature of LED technology allows straightforward implementation of multispectral PAI, which enables functional imaging such as oxygen saturation (sO_2) and hemoglobin concentration mapping. Arrays composed of multiple wavelength bands, commonly 470, 590, 630, 850, and 940 nm can sequentially illuminate the sample. By switching between diode banks on successive pulses, spectral unmixing algorithms can be applied to reconstruct chromophore distributions [7, 9, 23]. Spectral tuning is accomplished by temperature control or by using multiple LED types on a common substrate. Dual or quad-wavelength boards are increasingly available commercially, simplifying integration. For accurate quantitative imaging, pulse-to-pulse energy variation between wavelengths should be <5%, and wavelength calibration must account for thermal drift.

LED arrays, when properly heat-sinked and operated at controlled duty cycles, maintain stable pulse output and spectral consistency even under extended operation. Their inherent multispectral capability further enables simultaneous or sequential illumination across several wavelengths, supporting functional and molecular imaging applications. Figure 2.4 illustrates a representative multispectral LED-based photoacoustic computed tomography (PACT) setup, demonstrating stable performance and effective spectral unmixing of key chromophores such as melanin, indocyanine green (ICG), and methylene blue (MB) in a tissue-mimicking phantom.

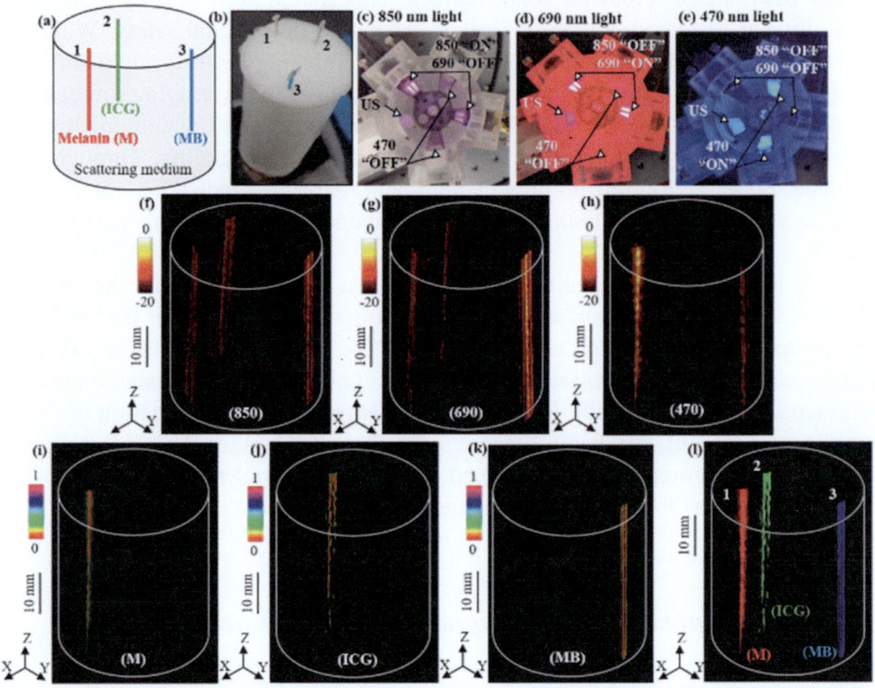

Fig. 2.4 LED-based multispectral photoacoustic computed tomography (PACT). **a** Schematic diagram and **b** photograph of a cylindrical tissue-mimicking phantom (35 mm diameter, 80 mm height) embedded with 0.5 mm polyethylene tubes filled with 1 mM solutions of melanin (M), indocyanine green (ICG), and methylene blue (MB). **c–e** Experimental configuration showing the ultrasound (US) transducer array, dual 850/690 nm LED arrays, and dual 470 nm LED arrays, with sequential illumination at 850, 690, and 470 nm. **f–h** Three-dimensional PACT reconstructions of the phantom under each illumination wavelength. **i–k** Spectrally unmixed volumetric images highlighting the spatial distributions of M, ICG, and MB, respectively. **l** Composite 3D rendering showing superimposed unmixed chromophore distributions. Scale bar = 10 mm (**f–l**). Reproduced with permission from [21]

2.2.8 Pulse-to-Pulse Energy Stability

Energy stability plays a critical role in ensuring uniform image intensity and reliable quantitative analysis. During overdriven operation, small fluctuations in driver current or junction temperature translate proportionally to variations in optical output. A pulse-to-pulse optical energy variation ($\Delta E/E$) below 2% is generally acceptable for functional imaging performance. Stability can be enhanced through constant-voltage pre-charging circuits, short electrical interconnects, and low-inductance PCB layouts. LEDs demonstrate excellent long-term stability when operated within their specified thermal limits. However, excessive overdriving, typically beyond ten times the nominal current can induce electromigration and bond-wire fatigue. Empirically, maintaining peak current below eight times the nominal value and a duty cycle under 1% provides operational lifetimes exceeding 10^8 pulses without measurable degradation [7–9, 23].

2.2.9 System-Level Integration and Illumination Architectures

LED-based PAI systems can employ either single-element or array-based illumination. Early prototypes used single high-power LEDs focused onto small regions, offering simplicity but limited optical fluence and field of view. In contrast, modern systems integrate multiple LEDs in linear or circumferential arrays around the ultrasound probe to achieve uniform, high-intensity illumination over larger areas. Array configurations also support multispectral imaging, with diodes of different wavelengths fired sequentially or in multiplexed mode. Each LED or sub-array is driven by a synchronized electronic channel, ensuring precise timing with ultrasound acquisition. Thermal management is achieved through shared metal substrates or compact heat sinks to maintain output stability. These integrated architectures combining compact arrays, synchronized driving, and efficient heat dissipation enable the compact, handheld LED-PAI probes now used in clinical and preclinical imaging.

2.2.10 Summary

LED illumination for PAI is primarily determined by the interplay between electrical drive conditions, thermal management, and safety constraints. Although LEDs deliver lower pulse energies than solid-state lasers, advances in packaging, driver design, high repetition rate and multispectral control have made them highly effective for shallow to intermediate-depth imaging. Their compactness, low cost, and excellent safety profile make LEDs ideal light sources for portable and clinical photoacoustic systems, paving the way toward broader accessibility and adoption in biomedical imaging.

2.3 Historical and Technological Developments

The evolution of LED-based PAI marks a crucial turning point in the democratization of photoacoustic technology. While conventional PAI relied heavily on bulky and expensive laser systems, the concept of using LEDs as excitation sources introduced an affordable, compact, and safer alternative for biomedical imaging. The following sections outline the chronological development of LED-based PAI—from its earliest proof-of-concept experiments to the emergence of clinically deployable systems.

2.3.1 Early Concept and Proof-of-Feasibility (2010–2013)

Around a decade ago, Hansen et al. first proposed and demonstrated the use of LEDs operating at a 627-nm wavelength as compact and inexpensive excitation sources for photoacoustic imaging [24]. This pioneering study marked the first experimental validation of LED-based excitation in PAI. The key principle underlying this development was the realization that low-power continuous-wave (CW) LEDs could be "overdriven" using high-current, short-duration electrical pulses. Under such pulsed overdrive conditions, LEDs can deliver optical output powers far exceeding their CW-rated specifications, without permanent damage, provided the duty cycle remains low [25–28].

In Hansen's proof-of-concept setup, a Luxeon LXHL_PD09 LED was used. When driven with a 1 A DC current, this LED produced approximately 250 mW of optical output. By applying 60-ns current pulses with peak currents up to 40 A using a MOSFET-based driver circuit derived from Alton and Raji's design [29], the optical pulse energy reached approximately 400 nJ per pulse. The focused light generated sufficient radiant energy to induce a detectable photoacoustic signal from a simple phantom. However, the experimental setup faced significant limitations: to obtain a usable photoacoustic signal, 50,000 A-line signals had to be averaged, and with a pulse repetition rate (PRR) of 200 Hz, the acquisition time for a single image exceeded 250 s. While inadequate for real-time imaging, this experiment successfully established the feasibility of using pulsed LEDs as excitation sources for PAI.

2.3.2 *Advancements in High-Power LED Excitation (2013–2016)*

Following Hansen's work, improvements in solid-state device fabrication and driver electronics enabled rapid increases in LED pulse energy and efficiency. In 2013, Allen et al. introduced high-power Luminus CBT-120 LEDs (operating in the 400–650 nm range) as illumination sources for biomedical PAI [30]. By overdriving the LEDs at ten times their rated current, they achieved a pulse energy of 22 μJ per pulse with a pulse duration of 500 ns. A MOSFET-based driver circuit similar to the one proposed by Chaney et al. [29] was used to deliver these high-current pulses. This substantial increase in optical energy allowed a dramatic reduction in frame averaging (by a factor of 1000) while maintaining the same 200 Hz PRR, marking a major step toward practical imaging speeds.

By 2016, Allen and colleagues had further refined their system using Luminus SST-90 LEDs (400–650 nm) and a commercial driver (PCO-7120, Directed Energy Inc.). This configuration produced 9 μJ per pulse with 200-ns pulse duration, 50 A peak current, and a PRR of 500 Hz, achieved through 20× overdriving of the LED elements [31]. Despite the extreme overdrive, the duty cycle remained below 0.01%, well within safe operational limits (far below the 1% damage threshold) [32]. In this work, the group successfully performed photoacoustic imaging of realistic tissue-mimicking phantoms, using wide-field illumination and averaging 5000 frames. The system achieved an imaging depth of approximately 15 mm in 1% intralipid, demonstrating for the first time that visible-light LEDs could produce useful photoacoustic signals in scattering media.

These pioneering studies collectively established the technical foundations of LED-based PAI, proving that visible-wavelength excitation impractical with many laser sources was both feasible and beneficial for functional imaging applications such as oxygen saturation mapping.

2.3.3 *Expansion to NIR Wavelengths and System Optimization (2016–2018)*

One of the most significant limitations of early LED-PAI systems was the restricted wavelength range. Oxygen saturation imaging, a key clinical application requires illumination near the absorption peaks of deoxyhemoglobin (~690 nm) and oxyhemoglobin (~850 nm). Recognizing this, researchers turned their focus to developing high-power near-infrared (NIR) LEDs in these spectral regions.

Between 2016 and 2017, Agano et al. introduced high-power 850-nm LED arrays capable of delivering ~200 μJ per pulse with 70–100 ns pulse durations [33–35]. To achieve this, they employed a double-stack LED design and optimized array geometry to enhance optical output. Each single LED element emitted 0.024 μJ per pulse at 1 A DC current, but by overdriving the stacked arrays at 20× rated current, they

Table 2.1 Chronological summary of key developments in LED-based PAI systems [7]

Year	Authors	Pulse width (ns)	Peak current (A)	Pulse energy (mJ)	Repetition rate (Hz)	Wavelength (nm)
2011	Hansen et al. [24]	60	40	0.0004	200	627
2013	Allen et al. [30]	500	200	0.0022	200	623
2016	Allen et al. [31]	200	50	0.0009	500	623
2016–2017	Agano et al. [33–35]	70*	15–20	0.15–0.2**	4000	850
2018	Zhu et al. [36]	70*	20	0.2**	4000–16,000	850

* It is tunable from 35–150 ns and measured with 70 ns; ** value are measured from the arrays of LED

achieved the desired energy output at high repetition rates. Complementing the optical advances, Agano's team developed low-noise, high-gain front-end electronics (~106 dB total gain) and acoustic matching between the LED pulse duration and ultrasound transducer bandwidth. The culmination of these innovations was the first commercial LED-based PAI system, AcousticX (CYBERDYNE Inc., Japan), operating at up to 4 kHz PRR, a significant leap toward real-time imaging. In 2018, Zhu et al. further advanced the technology, pushing the PRR to 16 kHz and demonstrating dynamic structural and functional imaging using the same system [36]. These developments collectively bridged the gap between laboratory prototypes and clinically relevant LED-PAI systems. Table 2.1 shows the chronological summary of key developments in LED-based PAI.

2.3.4 System Characterization and Clinical-Grade Performance (2018–2020)

The AcousticX system (Fig. 2.5) underwent extensive performance evaluation in both phantom and in vivo studies. Using a 9-MHz ultrasound probe paired with 850-nm LED arrays, Xia et al. reported axial and lateral resolutions of approximately 220 and 460 μm, respectively comparable to conventional ultrasound imaging [37]. The system achieved an imaging depth of 2.8 cm at an interleaved ultrasound/PA frame rate of 30 Hz, which could be extended to 3.8 cm by averaging more frames, albeit at a reduced frame rate (1.5 Hz). In this work, authors also reported system's feasibility in imaging superficial vasculature in vivo on human volunteers. Similarly, Hariri et al. reported axial and lateral resolutions of 268 and 570 μm, achieving 3.2

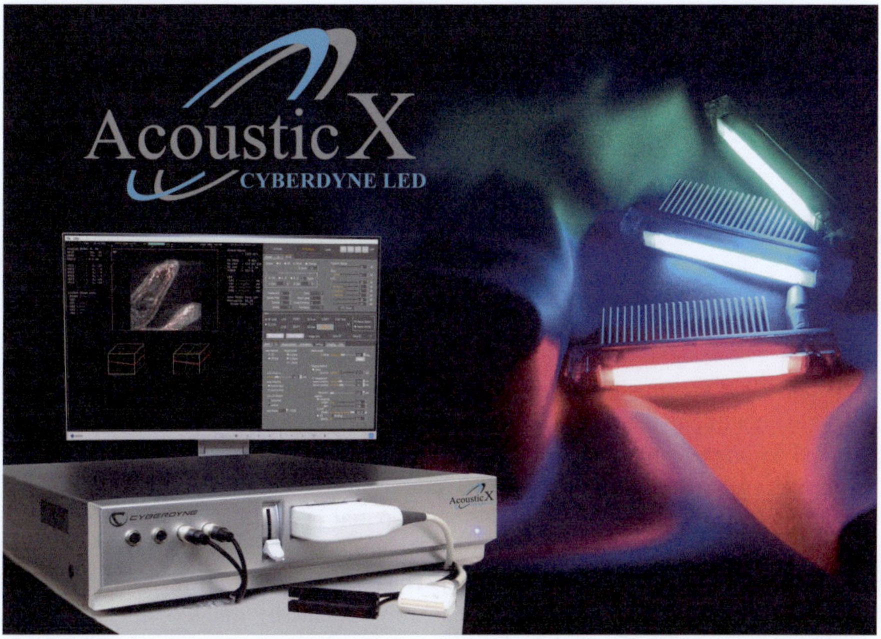

Fig. 2.5 Photograph of CYBERDYNE AcousticX system with real-time dual-mode ultrasound and PAI capability

cm imaging depth in a phantom at 15 Hz frame rate [38]. Importantly, this study also quantified the molecular sensitivity of LED-based PAI to contrast agents such as ICG (9 μM detection limit), methylene blue (0.75 mM), and DiR (68 μM)—establishing a baseline for molecular imaging performance.

Further investigations by Kalloor Joseph et al. utilized dual-wavelength (750/850 nm) LED excitation for oxygen saturation imaging in phantoms, small animals, and human volunteers [14, 16]. Using 28 in-vitro human-blood measurements, they reported a standard error of 8.4% between true and measured SO_2 values. In vivo mouse experiments demonstrated high repeatability and reproducibility, confirming the potential of LED-based systems for functional monitoring. In vivo imaging in human volunteers further demonstrated the clinical feasibility of LED-based photoacoustic systems for oxygen saturation mapping, successfully distinguishing veins from arteries with high sensitivity and specificity.

2.3.5 Summary

Over the past decade, LED-based PAI has evolved from a conceptual demonstration into a clinically evaluated modality. Continuous advances in semiconductor technology, driver design, and thermal management have enabled multi-wavelength, high-repetition-rate systems capable of real-time functional imaging. The development of commercial platforms such as AcousticX has firmly established LED illumination as a viable and safe alternative to laser-based excitation, paving the way for widespread clinical translation and point-of-care imaging applications.

2.4 Preclinical Applications

2.4.1 Phantom and Ex Vivo Studies

LED-based PAI systems have been extensively characterized on standardized phantoms to evaluate spatial resolution, depth, and quantitative accuracy. Joseph et al. [39] provided a comprehensive validation of a dual-wavelength LED/ultrasound system (AcousticX, CYBERDYNE Inc., Japan), confirming linear photoacoustic response, excellent signal stability, and sub-millimetre spatial resolution in tissue-mimicking and biological phantoms. Figure 2.6 further illustrates the high temporal precision of LED excitation demonstrated in their study. The PVCP phantom (Fig. 2.6a) contained multiple embedded absorptive targets that were repeatedly imaged at 750 and 850 nm. The corresponding intensity plots (Fig. 2.6b–e) show mean pixel values over continuous acquisitions of 180 min, repeated measurements over 30 days, and hourly sessions with and without phantom replacement. The near-zero gradients of the fitted lines indicate negligible signal drift, confirming outstanding long-term stability and reproducibility of LED-based PAI under both continuous and interrupted acquisition. This level of stability, achieved without recalibration or optical realignment demonstrates the robustness of LED excitation for quantitative and longitudinal imaging applications.

Hariri et al. [38] characterized the same platform for molecular sensitivity, reporting detection limits of ~9 μM for ICG, 0.75 mM for methylene blue, and 68 μM for DiR dye, with axial/lateral resolutions of 268 μm/570 μm and imaging depths approaching 3 cm. Francis et al. [12] implemented a tomographic configuration using four 850 nm LED arrays surrounding a 128-element ultrasound probe, showing that multi-angle illumination substantially improves reconstruction fidelity compared with planar illumination.

Agrawal et al. [13] directly compared LED and laser excitation on identical hardware, showing comparable signal-to-noise ratios up to ~3 cm depth and confirming that LEDs can achieve diagnostically useful contrast under safe fluence levels. Figure 2.7 summarizes their findings in a multilayer chicken tissue phantom containing four 0.5 mm pencil-lead targets embedded at depths up to 31 mm. Panels (b–d) show ultrasound, co-registered US + PA, and pure photoacoustic images obtained under laser illumination at 850 nm, while panels (e–h) present corresponding results acquired using an 850 nm LED array operating at frame rates from 6–30 Hz. Despite a three orders of magnitude lower per-pulse energy (0.4 mJ for LED versus 40 mJ for laser), LED-based illumination produced detectable signals from

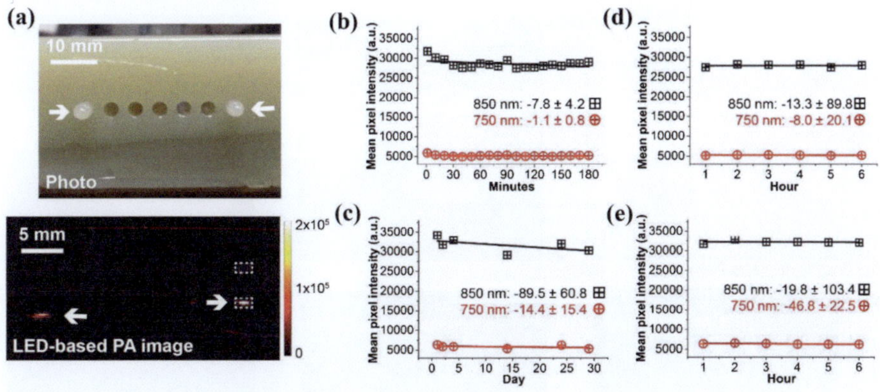

Fig. 2.6 LED-based PAI exhibits high temporal precision in phantom imaging. **a** Photo (top) and PA image (bottom) of the PVCP phantom used for stability studies. Target inclusions are indicated with arrows. Mean pixel intensity values extracted from regions of interest (exemplars shown in dotted lines in **a** were examined for continual acquisition over 180 min **b**, repeated acquisitions over 30 days (**c**) and over 6 h with (**d**) and without (**e**) replacement. Replacement indicates removal of the phantom from the imaging system in between measurements. Error bars are within the symbols. The gradient of the linear fits to the data for the 850 nm LED (black) and 750 nm LED (red) are given within each panel; none of the gradients are significantly non-zero. Reproduced with permission from [39]

all embedded targets, maintaining comparable SNR up to 3 cm depth (Fig. 2.7i). These results demonstrate that by leveraging higher repetition rates and frame averaging, LED-based PAI can achieve imaging performance on par with conventional laser systems for mesoscopic depths, while retaining compactness and inherent safety advantages (Fig. 2.7).

These studies collectively established that LED-PAI can deliver quantitative, repeatable imaging performance within the mesoscopic depth range (2–4 cm) with sub-millimetre resolution, achieving results comparable to conventional laser-based systems and forming a benchmark for subsequent in vivo experiments.

2.4.2 *Small-Animal Imaging*

Preclinical LED-PAI has advanced rapidly from benchtop validation to in vivo imaging of small animals. Zhu et al. [36] employed dual 850 nm LED bars integrated with a linear ultrasound probe to visualise murine vasculature and ocular microvessels, clearly resolving vessels smaller than 0.5 mm. Francis et al. [14] extended this concept to joint imaging in rodent models of rheumatoid arthritis, demonstrating LED-based detection of inflammatory hyper-vascularisation while adhering to 3Rs (Replacement, Reduction, Refinement) principles.

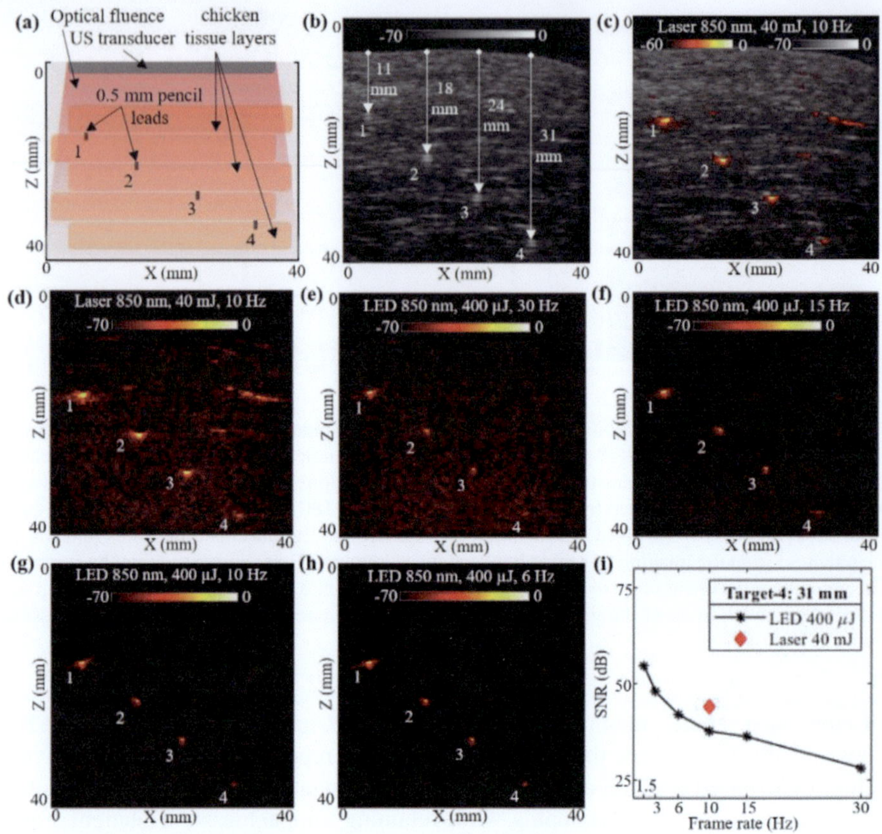

Fig. 2.7 Comparison of LED- and laser-based photoacoustic imaging in a multilayer chicken tissue phantom. Schematic **a** shows target positions (pencil leads) embedded between stacked chicken layers. Panels **b–d** display ultrasound, co-registered US + PA, and PA images acquired with laser illumination at 850 nm; panels **e–h** show corresponding results using 850 nm LED arrays at various frame rates. **i** Signal-to-noise ratio (SNR) comparison for the deepest target demonstrates that LED illumination achieves comparable performance to laser excitation at depths up to ~30 mm. Reproduced with permission from [13]

Xavierselvan et al. [40] applied the AcousticX system to image in vivo subcutaneous tumours in small-animal models. By combining photoacoustic and ultrasound modes, they obtained high-contrast vascular maps up to ~10 mm depth and revealed distinct vascular heterogeneity between tumour and surrounding tissue, corroborated by histological analysis. In addition to structural vascular mapping, Xavierselvan et al. [41] demonstrated that LED-based PAI can sensitively monitor tumour contrast enhancement following administration of exogenous dyes. Figure 2.8 shows representative photoacoustic/ultrasound overlay images of a subcutaneous FaDu tumour in a mouse before and after injection of a near-infrared (NC) dye. Post-administration images exhibit markedly increased photoacoustic signal within the tumour core

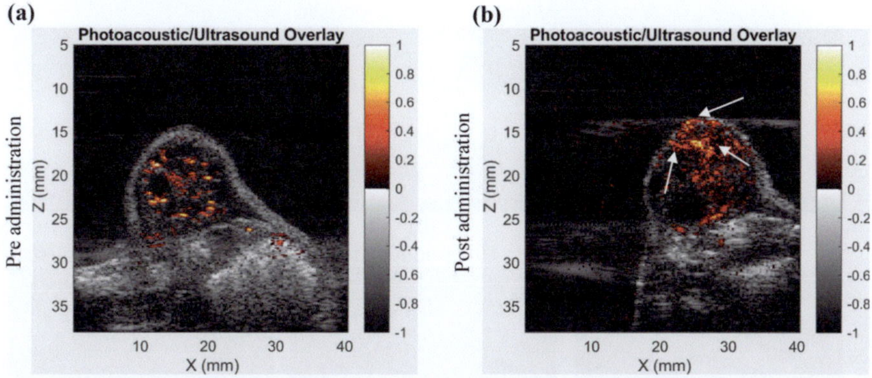

Fig. 2.8 LED-based photoacoustic/ultrasound imaging of a subcutaneous tumour in a mouse before and after contrast administration. Composite PA/US overlays acquired at 850 nm using the AcousticX system show signal enhancement (arrows) within the tumour region following injection of a near-infrared dye, illustrating LED-PAI's capacity for contrast-agent-based molecular imaging. Reproduced with permission from Xavierselvan et al. [41]

(arrows), confirming enhanced optical absorption and the system's ability to detect dynamic contrast uptake using 850 nm LED illumination. This result underscores the capability of LED-PAI to perform molecular and functional imaging in small-animal oncology models, complementing its structural and vascular applications (Fig. 2.8).

The tomographic system initially developed by Francis et al. [12] was subsequently employed to perform volumetric LED/ultrasound imaging of the mouse abdomen [42]. Using four 850 nm LED arrays positioned around a linear ultrasound probe, the setup enabled tomographic photoacoustic reconstruction of multiple anatomical planes. As illustrated in Fig. 2.9, multi-angle B-scan acquisitions (a–c) were combined to generate complete tomographic volumes (d–f), clearly visualizing key abdominal organs such as the liver, spleen, kidneys, and spinal cord. This multi-illumination configuration improved fluence uniformity and substantially reduced limited-view artefacts compared with single-angle imaging, demonstrating that LED excitation can effectively support true three-dimensional photoacoustic tomography in small-animal models.

Recent developments in full-view tomography have pushed these capabilities further: Liu et al. [23] introduced the full-view LED-based optoacoustic tomography (FLOAT) system, in which circumferential LED arrays (~0.5 mJ per pulse, 100 ns) surround a ring-shaped ultrasound detector to achieve volumetric imaging of mouse vasculature to depths exceeding 20 mm while minimizing limited-view artefacts. Building upon earlier LED tomographic efforts, the latest 2025 FLOAT implementation offers full-view volumetric imaging of the mouse body with enhanced depth and anatomical fidelity [43]. As shown in Fig. 2.10, cross-sectional reconstructions at multiple levels reveal structures such as skin, spinal cord, liver, spleen, kidneys, intestine, femoral and tail veins, and major vessels. This full-circumferential LED illumination geometry significantly reduces limited-view artefacts and ensures uniform

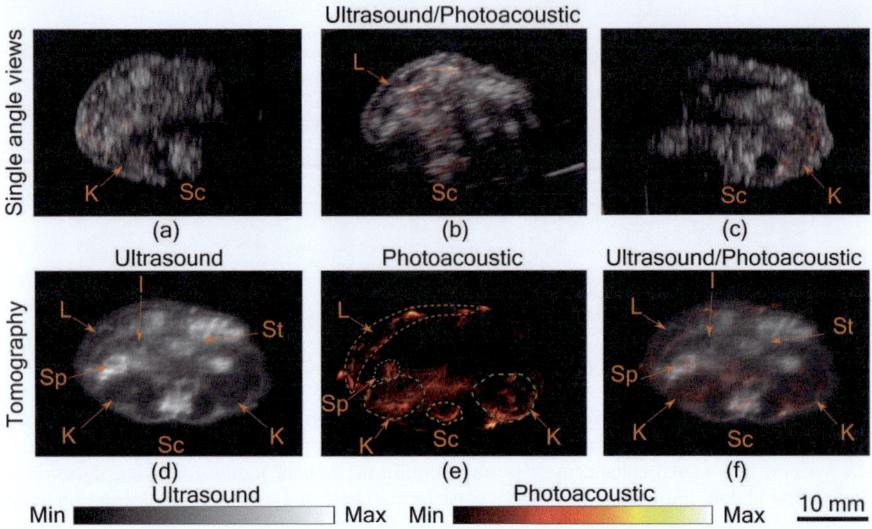

Fig. 2.9 Tomographic LED-based photoacoustic and ultrasound imaging of a mouse abdomen. Multi-angle co-registered ultrasound and photoacoustic B-scans (**a–c**) were combined to generate tomographic volumes (**d–f**), clearly depicting abdominal organs including the liver, spleen, kidneys, spleen, and spinal cord. The multi-illumination design using four 850 nm LED arrays improved fluence uniformity and reduced limited-view artefacts. Reproduced with permission from [42]

excitation, showcasing that LED-based photoacoustic tomography can now rival traditional laser-PA systems in comprehensive small-animal imaging.

Collectively, these investigations demonstrate that compact LED-based systems can effectively support both planar and tomographic in vivo imaging in small-animal models, delivering high-contrast visualization of vascular, tumoral, and functional structures. While preclinical laser-based photoacoustic platforms offer higher pulse energies and flexible wavelength tuning, their high cost, bulk, and safety requirements continue to limit accessibility and throughput in routine research. In contrast, LED-based PAI provides comparable image quality for mesoscopic applications at a fraction of the cost and system complexity. Its compact architecture, electrical efficiency, and inherent eye safety make it ideally suited for longitudinal and large-cohort preclinical studies, bridging the gap between benchtop experimentation and scalable translational imaging.

2.4.3 Functional Imaging

Beyond structural imaging, LED-based photoacoustic systems have proven capable of capturing functional and physiological information such as oxygen saturation (sO_2) and lymphatic vessel dynamics. These capabilities broaden the biomedical

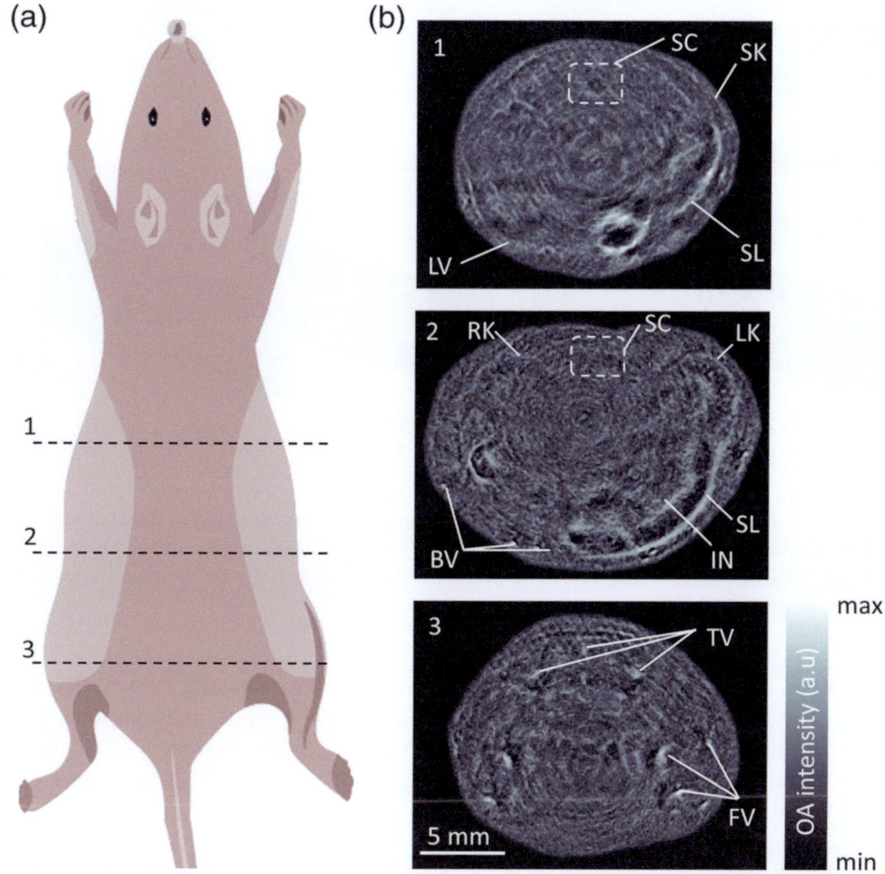

Fig. 2.10 Whole-body in vivo LED-based optoacoustic tomography (FLOAT) in mice **a** Schematic showing the cross-sectional planes used for volumetric imaging. **b** Reconstructed optoacoustic images at three anatomical levels demonstrate clear visualization of major internal structures including the spinal cord (SC), skin (SK), spleen (SL), liver (LV), kidneys (RK, LK), intestine (IN), blood vessels (BV), and femoral and tail veins (FV, TV). The full-view LED configuration provided uniform illumination and reduced limited-view artefacts across the entire cross-section. Reproduced with permission from [43]

relevance of LED-PAI, positioning it as a low-cost, safe, and quantitative tool for tissue characterization.

Using dual-wavelength LED arrays (typically 750 and 850 nm), Bulsink et al. [16] achieved quantitative mapping of oxygen saturation in tissue-mimicking phantoms, small animals and human subjects. Their study demonstrated high repeatability and less than 10% deviation from reference oximetry, confirming the potential of LED-PAI for reliable sO_2 imaging despite its lower optical fluence compared with laser

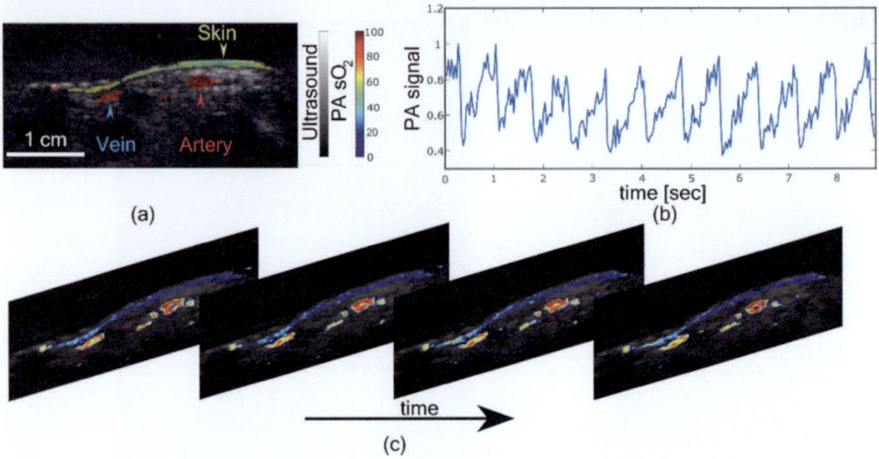

Fig. 2.11 In vivo LED-based photoacoustic imaging of the human wrist. **a** Fluence-compensated photoacoustic oxygen saturation (PA sO_2) map overlaid on an ultrasound image showing differentiation of artery and vein. **b** Normalized photoacoustic signal from the artery illustrating pulsation over time. **c** Sequential PA sO_2 frames visualizing arterial pulsation during imaging. Reproduced with permission from [16]

excitation. Figure 2.11 shows an example of human wrist imaging, where fluence-compensated sO_2 maps clearly differentiate arteries and veins, and temporal variations in the photoacoustic signal correspond to the cardiac pulse waveform. Such results confirm that LED-PAI can capture physiological dynamics in real time with excellent stability and safety.

Joseph et al. [24] further validated the dynamic performance of LED-based systems through repeated phantom and in vivo experiments, showing that hemoglobin contrast and oxygen-dependent responses could be reproduced over multiple sessions without significant signal drift. These studies established LED-based PAI as a stable and quantitative modality for monitoring functional hemodynamics.

Functional imaging has also extended into the lymphatic domain, demonstrating that LED-PAI can go beyond diagnostic visualization to provide real-time intraoperative guidance. In a key study, Singh et al. [46] presented LED-based photoacoustic lymphangiography for potentially guiding lymphaticovenous anastomosis (LVA) during microsurgical procedures. The system employed near-infrared LED arrays (820 and 940 nm) co-registered with real-time ultrasound to visualize functional lymphatic vessels and adjacent veins intraoperatively. As shown in Fig. 2.12, indocyanine green (ICG) contrast was used to trace lymphatic flow, with dual-wavelength imaging enabling spectral differentiation between veins and lymphatic vessels. This approach provided surgeons with precise feedback for identifying patent lymphatics and suitable recipient veins, demonstrating the feasibility of real-time LED-based photoacoustic guidance. Although conducted in a translational surgical setting, this work underscores that LED excitation can now support complex functional and interventional imaging tasks previously achievable only with laser-based systems.

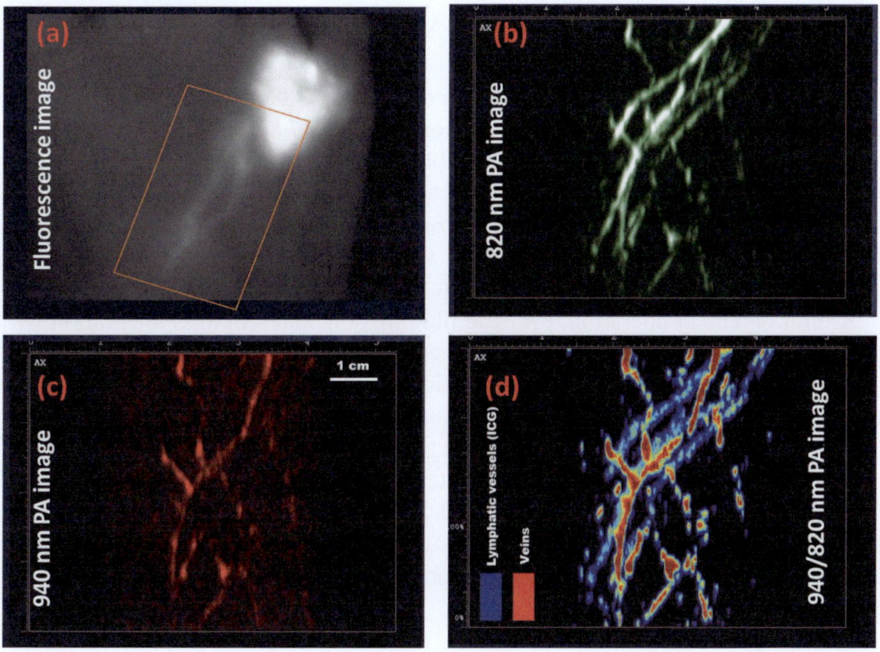

Fig. 2.12 LED-based photoacoustic lymphangiography for intraoperative visualization of lymphatic vessels. **a** Fluorescence image showing ICG uptake and flow through lymphatic vessels (orange rectangle marks the PA/US imaging plane). **b** 3D 820 nm photoacoustic maximum-intensity projection (MIP) image, **c** 3D 940 nm MIP image, and **d** combined 940/820 nm MIP image distinguishing lymphatic vessels (red) and veins (blue). Reproduced with permission from [44]

While laser-based systems still provide higher per-pulse energy and greater penetration depth, LED-based PAI achieves comparable oxygenation and vascular contrast for superficial and intermediate tissue depths (~1.5 cm). More importantly, LED systems combine low cost, compact architecture, eye-safe operation, and minimal maintenance, enabling high-throughput longitudinal imaging and even intraoperative use.

These attributes make LED-PAI uniquely suited for routine preclinical studies, where scalability, reproducibility, and ease of operation are critical. Collectively, the growing body of preclinical evidence confirms that LED-based PAI can deliver quantitative, reproducible, and functionally rich information at a fraction of the cost and complexity of laser-driven systems. By bridging performance and practicality, this technology has now matured to a point where its clinical potential is being actively realized. The following section explores how these advances have translated into clinical applications, including vascular imaging, dermatology, rheumatology, and surgical guidance.

2.5 Clinical and Translational Applications

The maturation of LED-based PAI from preclinical validation to clinical feasibility marks a major milestone in the translation of this technology. Building upon the safety, portability, and affordability established in animal studies, several groups have now demonstrated that LED-PAI can be deployed directly in clinical pilot studies for real-time, noninvasive visualization of vascular and functional features. Unlike conventional laser-based clinical systems which remain expensive, bulky, and restricted to specialized imaging suites, LED-driven platforms such as AcousticX offer handheld operation, eye-safe illumination, and seamless integration with ultrasound probes. These attributes have enabled a broad range of clinical investigations, including image-guided lymphatic microsurgery, angiogenesis monitoring during wound healing, rheumatoid arthritis diagnosis, port-wine stain assessment, dialysis access perfusion monitoring, and peripheral vascular disease evaluation.

Collectively, these studies demonstrate that LED-based PAI can deliver diagnostically useful contrast and functional information across multiple anatomical and pathological contexts, while retaining the advantages of cost efficiency and operational simplicity that are essential for widespread clinical adoption.

2.5.1 *Lymphangiography for Surgical Guidance*

The first European clinical pilot study employing LED-PAI for human lymphatic visualization was conducted at the Erasmus Medical Center, Rotterdam, The Netherlands, by Van Heumen et al. [19]. In this investigation involving 15 patients with secondary limb lymphedema, the feasibility of preoperative lymphatic vessel mapping was evaluated using a handheld AcousticX system equipped with dual-wavelength (820 and 940 nm) LED arrays co-registered with ultrasound. Indocyanine green (ICG) was administered to enhance optical absorption, and imaging regions were selected based on near-infrared fluorescence lymphography (NIRF-L). By computing the photoacoustic signal ratio (940/820 nm), lymphatic vessels showing stronger absorption at 820 nm were successfully differentiated from adjacent veins, which exhibited comparable signal intensities at both wavelengths.

The study demonstrated clear visualization of lymphatic and venous structures at depths up to 8–9 mm, including in regions of dermal backflow where NIRF-L failed to resolve distinct vessels. The handheld LED-PAI/ultrasound configuration provided real-time anatomical and functional feedback, supporting preoperative assessment and intraoperative orientation for lymphaticovenous anastomosis (LVA) procedures. Although limited to superficial depths, this first European clinical translation confirmed that compact, eye-safe, and low-cost LED-PAI can yield clinically actionable lymphatic imaging comparable to laser-based systems. Representative results are shown in Fig. 2.13, highlighting the ability of PAI ratio imaging

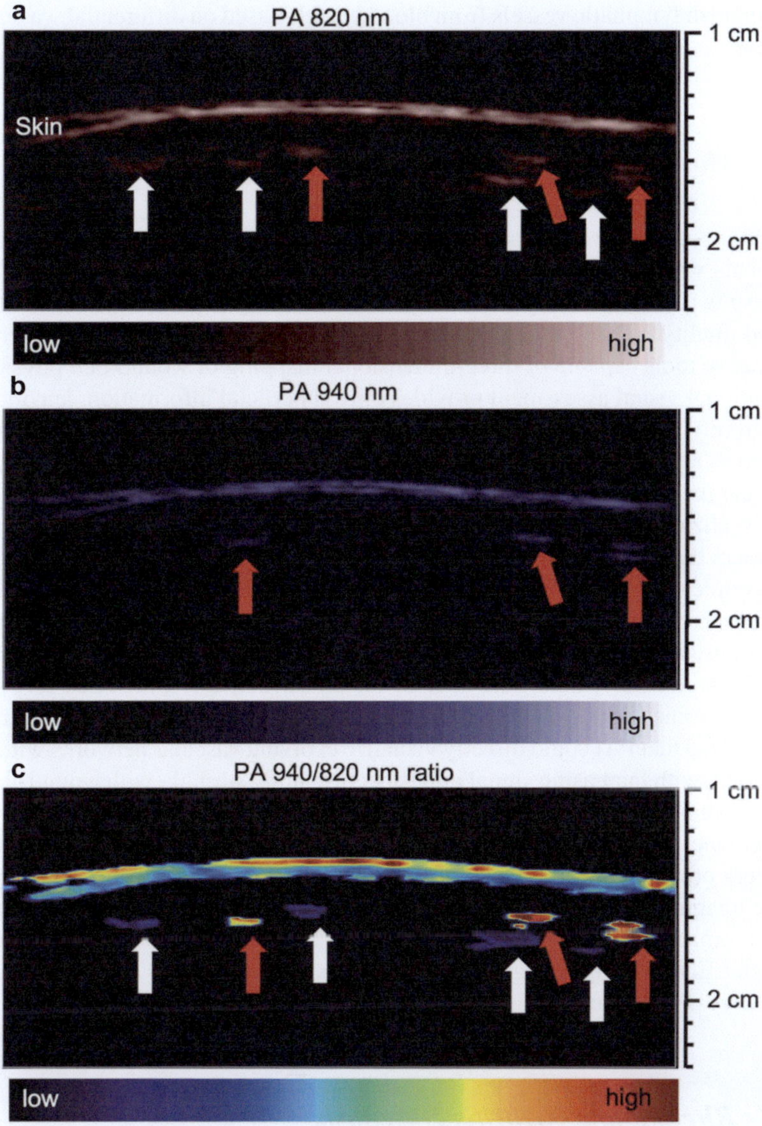

Fig. 2.13 Photoacoustic ratio imaging for lymphatic differentiation in a patient with secondary arm lymphedema. **a** PA image at 820 nm illumination; **b** PA image at 940 nm illumination; **c** 940/820 nm ratio image. White arrows denote lymphatic vessels, absent at 940 nm and appearing dark blue in the ratio map, while red arrows indicate blood vessels with signals at both wavelengths (ratio $\approx$ 1). The continuous band between 13 and 15 mm depth corresponds to the skin. Reproduced with permission from [19]

to distinguish lymphatic vessels from blood vessels based on differential wavelength responses.

2.5.2 *Monitoring Angiogenesis in Wound Healing*

Chronic wounds represent a major and growing healthcare burden, affecting millions of patients worldwide and costing billions of dollars annually. They often fail to heal despite advanced therapeutic interventions, leading to prolonged morbidity and reduced quality of life. A key limitation in wound management is the absence of noninvasive tools capable of three-dimensional mapping of wound-bed physiology. Conventional visual assessment provides only superficial information, leaving clinicians to rely heavily on experience when predicting therapeutic response. PAI has emerged as a powerful hybrid modality capable of overcoming these limitations. By exploiting the intrinsic optical absorption of hemoglobin, PA imaging enables real-time visualization of angiogenesis, microvascular perfusion, and oxygen saturation parameters that are directly linked to wound healing dynamics.

In a clinical pilot study, Mantri et al. [45] investigated the use of combined LED-based photoacoustic and ultrasound imaging to monitor angiogenesis and stratify patients based on their healing response. The study, conducted at the UCSD Hyperbaric Medicine and Wound Care Center, Encinitas, CA, USA, included 19 patients with 22 chronic wounds, imaged weekly over a three-week period. The results demonstrated that PAI could directly visualize evolving vascular networks within the wound bed, with increasing signal intensity correlating strongly with wound closure. Patients exhibiting early and sustained rises in photoacoustic intensity showed faster healing, whereas minimal vascular response was associated with delayed recovery. This work provided one of the first demonstrations of LED-based functional photoacoustic imaging for clinical wound assessment, confirming its potential as a noninvasive biomarker for therapy response. Figure 2.14 illustrates a representative case in which LED-based PA imaging was used to track angiogenesis and wound healing in an 82-year-old female with a chronic posterolateral ankle ulcer.

2.5.3 *Rheumatoid Arthritis Diagnosis*

PAI, owing to its sensitivity to hypervascularization and hypoxia—key early biomarkers of rheumatoid arthritis (RA)—has been extensively explored in both preclinical and early clinical studies. Inflammatory arthritis typically affects small joints such as those of the hands and wrists, where early detection of synovial inflammation is crucial for timely intervention. The combination of PAI with conventional ultrasound offers complementary structural and functional information, providing simultaneous visualization of joint anatomy, blood volume, and tissue oxygenation. Early findings from animal and pilot human studies suggest that PA imaging is

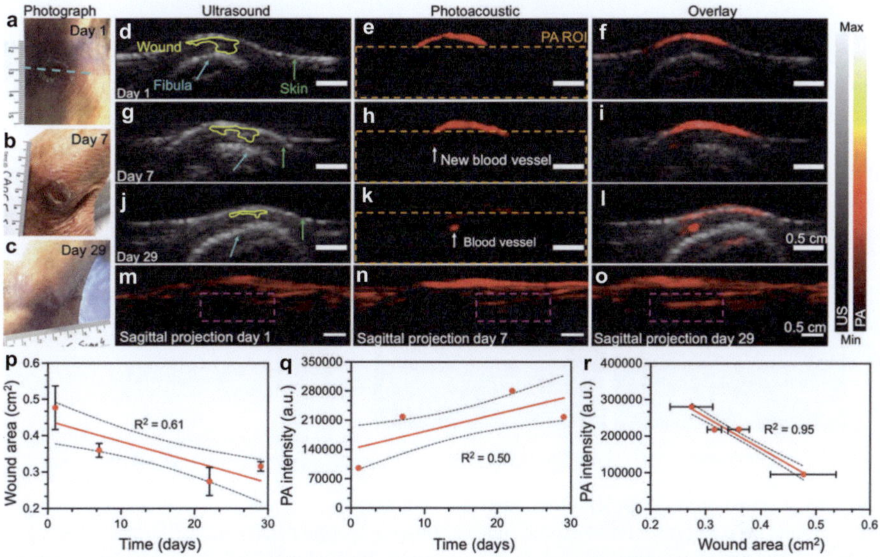

Fig. 2.14 Photoacoustic imaging tracks angiogenesis in a healing ankle wound over 29 days. Panels **a–c** show wound photographs on days 1, 7, and 29. Yellow regions of interest (ROI) in the ultrasound images outline the wound, with green and blue arrows indicating the skin surface and fibula, respectively. Panels **d–f**, **g–i**, and **j–l** display ultrasound, photoacoustic, and co-registered overlay images at corresponding time points; white arrows mark sites of new vessel formation. Sagittal maximum intensity projections **m–O** reveal progressive vascular ingrowth within the wound bed. Graphs **p–r** demonstrate a negative correlation between wound area and time ($R^2 = 0.61$), a positive correlation between photoacoustic intensity and time ($R^2 = 0.50$), and a strong linear relationship between photoacoustic intensity and wound closure ($R^2 = 0.95$), confirming that angiogenesis directly parallels healing. Scale bars represent 0.5 cm. Reproduced with permission from [45]

nearing clinical translation as a low-cost, point-of-care modality for early inflammatory disease assessment. In a seminal clinical pilot study, Janggun et al. [46] investigated the feasibility of using a handheld LED-based PA/US system to detect inflammatory changes in soft tissues surrounding joints. The study aimed to differentiate arthritic joints from healthy controls by assessing blood-volume-related PA signals within the synovium. Patients with clinically active arthritis, subclinical arthritis, and healthy volunteers were imaged at the metacarpophalangeal (MCP) joints, with LED-photoacoustic results compared directly against Doppler ultrasound findings.

As shown in Fig. 2.15, joints with clinically active arthritis displayed marked hyperemia in both Doppler and PA images. In subclinical cases, however, hyperemia was visible only in the PA image, undetected by Doppler ultrasound, demonstrating the superior sensitivity of PA imaging to microvascular changes. In contrast, healthy joints exhibited no significant hyperemia in either modality. Quantitative analysis confirmed statistically significant differences in PA signal intensity between clinically active, subclinical, and normal joints ($p < 0.05$).

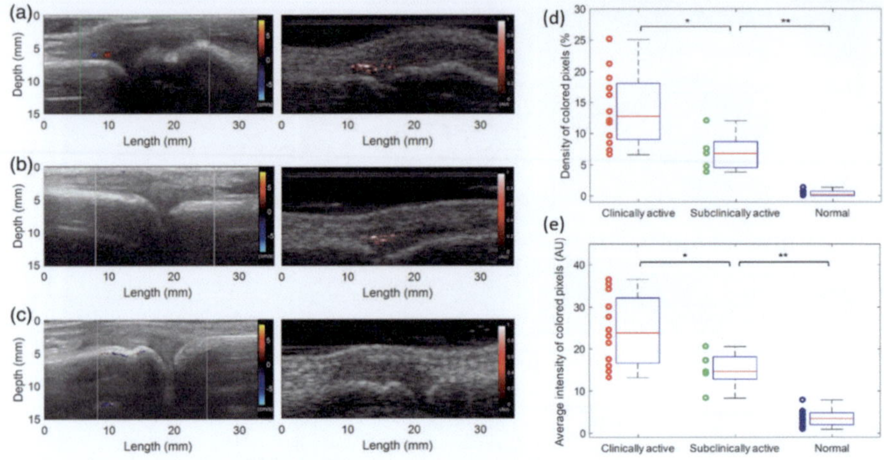

Fig. 2.15 Comparative Doppler ultrasound and photoacoustic images of metacarpophalangeal (MCP) joints. **a–c** Representative paired images showing Doppler (left) and LED-based PA (right) scans for three groups: **a** clinically active arthritis joints, **b** subclinical arthritis joints, and **c** normal healthy joints. Hyperemia is visible in both modalities for active arthritis but only in the PA image for subclinical cases, demonstrating enhanced sensitivity to microvascular changes. **d–e** Quantitative analysis of colored pixel density and mean PA intensity across groups. Statistical significance indicated by $^{*}p < 0.05$, $^{**}p < 0.005$. Adapted with permission from [46]

These results provide direct clinical evidence that LED-based PA imaging can detect early synovial angiogenesis with higher sensitivity than Doppler ultrasound, the current clinical standard. The study highlights the promise of LED-PA/US systems as affordable and noninvasive tools for the early detection and monitoring of rheumatoid arthritis activity in outpatient and point-of-care settings [47].

2.5.4 Port Wine Stain Imaging and Treatment Monitoring

Port-wine stain (PWS) is a congenital vascular malformation that can cause significant psychosocial distress due to its visible appearance. Characterized by ectatic dermal capillaries, PWS lesions typically appear on the face but may occur elsewhere on the body. With age, the affected skin may become thickened, nodular, or pebbled, further impacting cosmetic outcomes and mental well-being. Despite various therapeutic approaches, most complete clearance remains challenging, emphasizing the need for objective imaging tools that can assess vascular architecture and treatment response in depth.

In a clinical pilot study, Cheng et al. [48, 49] demonstrated the feasibility of LED-based PAI as a point-of-care technique for clinical evaluation and photodynamic therapy (PDT) monitoring in PWS patients. The study enrolled 22 patients and divided them into two age groups: 3–6 years (Group 1) and above 6 years (Group

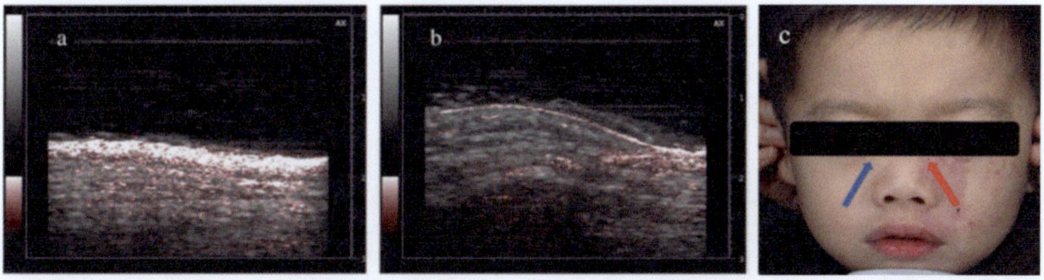

Fig. 2.16 LED-PA and ultrasound imaging of port-wine stain lesions. **a** Co-registered PA/US overlay image of a PWS region of interest (ROI). **b** Corresponding PA/US image of the contralateral normal skin region (HR). **c** Clinical photograph of a PWS patient with markings indicating the imaged regions. The photoacoustic images reveal enhanced subsurface vascular contrast within the lesion compared to surrounding skin, demonstrating the capability of LED-based PA imaging to map lesion depth and vascular density. Adapted with permission from [49]

2). Using a dual-mode LED-PA/ultrasound imaging system (AcousticX), the authors obtained cross-sectional maps of vascular structures within the PWS lesions and adjacent normal skin. The PA/ultrasound overlay images clearly revealed enhanced vascular contrast in the lesion area compared with normal tissue, with measurable differences in vascular density and depth distribution between the two age groups. These age-dependent variations corresponded well with established histopathological knowledge that older PWS lesions exhibit deeper and larger-caliber vasculature, often contributing to diminished treatment efficacy. The study thus provided the first clinical evidence that LED-based PA imaging can serve as a noninvasive, real-time tool for both diagnosis and treatment monitoring of PWS. Representative results are shown in Fig. 2.16, illustrating clear vascular contrast between affected and control regions.

2.5.5 *Monitoring Perfusion in Dialysis Patients*

Hemodialysis (HD) patients often experience recurrent drops in blood pressure (intradialytic hypotension), which can compromise peripheral perfusion and contribute to chronic limb ischemia, delayed wound healing, and increased amputation risk. Monitoring such hemodynamic fluctuations noninvasively remains a significant clinical challenge.

In a clinical study by Mantri et al. [50], an LED-based PA/US imaging system was employed to evaluate peripheral blood perfusion in 20 HD patients and 11 healthy controls. Imaging was performed on the plantar and heel regions of the feet before and after dialysis sessions in HD patients, and at rest and after mild exercise in controls. The study found a significant correlation between photoacoustic signal intensity and systolic blood pressure ($r = 0.70$ in controls; $r = 0.43$ in HD patients), demonstrating that PA intensity reflects real-time perfusion changes. However, HD patients exhibited a markedly reduced PA response post-dialysis ($p < 0.01$), suggesting impaired

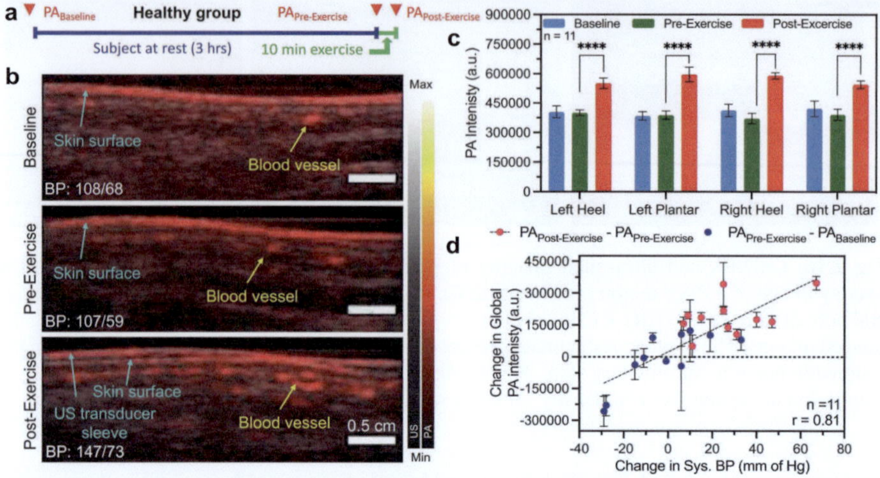

Fig. 2.17 Changes in peripheral tissue perfusion during hemodialysis monitored using LED-based photoacoustic imaging. **a** Photoacoustic and ultrasound overlay of the left heel showing reduced perfusion after dialysis when BP decreases. Scale bar: 0.5 cm. **b** HD patients were imaged before and after dialysis. **c** Global PA signal significantly decreased post-dialysis ($p < 0.01$). **d** Regional PA intensities decreased in all imaged areas. **e** A positive correlation between PA intensity and blood pressure ($r = 0.43$, $p < 0.01$) confirmed perfusion dependence. Reprinted with permission from [50]

vascular reactivity linked to hemodynamic instability. These results confirm that LED-based PA imaging can sensitively detect microvascular perfusion changes associated with blood pressure variations, enabling quantitative, bedside monitoring of peripheral vascular health in patients with end-stage renal disease. The portability and safety of LED systems further highlight their suitability for routine vascular surveillance in dialysis clinics. Figure 2.17 shows representative PA/US overlays of the heel region before and after dialysis, demonstrating reduced perfusion corresponding to blood pressure decline, along with quantitative correlations between PA intensity and hemodynamic changes.

2.5.6 *Translational 3D Peripheral Vascular Imaging*

Beyond established clinical studies, several translational demonstrations have explored the capabilities of LED-based PAI for visualizing microvascular networks in peripheral tissues. High-resolution mapping of dermal and subdermal vasculature is particularly relevant for disorders such as diabetes, peripheral arterial disease, and connective tissue disorders, where early vascular compromise precedes visible symptoms. Using compact LED arrays operating at near-infrared wavelength, volumetric imaging of extremities including the foot and fingers has been achieved with sub-millimeter resolution [51, 52]. The combination of LED excitation and real-time

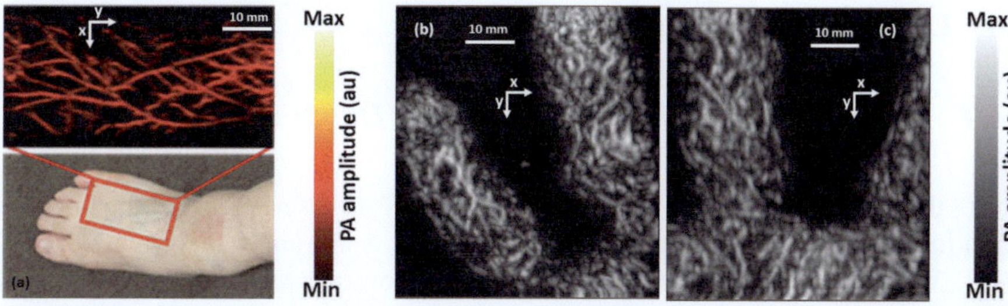

Fig. 2.18 Translational demonstration of 3D peripheral vascular imaging using LED-based photoacoustics. **a** Imaging site on the dorsum of the human foot with the corresponding 3D LED-based photoacoustic reconstruction rendered using an enhanced hot colormap. **b**, **c** Three-dimensional maximum-intensity-projection (MIP) images of human fingers acquired at different locations using the AcousticX system and displayed with a filtered gray colormap. Fine microvascular structures are clearly visualized, achieving an estimated spatial resolution of approximately 200 μm. Reprinted with permission from [47]

ultrasound detection enables detailed visualization of capillary-level vessels in three dimensions without the need for laser-classified light sources. These results highlight the translational potential of LED-PAI to serve as a noninvasive bedside imaging tool for assessing peripheral vascular health and microcirculatory impairment.

Figure 2.18 illustrates representative three-dimensional LED-based PAI results from human subjects. Panel (a) shows the imaging region on the dorsum of the human foot alongside a corresponding 3D photoacoustic reconstruction rendered using the enhanced hot colormap. Panels (b) and (c) display three-dimensional maximum-intensity-projection (MIP) images of human fingers obtained at different locations with the AcousticX system. Fine vascular structures are clearly visualized, with an estimated spatial resolution of ~200 μm, confirming the feasibility of LED-based imaging for peripheral microvascular assessment.

Summary of Clinical and Translational Applications

To date, LED-based PAI has been evaluated in more than 100 patients worldwide, spanning applications in vascular diagnostics, wound healing, arthritis assessment, and lymphatic surgery. These studies collectively demonstrate that compact, eye-safe LED systems can deliver clinically meaningful, quantitative imaging performance comparable to laser-based platforms, but at a fraction of the cost and complexity. Importantly, their portability and safety have enabled deployment in preoperative and resource-limited settings, expanding the accessibility of functional and vascular imaging beyond traditional hospital environments. Together, these clinical and translational efforts firmly establish LED-based PAI as a scalable, point-of-care technology poised for widespread clinical integration.

2.6 Latest Developments

2.6.1 Frequency-Matched LED–PAI for Depth Extension

One of the most persistent challenges in LED-based PAI has been its limited imaging depth in human tissue, typically restricted to about 1 cm under in vivo conditions [7–9]. This limitation primarily arises from the low per-pulse optical energy of LEDs, which is usually one to two orders of magnitude below that of laser sources. Consequently, the generated acoustic signals are weak at depth, and most clinical implementations have focused on superficial vasculature and lymphatic structures.

A recent study by Thomas et al. introduced a frequency-matching approach to extend the effective penetration depth of LED-PAI by optimizing both optical and acoustic parameters [53]. Specifically, the researchers investigated how modifying LED pulse width and matching it to the ultrasound probe's detection bandwidth could enhance sensitivity at depth. By increasing the pulse width from 30 to 100 ns, the LED arrays achieved a four-fold increase in pulse energy from approximately 118–498 μJ per array, while maintaining thermal stability within safe limits. This longer optical pulse generated lower-frequency acoustic components that were better suited for detection with low-frequency (5 MHz) ultrasound probes, thereby improving sensitivity and signal-to-noise ratio in deeper tissue layers.

Using dual 850 nm LED arrays integrated into an LED-PA/US platform, the study systematically compared 5, 7, and 10 MHz probes under matched conditions. The 5 MHz probe combined with 100 ns pulses showed the greatest overlap between the generated photoacoustic spectrum (3–6 MHz) and the detector sensitivity band, resulting in 8–10× higher PA signal amplitude at depth compared with higher-frequency probes. Phantom experiments confirmed detectable targets at 29 mm, while in vivo imaging visualized the carotid artery (~9 mm) and breast vasculature up to 18 mm beneath the skin surface. Despite the expected reduction in axial resolution (~470 μm axial, 700 μm lateral), the configuration delivered a clinically useful balance between depth and image clarity, expanding LED-PAI beyond the traditional mesoscopic regime.

Figure 2.19 shows representative time- and frequency-domain responses of photoacoustic signals from an absorber located 2.8 cm deep in an ex vivo chicken tissue phantom. The data highlight how varying LED pulse duration (30–100 ns) and ultrasound transducer frequency (5–10 MHz) influence signal amplitude and spectral bandwidth. Longer pulses substantially increase signal strength while maintaining good spectral overlap with clinical transducer bandwidths. Building on this optimization, Fig. 2.20 presents in vivo LED-based PAI of the human carotid and breast, demonstrating high-quality vascular visualization under realistic clinical conditions. The system clearly resolved carotid bifurcation vessels and achieved enhanced vascular contrast in breast tissue up to a depth of 1.8 cm, exceeding previously reported in vivo penetration depths for LED-based imaging. These findings underscore the potential of optimized LED excitation for extending imaging depth while maintaining clinical image quality for vascular assessment.

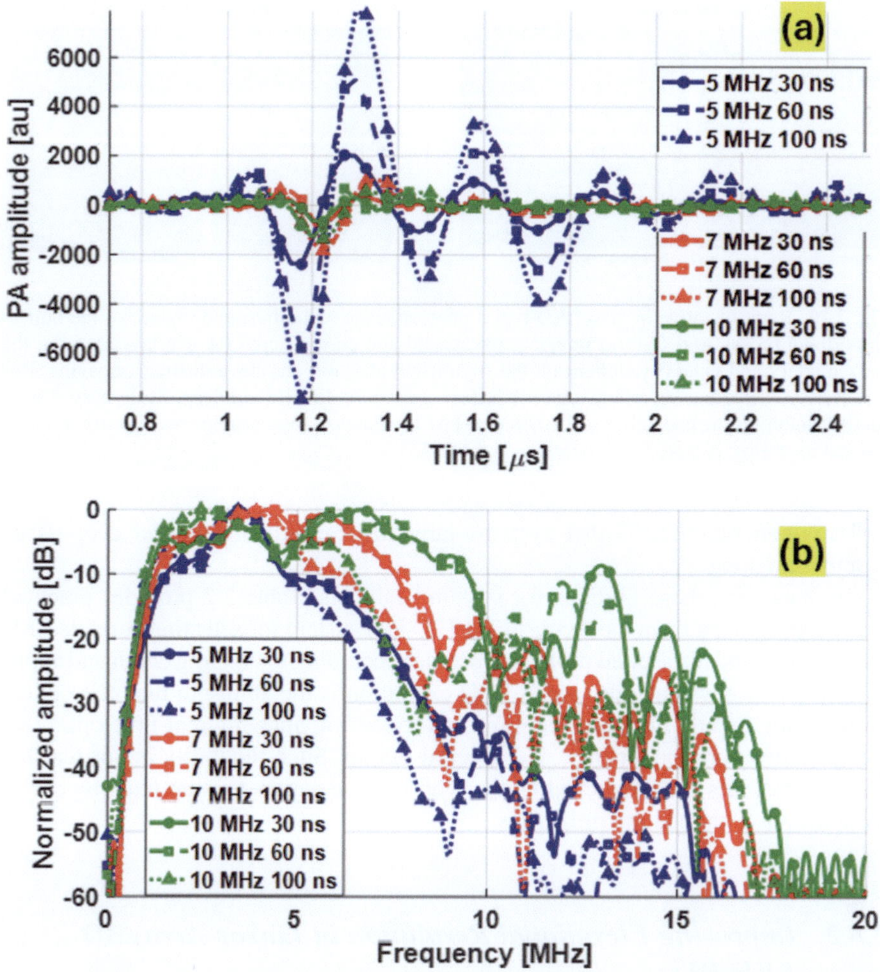

Fig. 2.19 Effect of pulse duration and transducer frequency on LED-based PA signal characteristics. **a** Time-domain photoacoustic waveforms recorded from an absorber positioned 2.8 cm deep within a chicken-tissue phantom, measured using transducers of 5, 7, and 10 MHz center frequencies. **b** Corresponding normalized frequency spectra showing improved amplitude and broader spectral response for longer LED pulse widths (60–100 ns). Results show that tuning LED pulse width to the transducer bandwidth enhances penetration and sensitivity. Adapted with permission from [53]

This work reframes the depth limitation of LED-PAI as a spectral-matching problem rather than solely an optical power constraint. By combining longer optical pulses within confinement limits with low-frequency, high-sensitivity ultrasound detection, it establishes a generalizable strategy for enhancing penetration while retaining LED safety and cost advantages. The approach offers a clear design pathway

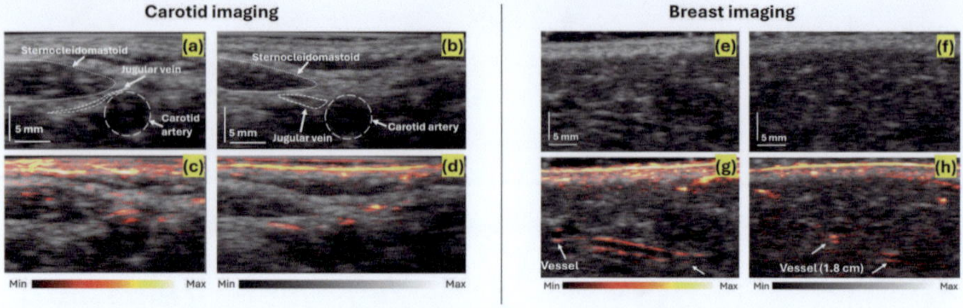

Fig. 2.20 Representative in vivo LED-based photoacoustic and ultrasound images of the human carotid and breast. **a–d** Carotid imaging: ultrasound and co-registered PA overlays showing the carotid artery and jugular vein beneath the sternocleidomastoid muscle. **e–h** Breast imaging: PA/US overlays revealing superficial and subsurface vessels up to ~1.8 cm deep. Scale bars: 5 mm. Images highlight the capability of LED-based PAI to visualize mesoscopic vascular structures in clinical scenarios. Adapted with permission from [53]

for next-generation LED-PAI systems targeting carotid, breast, and deep-tissue vascular imaging.

To generalize these findings for system designers, Table 2.2 provides practical guidelines derived from the AcousticX LED–PAI system for selecting suitable LED pulse widths and ultrasound probe frequencies according to imaging depth and application. A key advantage unique to LED excitation is the ability to tune the optical pulse width directly through electronic driver control, allowing real-time optimization of pulse duration, energy, and spectral output. Such flexibility is not feasible with laser diodes or Q-switched lasers, whose pulse parameters are fixed by cavity dynamics or gain medium properties.

2.6.2 *Improving Elevational Resolution in Linear-Array 3D LED-PAI*

While pulse-width optimization and spectral matching address sensitivity and penetration limits, spatial resolution—particularly in the elevational dimension—remains a critical bottleneck for linear-array-based LED-PA systems. The inherent out-of-plane blurring in conventional 2D acquisitions restricts accurate 3D vascular reconstruction, especially in mesoscopic human imaging where isotropic detail is essential for diagnostic interpretation. To overcome this limitation, Miao et al. developed a multiview LED-based photoacoustic computed tomography (MvLPACT) system using the AcousticX platform [54]. The approach acquires multiple angular views around the imaging target and reconstructs a 3D composite volume to enhance isotropy and reduce elevational artefacts. This strategy effectively improves axial–lateral–elevational balance while maintaining real-time imaging capability and compact form factor.

Table 2.2 Recommended LED pulse width and ultrasound probe configurations for depth-optimized PAI

Ultrasound probe (center frequency) (MHz)	Recommended LED pulse width (ns)	Typical optical pulse energy per array (μJ)	Expected imaging depth	Application domain	Key remarks
10	30–50	100–200	≤10 mm	Superficial microvasculature, dermatology	High axial resolution (~300 μm), shallow depth
7	60–80	150–250	10–20 mm	Musculoskeletal, lymphatic, small-animal	Balanced depth and resolution
5	90–100	450–500	15–30 mm (phantom), ≤20 mm (in vivo)	Carotid, breast, peripheral vasculature	Optimal depth–SNR match; frequency-matched illumination

Figure 2.21 presents representative in vivo results from the human palm obtained with the MvLPACT configuration. Compared to single-view linear-array PACT (SvLPACT), the multiview reconstruction delivers markedly sharper vascular delineation and continuity across depth. Quantitative analysis revealed a >60% improvement in full-width-at-half-maximum (FWHM) for representative vessels, confirming enhanced elevational resolution without compromising signal intensity. This work demonstrates that advanced acquisition geometries can complement LED excitation, paving the way for clinically relevant 3D imaging at significantly reduced system cost and complexity.

2.6.3 Hybrid Illumination Strategies for Enhanced Performance

Recent work by Periyasamy et al. [55] introduced a hybrid LED–pulsed laser diode (PLD) excitation approach integrated with the AcousticX system. The design interleaved LED and PLD pulses with microsecond-scale temporal delays (Fig. 2.22a), enabling rapid sequential acquisition from both illumination sources. As shown in Fig. 2.22b, this configuration achieved substantial improvement in temporal resolution, capturing dynamic flow at rates up to 1.2 mL/min in real time. Beyond faster frame rates, such hybrid designs also improve signal-to-noise ratio (SNR) and offer multispectral flexibility while maintaining the compactness, low cost, and safety

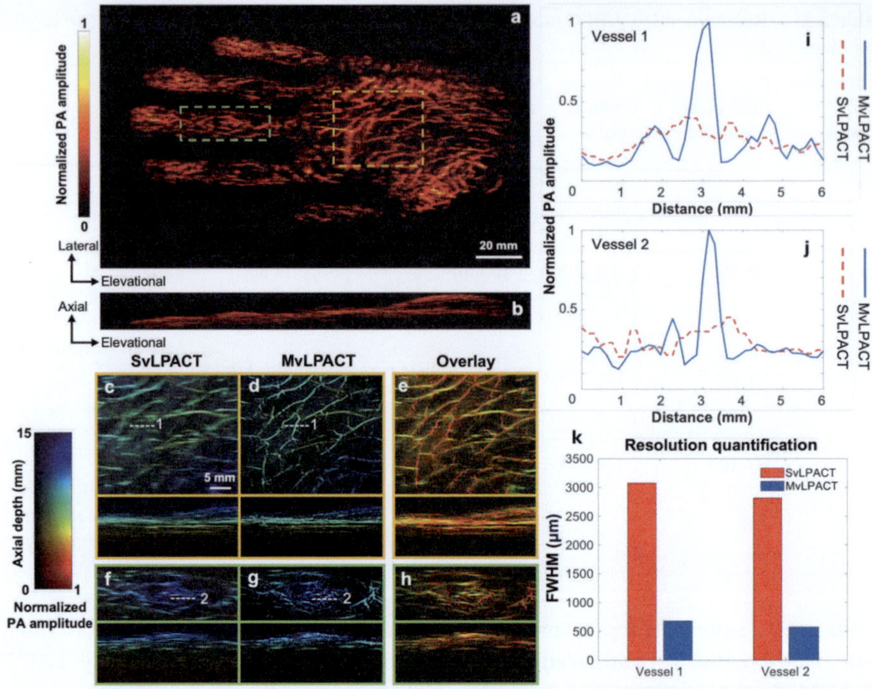

Fig. 2.21 Multiview linear-array LED-based photoacoustic computed tomography (MvLPACT) improves elevational resolution in in vivo human palm imaging. **a**–**b** Full-view maximum amplitude projections (MAPs) in lateral–elevational and axial–elevational planes. **c**–**h** Comparison of selected palm regions showing single-view (SvLPACT) and multiview reconstructions, with overlays highlighting enhanced vessel delineation. **i**–**j** Line profiles of two vessels demonstrate sharper peaks in MvLPACT. **k** Quantitative resolution analysis confirms a >60% improvement in full-width-at-half-maximum (FWHM) compared with single-view imaging. Reproduced with permission from [54]

advantages of LED-based platforms. This development illustrates how modest hardware augmentation, without resorting to expensive laser systems, can extend the functional reach of LED-PAI toward high-speed physiological imaging and quantitative spectroscopy.

2.6.4 LED-Based Photoacoustic Microscopy

While most LED-based photoacoustic systems are optimized for mesoscopic or tomographic imaging [7–9], exploratory studies have examined their potential for photoacoustic microscopy (PAM)—targeting high-resolution visualization of superficial microvasculature. Optical-resolution PAM (OR-PAM) typically requires ultrashort optical pulses (<10 ns) to preserve the broadband acoustic spectra necessary for

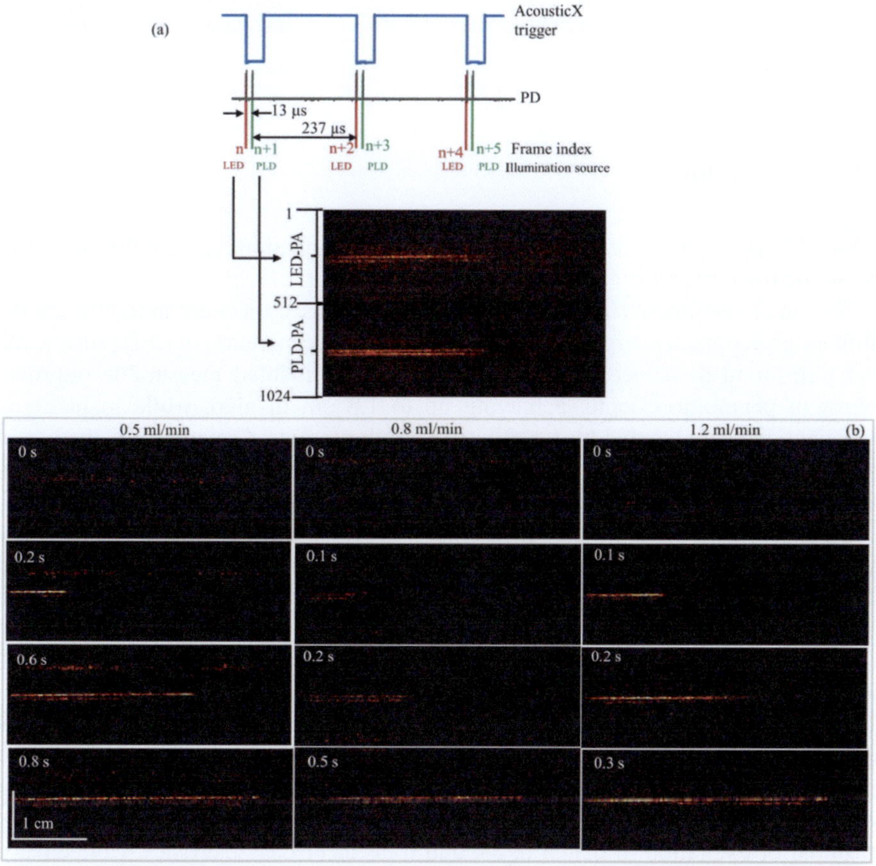

Fig. 2.22 Hybrid LED–PLD illumination enhances temporal resolution in flow imaging. **a** Interleaved LED and pulsed laser diode (PLD) excitations within the AcousticX system. **b** Sequential LED-PA frames during ink flow at different rates, demonstrating improved temporal resolution and motion tracking. This hybrid design also enhances SNR and multispectral flexibility while retaining LED affordability. Reproduced with permission from [55]

micron-scale resolution. In contrast, the longer pulse widths (50–200 ns) typical of LEDs limit the achievable axial resolution and reduce image sharpness, even though the available optical fluence is generally adequate for millimeter-scale penetration.

Despite this limitation, early feasibility studies demonstrated that overdriven visible LEDs operating at kilohertz repetition rates can generate sufficient photoacoustic contrast to resolve vascular networks in small-animal tissues such as the mouse ear [56]. These efforts confirm the viability of LED excitation for microimaging, motivating continued research into reducing pulse duration and improving driver synchronization. With emerging advances in nanosecond-scale GaN LED emitters and high-speed driver electronics, LED-based PAM could evolve into a compact, low-cost alternative for dermatologic and microvascular imaging, bridging the gap between benchtop laser microscopy and handheld photoacoustic tomography.

2.6.5 Summary

Table 2.3 summarize preclinical and clinical applications of LED-based PAI, including the most recent developments.

Recent developments in LED-based PAI have focused on overcoming traditional limitations in optical energy and spatial resolution. Optimization of LED pulse width and ultrasound transducer frequency matching has enabled measurable improvements in penetration depth, achieving up to 1.8 cm in vivo, while maintaining sufficient acoustic bandwidth for clinical vascular targets. Concurrently, the use of multi-view linear-array configurations has enhanced elevational resolution and volumetric uniformity, yielding near-isotropic 3D reconstructions of human vasculature. Complementing these architectural advances, hybrid illumination strategies that integrate LEDs with pulsed laser diodes have demonstrated substantial gains in temporal resolution, signal-to-noise ratio, and multispectral capability, all within compact and low-cost system designs.

Looking ahead, progress in the solid-state lighting industry, including multi-stack LED architectures, advanced thermal packaging, and high-efficiency phosphor formulations will further boost the brightness and lifetime of photoacoustic light sources. When combined with time-multiplexed multispectral arrays, these advances enable near-simultaneous functional imaging across multiple wavelengths. Emerging ideas such as coded excitation and adaptive pulse modulation promise dynamic optimization of excitation to match tissue and transducer characteristics. Collectively, these trends signal a shift toward smart, energy-efficient, and depth-adaptive LED-based photoacoustic systems, bridging laboratory innovation with scalable clinical and point-of-care applications.

2.7 Future Directions and Opportunities

2.7.1 Standardization and Benchmarking

As LED-based PAI progresses toward broader clinical adoption, establishing clear and reproducible performance standards becomes increasingly essential. While numerous proof-of-concept and application-driven studies have demonstrated feasibility, there remains considerable variability in reported parameters such as optical fluence, frame averaging, and signal-to-noise ratio (SNR). The absence of a unified benchmarking framework complicates direct comparison between systems and hinders the development of standardized safety and performance criteria.

Table 2.3 Preclinical and clinical applications of LED-PAI (Adapted with permission from [7, 47])

Target	Application		Depth (mm)	Contrast agent	Wavelength (nm)
Medical needles, vasculature	Guidance of minimally invasive procedures with peripheral tissue targets [37]	Phantom and ex vivo studies	38	N/A	850
Vasculature	Imaging of human placental vasculature [18]		7	N/A	850
Tumor	Imaging of intraocular tumors [36]		10	N/A	850
Vasculature	Non-invasive monitoring of angiogenesis [57]	Animal in vivo	10	N/A	850
Ulcer	Noninvasive imaging of pressure ulcers [58]		10	N/A	690
Oxygen saturation	Oxygen saturation imaging in Rheumatoid arthritis [14]		5	N/A	750/850
Molecular	Detection and monitoring of reactive oxygen and nitrogen species [59]		10	CyBA	850
Tumor/ contrast agents	Imaging of tumor using contrast enhancement [41]		10	NC	850
Cells/ contrast agents	Imaging of molecular-labelled cells [38]		10	DiR	850
Whole body	Tomographic imaging of organs [42]		5 mm	N/A	850
Abdomen	Liver imaging in fibrosis model (tomographic) [42]		10 mm	N/A	850
Vasculature	Imaging of peripheral microvasculature and function [17]	Healthy human in vivo	10	N/A	690/850
Vasculature	Simultaneous imaging of veins and lymphatic vessels [44]		10	ICG	940/820

(continued)

Table 2.3 (continued)

Target	Application		Depth (mm)	Contrast agent	Wavelength (nm)
Finger joints	Full view tomography of finger joints [12]		5	N/A	850
Carotid, breast	Frequency-matched imaging of deep-vasculature [53]		18	N/A	850
Finger joints	Imaging of inflammatory arthritis [46]	Patient in vivo	5	N/A	850
Skin	Imaging of port wine stain [48]		10	N/A	850
Lymphatic vessels	Lymphovenous anastomosis [19]		8	N/A	940/820
Vasculature	Wound healing monitoring [45]		5	N/A	850
Vasculature	Imaging perfusion in dialysis patients [50]		5	N/A	850

Early comparative works, such as those by Agrawal et al. and Zhu et al., provided initial evaluations of LED- and laser-based systems under controlled phantom conditions [13, 36]. Tissue-mimicking phantoms made of intralipid–agar or India-ink composites have been used to characterize SNR versus depth and spatial resolution, yet most of these experiments have remained single-site studies. More recently, the International Photoacoustic Standardization Consortium (IPASC) has coordinated multicenter phantom studies and open-data initiatives encompassing a broad range of photoacoustic systems, including commercial LED-based platforms such as AcousticX [60]. These collaborative efforts mark an important step toward harmonized benchmarking.

However, the currently available IPASC datasets include only limited system-level metadata, typically ultrasound probe type and illumination wavelength, while crucial LED-specific parameters such as pulse width, repetition rate, delivered fluence, and array geometry remain proprietary or undocumented. As a result, quantitative cross-platform comparisons and reproducibility analyses are still difficult. Future standardization initiatives should therefore prioritize transparent documentation of excitation conditions, optical coupling geometry, and acquisition protocols, enabling reproducible performance evaluation and equitable comparison between high- and low-fluence sources.

To facilitate this, LED-PAI researchers could adopt a concise reporting checklist encompassing:

- per-pulse optical energy and surface fluence, with measurement method
- pulse width, repetition rate, duty cycle, and averaging strategy

- array geometry, illumination coverage, and stand-off distance
- thermal stabilization and drift metrics
- transducer frequency/bandwidth and acoustic coupling details
- reconstruction settings and fluence-compensation method if any
- compliance with IEC 62,471 photobiological safety classification

Establishing such a consistent reporting framework would significantly improve transparency and comparability, laying the groundwork for future inter-laboratory validation and eventual clinical standardization of LED-based PAI.

2.7.2 Global Health Applications

LED-based PAI holds remarkable promise for global and resource-limited healthcare, driven by its affordability, intrinsic safety, and minimal infrastructure requirements. In contrast to laser-based systems that demand high-voltage power supplies, cooling mechanisms, and specialized safety infrastructure, LED-PAI platforms can operate using compact solid-state illumination and conventional ultrasound front-ends. This combination allows imaging to be performed safely in basic clinical environments, rural health centers, or mobile screening units without extensive technical training or facility modifications [7, 9, 36, 61].

In this context, portable and battery-powered LED-PAI probes represent an emerging step toward truly decentralized diagnostics. Early prototypes integrating LED arrays with handheld ultrasound probes have already demonstrated practicality for vascular and soft-tissue imaging [9, 17]. Recent laboratory-scale demonstrations and conference reports have described flexible LED array designs and preliminary battery-operated handheld systems, highlighting the engineering feasibility of wireless, untethered PAI probes for point-of-care use [61]. Although these developments remain at the prototype stage, they signify an important step toward future deployment in field and home-care environments where access to conventional laser-based systems is limited.

By uniting low cost, portability, and inherent eye safety, LED-PAI embodies the principles of frugal innovation and equitable access to diagnostic imaging. Its ability to visualize blood oxygenation, perfusion, and inflammatory changes without exogenous contrast agents positions it as an ideal candidate for population-scale health screening and long-term monitoring programs in developing regions [7, 9, 62, 63].

2.7.3 Integration with AI and Multimodal Imaging

Artificial intelligence (AI) and machine learning are increasingly transforming photoacoustic image reconstruction, interpretation, and system optimization. For

LED-PAI, where signal strength is inherently limited by lower optical fluence, AI-driven algorithms offer powerful tools to enhance SNR, suppress noise, and achieve real-time visualization. Deep learning frameworks such as convolutional neural networks (CNNs) and generative adversarial networks (GANs) have been successfully applied to address the challenges of limited-view, bandlimited, and low-SNR data common to LED-based systems [64–67].

Recent work by Francis et al. [68] demonstrated GAN-based photoacoustic image reconstruction capable of recovering high-quality images from bandlimited and limited-view datasets, a scenario highly representative of LED-PAI systems. Building on this, Paul and Mallidi [69] introduced DenP2P, a conditional GAN framework enabling real-time LED-based PAI by denoising and reconstructing data in a single step. These studies highlight how AI integration can directly compensate for the limited fluence and bandwidth in LED-driven systems, potentially matching or surpassing the perceptual quality of laser-based reconstructions.

Beyond signal recovery, AI algorithms are increasingly being used to fuse multimodal data from co-registered ultrasound (US) and photoacoustic (PA) channels, improving vascular segmentation, oxygenation mapping, and functional quantification [3, 70]. The integration of AI-based feature extraction with LED-PAI's inherent multispectral capability could eventually enable intelligent, low-cost diagnostic devices capable of automated tissue characterization and disease detection. Together, these trends suggest that the future of LED-based PAI will be strongly shaped by AI-assisted reconstruction and multimodal fusion, bridging the gap between compact hardware and intelligent clinical decision-making.

2.7.4 Regulatory Pathways

The regulatory trajectory for LED-PAI is uniquely favorable compared with conventional laser-driven systems. Because LEDs are categorized under *non-laser optical radiation sources* (IEC 62471) rather than high-risk laser classes (IEC 60,825–1), they are considered inherently eye-safe and skin-safe. This classification substantially reduces the regulatory burden typically associated with optical medical devices, which often require extensive safety validation and restricted operating environments. The low-risk status of LEDs also simplifies integration into handheld and portable systems, allowing device developers to focus on clinical performance and usability rather than complex safety controls or interlock mechanisms.

From a translational standpoint, the most practical pathway for clinical adoption lies in integration with ultrasound, a modality already established, regulated, and ubiquitous across healthcare systems worldwide. By coupling LED arrays with standard linear ultrasound probes, manufacturers can create dual-mode photoacoustic/ultrasound systems that operate within the existing safety, sterilization, and ergonomic frameworks of ultrasound imaging. Such hybrid configurations effectively introduce photoacoustic vascular and functional contrast without altering the device's base classification. Because the ultrasound front-end already complies with FDA

510(k) or CE-mark requirements, only the optical module and combined software components require supplementary evaluation. This hybrid strategy significantly accelerates translational approval timelines, particularly for diagnostic applications such as vascular assessment, perfusion monitoring, and intraoperative visualization.

Importantly, LEDs make this integration both technically and economically feasible. Their compact form factor, low cost, and minimal heat generation enable direct attachment to transducers or probe housings without compromising ergonomics or sterility. In contrast, integrating lasers into clinical scanners remains challenging due to eye-safety considerations, optical alignment requirements, and high cost. Consequently, LED-PAI holds a realistic path toward inclusion in routine point-of-care ultrasound platforms, potentially as a software-activated vascular imaging mode or "PA assist" function. As regulatory agencies continue to modernize frameworks for hybrid optical–acoustic systems, LED-based excitation stands out as the safest and most scalable route toward widespread clinical deployment of PAI.

Collectively, these future directions highlight the rapid convergence of safety, accessibility, and intelligent automation in LED-based PAI. Continued progress in performance benchmarking, low-cost handheld integration, AI-assisted reconstruction, and regulatory harmonization will determine how quickly LED-PAI transitions from pilot studies to routine clinical workflows. The unique safety classification and seamless compatibility with ultrasound platforms position LED illumination as the most practical route for scalable, affordable photoacoustic imaging—bridging laboratory innovation with global point-of-care healthcare delivery.

2.8 Summary and Outlook

Over the past decade, LEDs have transformed PAI from a laser-dependent laboratory technique into a safe, compact, and affordable clinical modality. The unique combination of eye-safe operation, low cost, and compatibility with standard ultrasound systems has positioned LED-based PAI as the most pragmatic pathway toward real-world adoption. Continuous advances in semiconductor fabrication, thermal management, and driver design have improved optical pulse energy and stability, enabling real-time imaging at clinically relevant depths.

Historically, the field progressed from early proof-of-concept demonstrations using single LED elements to fully integrated handheld and tomographic systems capable of vascular, lymphatic, and functional imaging. Preclinical and clinical studies have validated LED-based PAI for diverse applications from monitoring wound healing and rheumatoid arthritis to guiding microsurgery and assessing peripheral perfusion. These investigations have collectively demonstrated that LED-PAI can deliver diagnostic information comparable to laser-based systems within the mesoscopic regime, while offering a fraction of the cost and complexity.

Recent developments continue to expand the modality's capabilities. Optimization of LED pulse parameters for specific ultrasound probes has extended penetration depth, while innovations in multi-view acquisition and pulse modulation have improved spatial resolution and signal-to-noise ratio. Hybrid designs combining LEDs with laser diodes, and the emergence of portable or battery-operated handheld probes, point toward a new generation of autonomous, AI-enabled imaging systems suited for global health deployment.

In summary, LED-based PAI now represents a mature, clinically relevant technology with unique advantages in safety, portability, and affordability. Its integration with ultrasound platforms provides a natural bridge between optical and acoustic imaging, fostering the transition from experimental research to everyday clinical practice. Continued collaboration across optics, electronics, and medicine will be key to realizing the full translational potential of LED-based photoacoustics in precision and point-of-care healthcare.

Key Takeaways

- **LEDs have democratized photoacoustic imaging**, transforming it from a laser-based research technique into a compact, safe, and affordable clinical modality.
- **Advances in semiconductor and driver technology** have increased optical pulse energy, thermal stability, and emission consistency, enabling real-time imaging at clinically relevant depths.
- **Clinical validation is well established**—LED-PAI has demonstrated diagnostic performance comparable to laser systems for vascular, lymphatic, rheumatologic, and perfusion monitoring applications.
- **System integration with standard ultrasound platforms** has made hybrid LED-PAI/US scanners practical, cost-efficient, and easily deployable in point-of-care environments.
- **Ongoing innovations**, including optimized pulse shaping, multi-view acquisition, and hybrid LED/LD architectures, continue to improve depth, resolution, and signal-to-noise ratio.
- **Future LED-PAI systems will be portable and intelligent**, combining battery-powered probes with embedded AI processing, expanding access to precision and global health imaging.

References

1. M. Xu, L.V. Wang, Photoacoustic imaging in biomedicine. Rev. Sci. Instrum. **77**, 041101 (2006)
2. P. Beard, Biomedical photoacoustic imaging. Interface Focus **1**, 602–631 (2011)
3. A.B.E. Attia, G. Balasundaram, M. Moothanchery et al., A review of clinical photoacoustic imaging: current and future trends. Photoacoustics **16**, 100144 (2019)
4. I. Steinberg, D.M. Huland, O. Vermesh et al., Photoacoustic clinical imaging. Photoacoustics **14**, 77–98 (2019)
5. L.V. Wang, J. Yao, A practical guide to photoacoustic tomography in the life sciences. Nat. Methods **13**, 627–638 (2016)
6. S. Manohar, M. Dantuma, Current and future trends in photoacoustic breast imaging. Photoacoustics **16**, 100134 (2019)

7. Y. Zhu, T. Feng, Q. Cheng, X. Wang, S. Du, N. Sato, J. Yuan, M.K.A. Singh, Towards clinical translation of LED-based photoacoustic imaging: a review. Sensors **20**(9), 2484 (2020)
8. T.J. Allen, High-power light emitting diodes; an alternative excitation source for photoacoustic tomography, in *LED-Based Photoacoustic Imaging. Progress in Optical Science and Photonics*, vol. 7, ed. by M. Kuniyil Ajith Singh (Springer, Singapore, 2020)
9. M.K.A. Singh, *LED-Based Photoacoustic Imaging. Progress in Optical Science and Photonics*, vol. 7 (Springer, Singapore, 2020)
10. M.K.A. Singh, T. Agano, N. Sato, Y. Shigeta, T. Uemura, Real-time in vivo imaging of human lymphatic system using an LED-based photoacoustic/ultrasound imaging system, in *Proceedings of SPIE Photons Plus Ultrasound: Imaging and Sensing*, vol. 10494 (2018), p. 1049404
11. M.K.A. Singh, Y. Shigeta, T. Hanaoka, T. Agano, N. Sato, High-speed photoacoustic imaging using an LED-based photoacoustic imaging system, in *Proceedings of SPIE Photons Plus Ultrasound: Imaging and Sensing*, vol. 10494 (2018), p. 104943N
12. K.J. Francis, Y.E. Boink, M. Dantuma, M.K.A. Singh, S. Manohar, W. Steenbergen, Tomographic imaging with an ultrasound and LED-based photoacoustic system. Biomed. Opt. Express **11**, 2152–2165 (2020)
13. S. Agrawal, M.K.A. Singh, K. Johnstonbaugh, D.C. Han, C.R. Pameijer, S.-R. Kothapalli, Photoacoustic imaging of human vasculature using LED versus laser illumination: a comparison study on tissue phantoms and in vivo humans. Sensors **21**, 424 (2021)
14. K.J. Francis, M. Xavierselvan, M.K.A. Singh, S. Mallidi, C. Van Der Laken, F. Van De Loo, W. Steenbergen, LED-based photoacoustic imaging for early detection of joint inflammation in rodents: towards achieving 3Rs in rheumatoid arthritis research, in *Proceedings of SPIE Photons Plus Ultrasound: Imaging and Sensing*, vol. 11240 (2020), p. 112400M
15. S.L. Jacques, Optical properties of biological tissues: a review. Phys. Med. Biol. **58**, R37–R61 (2013)
16. R. Bulsink, M.K.A. Singh, M. Xavierselvan, S. Mallidi, W. Steenbergen, K.J. Francis, Oxygen saturation imaging using LED-based photoacoustic system. Sensors **21**, 283 (2021)
17. W. Xia, E. Maneas, N.T. Huynh, M.K.A. Singh, N.M. Brown, S. Ourselin, E. Gilbert-Kawai, S.J. West, A.E. Desjardins, Imaging of human peripheral blood vessels during cuff occlusion with a compact LED-based photoacoustic and ultrasound system, in *Proceedings of SPIE Photons Plus Ultrasound: Imaging and Sensing*, vol. 10878 (2019), p. 1087804
18. E. Maneas, R. Aughwane, N. Huynh, W. Xia, R. Ansari, M.K.A. Singh, J.C. Hutchinson, N.J. Sebire, O.J. Arthurs, J. Deprest et al., Photoacoustic imaging of the human placental vasculature. J. Biophotonics **13**, e201900167 (2020)
19. S. Van Heumen, J.J.M. Riksen, M.K.A. Singh, G. Van Soest, D. Vasilic, LED-based photoacoustic imaging for preoperative visualization of lymphatic vessels in patients with secondary limb lymphedema. Photoacoustics **29**, 100446 (2022)
20. International Electrotechnical Commission (IEC), *IEC 60825-1: Safety of Laser Products—Part 1: Equipment Classification and Requirements* (IEC, Geneva, 2014)
21. S. Agrawal, C. Fadden, A. Dangi, X. Yang, H. Albahrani, N. Frings, S. Heidari Zadi, S.-R. Kothapalli, Light-emitting-diode-based multispectral photoacoustic computed tomography system. Sensors **19**(22), 4861 (2019)
22. H. Zhong, T. Duan, H. Lan, M. Zhou, F. Gao, Review of low-cost photoacoustic sensing and imaging based on laser diode and light-emitting diode. Sensors **18**(7), 2264 (2018)
23. X. Liu, S.K. Kalva, B. Lafci, D. Nozdriukhin, X.L. Deán-Ben, D. Razansky, Full-view LED-based optoacoustic tomography. Photoacoustics **31**, 100521 (2023)
24. R.S. Hansen, Using high-power light emitting diodes for photoacoustic imaging. Proc. SPIE 7968 (Medical Imaging 2011: Ultrasonic Imaging, Tomography, and Therapy) (2011)
25. T. Ikuta, R. Shimizu, The dynamic response of magnetic domain walls to applied fields in yttrium orthoferrite, observed by a stroboscopic technique. J. Phys. D Appl. Phys. **7**, 726 (1974)
26. W. McFarlane, An inexpensive nanosecond light pulser for use in photomultiplier system testing. Rev. Sci. Instrum. **45**, 286–289 (1974)

27. T. Araki, H. Misawa, Light emitting diode-based nanosecond ultraviolet light source for fluorescence lifetime measurements. Rev. Sci. Instrum. **66**, 5469–5472 (1995)
28. T. Araki, Y. Fujisawa, M. Hashimoto, An ultraviolet nanosecond light pulse generator using a light emitting diode for test of photodetectors. Rev. Sci. Instrum. **68**, 1365–1368 (1997)
29. A. Chaney, R. Sundararajan, Simple MOSFET-based high-voltage nanosecond pulse circuit. IEEE Trans. Plasma Sci. **32**, 1919–1924 (2004)
30. T.J. Allen, P.C. Beard, Light emitting diodes as an excitation source for biomedical photoacoustics, in *Photons Plus Ultrasound: Imaging and Sensing*, vol. 8581 (2013)
31. T.J. Allen, P.C. Beard, High power visible light emitting diodes as pulsed excitation sources for biomedical photoacoustics. Biomed. Opt. Express **7**, 1260–1270 (2016)
32. C. Willert, B. Stasicki, J. Klinner, S. Moessner, Pulsed operation of high-power light emitting diodes for imaging flow velocimetry. Meas. Sci. Technol. **21**, 075402 (2010)
33. T. Agano, N. Sato, H. Nakatsuka, K. Kitagawa, T. Hanaoka, K. Morisono, Y. Shigeta, Attempts to increase penetration of photoacoustic system using LED array light source, in *Photons Plus Ultrasound: Imaging and Sensing*, vol. 9323 (2015)
34. T. Agano, N. Sato, H. Nakatsuka, K. Kitagawa, T. Hanaoka, K. Morisono, Y. Shigeta, Comparative experiments of photoacoustic system using laser light source and LED array light source, in *Photons Plus Ultrasound: Imaging and Sensing*, vol. 9323 (2015)
35. T. Agano, N. Sato, Photoacoustic imaging system using LED light source, in *Proceedings of the Conference on Lasers and Electro-Optics (CLEO)* (San Jose, CA, USA, 2016), pp. 1–2
36. Y. Zhu, G. Xu, J. Yuan, J. Jo, G. Gandikota, H. Demirci, T. Agano, N. Sato, Y. Shigeta, X. Wang, Light emitting diodes based photoacoustic imaging and potential clinical applications. Sci. Rep. **8**, 1–12 (2018)
37. W. Xia, M. Kuniyil Ajith Singh, E. Maneas, N. Sato, Y. Shigeta, T. Agano, S. Ourselin, S.J. West, A.E. Desjardins, Handheld real-time LED-based photoacoustic and ultrasound imaging system for accurate visualization of clinical metal needles and superficial vasculature to guide minimally invasive procedures. Sensors **18**, 1394 (2018)
38. A. Hariri, J. Lemaster, J. Wang, A.S. Jeevarathinam, D.L. Chao, J.V. Jokerst, The characterization of an economic and portable LED-based photoacoustic imaging system to facilitate molecular imaging. Photoacoustics **9**, 10–20 (2018)
39. J. Joseph, M.K.A. Singh, N. Sato, S.E. Bohndiek, Technical validation studies of a dual-wavelength LED-based photoacoustic and ultrasound imaging system. Photoacoustics **22**, 100267 (2021)
40. M. Xavierselvan, M.K.A. Singh, S. Mallidi, In vivo tumor vascular imaging with light emitting diode-based photoacoustic imaging system. Sensors **20**(16), 4503 (2020)
41. M. Xavierselvan, S. Mallidi, LED-based functional photoacoustics—portable and affordable solution for preclinical cancer imaging, in *LED-Based Photoacoustic Imaging: From Bench to Bedside*, ed. by M. Kuniyil Ajith Singh (Springer Singapore: Singapore, 2020), pp. 303–319
42. K.J. Francis, R. Booijink, R. Bansal, W. Steenbergen, Tomographic ultrasound and LED-based photoacoustic system for preclinical imaging. Sensors **20**(10), 2793 (2020)
43. S.K. Kalva, X. Liu, X.L. Deán-Ben, L. Tang, D. Razansky, LED-based optoacoustic tomography of mice. J. Biomed. Opt. **30**(4), 040501 (2025)
44. M.K.A. Singh, N. Sato, Y. Shigeta, T. Agano et al., Guidance of lymphaticovenous anastomosis using LED-based photoacoustic lymphangiography, in *Proceedings of SPIE Photons Plus Ultrasound: Imaging and Sensing*, vol. 11960 (2022), p. 119600R
45. Y. Mantri, J. Tsujimoto, B. Donovan et al., Photoacoustic monitoring of angiogenesis predicts response to therapy in healing wounds. Wound Repair Regen **30**(2), 258–267 (2022)
46. J. Jo, G. Xu, Y. Zhu, M. Burton, J. Sarazin, E. Schiopu, G. Gandikota, X. Wang, Detecting joint inflammation by an LED-based photoacoustic imaging system: a feasibility study. J. Biomed. Opt. **23**(11), 110501 (2018)
47. M.K.A. Singh, N. Sato, F. Ichihashi, W. Xia, Y. Sankai, The evolution of LED-based photoacoustic imaging: from labs to clinics, in *Biomedical Photoacoustics: Technology and Applications*, ed. by W. Xia, 1st edn. (Springer Nature, 2024), pp. 573–608

48. H. Zhang, Q. Cheng, T. Feng et al., A novel light-emitting diode–based photoacoustic evaluation method for port-wine stain and its clinical trial (Conference Presentation), in *Photons Plus Ultrasound: Imaging and Sensing. International Society for Optics and Photonics* (2019)
49. Q. Cheng, T. Feng, H. Zhang, Y. Zhu, J. Yuan, X. Wang, Diagnosis and treatment monitoring of port-wine stain using LED-based photoacoustics: theoretical aspects and first in-human clinical pilot study, in *LED-Based Photoacoustic Imaging: From Bench to Bedside* (Springer, Singapore, 2020), pp. 351–377
50. Y. Mantri, T.R. Dorobek, J. Tsujimoto, W.F. Penny, P.S. Garimella, J.V. Jokerst, Monitoring peripheral hemodynamic response to changes in blood pressure via photoacoustic imaging. Photoacoustics **26**, 100345 (2022)
51. M.K.A. Singh, N. Sato, F. Ichihashi, Y. Sankai, Improving 3D imaging capability of LED-based photoacoustic imaging for high-resolution human vascular imaging applications, in *Proceedings of SPIE Photons Plus Ultrasound: Imaging and Sensing*, vol. 12379 (2023)
52. M.K.A. Singh, N. Sato, F. Ichihashi, Y. Sankai, LED-based photoacoustic imaging: a point-of-care solution for microvascular health assessment, in *Proceedings of SPIE Photons Plus Ultrasound: Imaging and Sensing*, vol. 12842 (2024)
53. A. Thomas, M. Kuniyil Ajith Singh, N. Sato, F. Kalloor Joseph, Extending the depth of LED-based photoacoustic imaging for carotid and breast applications (medRxiv Preprint 2025.11.13.25339983). medRxiv (2025)
54. J. Miao, S. Zhang, G. Sun, L. Li, Multiview linear-array photoacoustic computed tomography with improved elevational resolution in 3D. Biomed. Opt. Express **16**, 4051–4062 (2025)
55. V. Periyasamy, A. Das, M. Pramanik, Temporally delayed, dual-pulse, dual-wavelength photoacoustic imaging (TD-DP-DW-PAI): advancing capabilities of a commercial light-emitting diode-based photoacoustic imaging system with a pulsed laser diode. Biomed. Opt. Express **16**, 2945–2958 (2025)
56. X. Dai, H. Yang, H. Jiang, In vivo photoacoustic imaging of vasculature with a low-cost miniature light-emitting diode excitation. Opt. Lett. **42**, 1456–1459 (2017)
57. Y. Zhu et al., LED-based photoacoustic imaging for monitoring angiogenesis in fibrin scaffolds. Tissue Eng. Part C Methods **25**(9), 523–531 (2019)
58. A. Hariri et al., Noninvasive staging of pressure ulcers using photoacoustic imaging. Wound Repair Regen. **27**(5), 488–496 (2019)
59. A. Hariri et al., Molecular imaging of oxidative stress using an LED-based photoacoustic imaging system. Sci. Rep. **9**(1), 1–10 (2019)
60. International Photoacoustic Standardisation Consortium (IPASC). Standardization initiatives and benchmarking datasets for quantitative photoacoustic imaging. Available at: https://ipasc.science
61. A. Das, M. Pramanik, Performance characterization of light-emitting diode array-based photoacoustic computed tomography (LED-PACT). Biomed. Opt. Express **16**, 3632–3647 (2025)
62. M. Kuniyil Ajith Singh, N. Sato, F. Ichihashi, Y. Sankai, In vivo demonstration of real-time oxygen saturation imaging using a portable and affordable LED-based multispectral photoacoustic and ultrasound imaging system, in *Proceedings of SPIE*, vol. 10878 (SPIE, 2019)
63. E. Maneas, W. Xia, M. Kuniyil Ajith Singh et al., Human placental vasculature imaging using an LED-based photoacoustic/ultrasound imaging system, in *Proceedings of SPIE*, vol. 10494 (SPIE, 2018), 104940Y
64. H. Lan, T. Duan, Z. Wang, F. Gao, Deep-learning-based limited-view photoacoustic reconstruction. Photoacoustics **18**, 100160 (2020)
65. R. Manwar, S.-R. Kothapalli, M.A. Anastasio, Deep learning in photoacoustic tomography. IEEE Trans. Med. Imaging **40**(3), 811–825 (2021)
66. M. Moothanchery, A.B.E. Attia, M. Pramanik, Deep-learning-enhanced photoacoustic imaging: recent trends and challenges. Photoacoustics **27**, 100374 (2022)
67. M. Kuniyil Ajith Singh, K. Sivasubramanian, N. Sato, F. Ichihashi, Y. Sankai, L. Xing, Deep learning-enhanced LED-based photoacoustic imaging, in *Photons Plus Ultrasound: Imaging and Sensing*, vol. 11240 (2020), p. 1124038

68. K.J. Francis, A. Arora, P. Kancharla, M.K.A. Singh, W. Steenbergen, S.S. Channappayya, Generative adversarial network-based photoacoustic image reconstruction from bandlimited and limited-view data, in *Proceedings of SPIE 11642, Photons Plus Ultrasound: Imaging and Sensing* (2021), p. 1164235
69. A. Paul, S. Mallidi, Facilitating real-time LED-based photoacoustic imaging with DenP2P: an optimized conditional generative adversarial deep learning solution. AIP Adv. **15**(5), 055227 (2025)
70. L. Hervé, B.T. Cox, S.R. Arridge, Multimodal photoacoustic and ultrasound image fusion using machine learning. Biomed. Opt. Express **13**(8), 4171–4187 (2022)

Chapter 3
Laser Diode-Based Photoacoustic Imaging

Learning objectives

After completing this chapter, the reader will be able to:

- **Explain the operating principles and technical characteristics** of pulsed laser diodes used in photoacoustic imaging.
- **Compare Laser diode-based excitation with LED and conventional laser sources**, focusing on differences in pulse energy, wavelength coverage, and system-level performance.
- **Identify the key elements of laser diode driver circuitry** and understand how they influence pulse width, repetition rate, and energy efficiency.
- **Recognize recent advances in laser diode packaging and integration**, enabling compact, portable, and handheld PAI systems.
- **Review preclinical and early clinical applications** of laser diode-based photoacoustic imaging in vascular, musculoskeletal, and dermatological imaging.

3.1 Introduction

Photoacoustic imaging (PAI) is a hybrid biomedical imaging modality that combines optical excitation with ultrasonic detection to provide high-resolution visualization of optical absorption within biological tissues. Upon absorption of a short light pulse, localized thermoelastic expansion generates broadband ultrasonic waves that are detected by an ultrasound transducer and reconstructed into a spatial map of tissue optical absorption. This approach enables label-free imaging of endogenous chromophores such as hemoglobin, melanin, and lipids, providing valuable functional and molecular information about vascular and metabolic activity in vivo [1–5].

Traditionally, PAI systems have relied on high-energy, nanosecond-pulsed laser sources such as Q-switched Nd:YAG and optical parametric oscillator (OPO) systems. While these lasers provide excellent wavelength tunability, high peak power,

M. Kuniyil Ajith Singh, *Photoacoustic Imaging Using Affordable Resources*, Progress in Optical Science and Photonics 40, https://doi.org/10.1007/978-981-95-7175-8_3

and short pulse duration ideal for satisfying stress and thermal confinement criteria, they are expensive, bulky, and require complex maintenance and cooling systems. In addition, their low pulse repetition rate (PRR), typically 10–50 Hz, limits frame averaging and hinders real-time imaging of dynamic physiological processes [6–9]. These limitations have motivated the exploration of compact, affordable semiconductor-based excitation sources such as light-emitting diodes (LEDs) and laser diodes (LDs) [10–13].

Both LEDs and laser diode (LD) convert electrical energy directly into light, offering notable energy efficiency, small form factor, and fast electrical modulation. LEDs, with their broad emission spectra and inherent safety, have already demonstrated promising results for shallow, high-frame-rate imaging applications [11, 13]. However, their relatively low pulse energy restricts penetration depth to the mesoscopic regime. Laser diodes, by contrast, bridge the performance gap between LEDs and conventional solid-state lasers. They emit coherent light with narrower spectral bandwidths and higher per-pulse energies (up to the millijoule range in stacked architectures), enabling deeper imaging and more efficient optical coupling into tissue [14–17].

Pulsed LDs, operating in the near-infrared (NIR) region (typically 750–900 nm), can achieve repetition rates up to several kilohertz while maintaining nanosecond pulse durations. These attributes allow extensive frame averaging without sacrificing temporal resolution, significantly improving the signal-to-noise ratio (SNR) and image quality. Recent advancements in driver electronics, diode stacking, and beam-combining optics have further enhanced the optical output and spatial uniformity of LD systems, paving the way for real-time, handheld, and even tomographic implementations. Integration of LDs within dual-modality ultrasound (US) probes has also been demonstrated [18–21], enabling simultaneous PA and US imaging within a single, compact housing [14–21].

Despite these advantages, LD-based PAI faces several technical challenges. The emission wavelengths of most high-power LDs lie within the NIR window, where the optical absorption of hemoglobin is weaker compared with the visible spectrum, reducing sensitivity to microvasculature [17]. Furthermore, the inherent beam divergence of LDs (typically ~10° (slow axis) to ~40° (fast axis)) necessitates beam-shaping optics such as cylindrical lenses or diffractive optical elements to ensure uniform illumination across the region of interest [15]. The pulse energy of even high-performance LD stacks remains lower than that of conventional solid-state lasers, constraining deep-tissue imaging; nonetheless, increasing PRR, optimizing illumination geometry, and improving driver efficiency continue to close this gap [14–21]. Table 3.1 compares different light sources in PAI in terms of pulse energy, repetition rate, pulse width, cost and various advantages and drawbacks.

Overall, LD-based PAI represents a versatile and practical compromise between LED-driven and laser-driven approaches. It combines the cost-effectiveness, portability, and high repetition rate of LEDs with the coherence, directionality, and energy efficiency of lasers. With demonstrated success in both preclinical and early clinical studies, including imaging of joints, vasculature, brain, and liver, LD-based systems continue to gain traction as a compact and affordable light source for real-time

Table 3.1 Comparison of different light sources used in PAI. Reproduced with permission from [11]

Light source	Energy (mJ)	PRR (Hz)	Pulse width (ns)	Cost (K)[a]	Advantages	Disadvantages
Laser	5–120	10–100	< 10	\$70–200	Powerful, ~5 cm penetration depth, tunable wavelength	Bulky size, eye protection and laser safe rooms needed
LD	0.5–2.5	~1–6 K	30–200	~\$10–25	Integration in a handheld probe feasible, high PRR, wide wavelength range	Limited penetration depth, eye protection and laser safe rooms needed, wavelength tuning not possible
LED	0.2	~200–16 K	30–100	\$10–15	Integration in a handheld probe feasible, high PRR, no need of laser-safe rooms and eye-safety goggles	Limited penetration depth, wavelength tuning not possible

[a] Cost includes the driving electronics and may vary based on different features, number of wavelengths etc. Integration to a US probe may also involve extra development cost

functional and molecular imaging [22–29]. This chapter will discuss the underlying technology, performance optimization strategies, and translational applications of LD-based PAI, highlighting its emerging role as a bridge between benchtop research systems and point-of-care clinical imaging.

3.2 Principles of Laser Diode Operation

LDs are electrically pumped semiconductor lasers that convert electrical energy directly into coherent optical radiation. The device consists of a p–i–n (positive–intrinsic–negative) junction structure where stimulated emission occurs within the active region sandwiched between cladding layers of lower refractive index. When a forward bias is applied, electrons and holes are injected into the intrinsic layer, where they recombine radiatively to produce photons. If the carrier injection exceeds the threshold current density, optical gain surpasses internal losses, and stimulated emission dominates spontaneous emission, leading to laser action [9–12]. A pair of cleaved, reflective facets at the ends of the semiconductor chip forms a Fabry–Pérot optical cavity, allowing light to oscillate back and forth to achieve amplification. The resulting laser output is highly directional, with a typical spectral linewidth of 1–3 nm and a well-defined emission mode. The optical power output is proportional to the input current above threshold, and the overall electrical-to-optical conversion efficiency of LDs is substantially higher (20–40%) than that of conventional flashlamp-pumped solid-state lasers, which are typically limited to <5% [14, 15].

Compared to traditional pulsed lasers such as Q-switched Nd:YAG or optical parametric oscillator (OPO) systems, LDs offer several practical advantages—compactness, robustness, low power consumption, and cost efficiency—making them suitable for portable and point-of-care PAI applications [13, 16]. However, due to their semiconductor nature, LDs exhibit asymmetric beam divergence (fast and slow axes) and limited tunability, which must be compensated by external optics or the use of diode stacks to achieve uniform illumination [15]. LDs can be operated in continuous-wave (CW) or pulsed mode. CW operation is widely used in optical communication, sensing, and industrial applications but is unsuitable for PAI because long exposure times violate the stress and thermal confinement conditions required to generate broadband photoacoustic signals [1, 5]. Pulsed LDs, by contrast, deliver nanosecond-scale optical pulses (typically 30–200 ns) that confine the thermal expansion to microscopic regions, producing efficient acoustic wave generation [13, 15, 16]. The short-pulse excitation ensures that heat does not diffuse significantly during the laser pulse, preserving high temporal fidelity and enabling quantitative imaging of chromophore distributions.

Modern LDs achieve high peak powers through improvements in epitaxial design, thermal management, and diode stacking, a technique in which multiple emitters are aligned and driven synchronously to increase total optical output while maintaining compact form factor [15–21]. Combined with advanced driver electronics capable of operating at kilohertz pulse repetition rates, these systems enable real-time, high-frame-rate imaging and integration into handheld or tomographic probes.

In summary, LDs combine the efficiency of solid-state lasers with the portability and affordability of semiconductor devices, offering a practical bridge between high-performance laboratory lasers and low-cost LED arrays. These characteristics make LDs particularly well suited for deep-tissue, high-speed PAI, where maintaining both penetration and temporal resolution is essential [22–29].

3.3 Technical Aspects of Pulsed Laser Diodes

3.3.1 Emission Wavelengths

LDs are available across a broad spectral range extending from the visible (≈400 nm) to the near-infrared (NIR, ≈1000 nm) region. However, high-power pulsed LDs used in PAI are predominantly limited to the NIR window (≈750–905 nm), where optical absorption by blood and tissue scattering are both relatively low, allowing improved photon penetration depth and reduced attenuation losses [9–12]. This spectral region coincides with the so-called "biological optical window," which balances tissue transparency and sufficient hemoglobin absorption to enable high-contrast vascular imaging. While the NIR range facilitates deeper imaging, it does not fully exploit the stronger optical absorption of hemoglobin in the visible spectrum, which would be advantageous for high-resolution microvascular applications. Continuous-wave LDs are widely available across visible and NIR wavelengths, but their pulsed counterparts, which are necessary for PAI to satisfy stress and thermal confinement, remain commercially restricted to a few near-infrared wavelengths such as 750, 803, 808, and 905 nm [10–12, 14, 15]. These emission lines coincide with spectral absorption peaks of oxy- and deoxy-hemoglobin, making them suitable for multispectral functional imaging such as oxygen saturation (sO_2) estimation [14, 15, 27].

Recent advances in semiconductor fabrication and epitaxial layer engineering have broadened the accessible wavelength range of pulsed LDs and improved their optical efficiency [16–19]. Multi-junction and diode-stack configurations have increased the achievable pulse energy while maintaining compact form factors. For instance, Canal et al. [14] demonstrated a portable multiwavelength LD source integrating 750 and 808 nm emitters for handheld PAI, while Daoudi et al. [15] employed an 805 nm LD in a dual-modality ultrasound-photoacoustic probe for in vivo vascular imaging. More recently, Liu et al. [19] presented a handheld embedded LD-based optoacoustic system operating near 850 nm capable of real-time 3D angiography of deep tissue, underscoring the progress toward clinical-grade performance.

Although wavelength tunability in LDs remains limited compared with OPO or dye-laser systems, multispectral operation can be achieved by multiplexing multiple diode stacks at distinct wavelengths [16, 19–21]. This approach has enabled functional imaging of oxygenation and hemoglobin distribution, and, when combined with advanced reconstruction and spectral unmixing algorithms, promises compact and cost-effective solutions for quantitative biomedical imaging.

3.3.2 Drivers and Electronics

A pulsed LD driver is a critical component that governs the temporal and energetic characteristics of the emitted light. Its main role is to deliver short, high-current electrical pulses, often tens of amperes in amplitude and tens of nanoseconds in width, while maintaining stability and protecting the LD from electrical overstress [10–12, 14, 16]. Two major architectures are typically employed: capacitor-discharge and direct-current modulation. In the capacitor-discharge design, a capacitor is charged to a predefined voltage and rapidly discharged through the LD using a fast switching transistor, enabling high peak current and precise control of pulse duration [12, 14, 28–30]. This approach is favored in PAI because of its excellent repeatability and controllable pulse shape. In contrast, direct-current modulation, where the driving current is switched directly through the diode, offers simplicity and very high repetition rates but lacks flexibility for pulse-width control. A typical LD driver circuit is shown in Fig. 3.1 and the basic idea of driving is similar to that of an LED driver [11].

A major design challenge is minimizing parasitic inductance and capacitance in the driver and interconnects—typically denoted as Lr, Lc, and Ld—as these limit the current rise time and hence optical output efficiency [16, 17]. Compact layouts and low-inductance PCB traces are therefore crucial. Canal et al. demonstrated one of the most efficient LD drivers for biomedical PAI, achieving 1.7 mJ per pulse with a pulse width of 40 ns and repetition rate up to 10 kHz, all within a 40 × 50 mm^2 board [14]. This compact design (Fig. 3.2) integrated the LD and its driver on a single board, paving the way toward handheld and portable PAI probes.

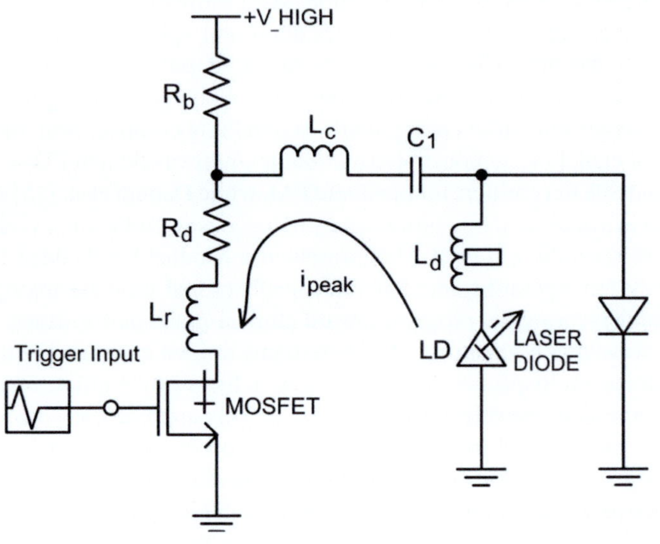

Fig. 3.1 Schematic of the laser diode driver circuit used for photoacoustic imaging

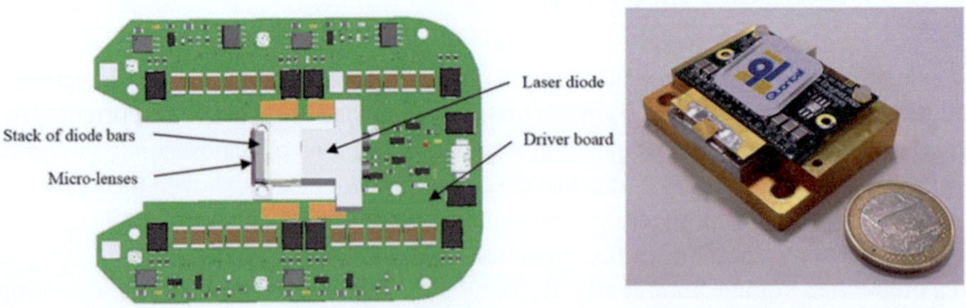

Fig. 3.2 Schematic diagram of the diode laser source showing implementation of the laser diode within the driver board (left). Ultra-short pulse laser source integrating a mini-laser diode providing 1.7 mJ in pulses as short as 40 ns from Quantel, France (right). Adapted with permission from [14]

Although the pulse-width tuning range of LD drivers remains narrower than that of LED systems, ongoing electronic design improvements are extending the flexibility of pulse generation. Further integration of the driver and LD module within the probe housing, together with thermoelectric cooling and impedance-matched interconnects, will continue to enhance portability and energy efficiency for clinical use [14–17, 29, 31].

3.3.3 Optical Power and Pulse Repetition Rate

Typically, pulsed LDs deliver pulse energies ranging from hundreds of nanojoules to a few microjoules per pulse, depending on the emitter architecture and driver configuration. Although these values remain below those of Q-switched or OPO-based lasers, their high PRR—often between 1 and 10 kHz—allows extensive frame averaging to compensate for reduced per-pulse energy while maintaining near-real-time imaging capability [12, 14, 16, 28]. The development of stacked and bar-array LD modules has substantially increased achievable optical power. Canal et al. [14] reported multi-wavelength diode stacks producing up to 1.7 mJ per pulse with pulse widths around 40 ns at 10 kHz PRR, enabling centimeter-scale tissue imaging. These stacking and beam-combining techniques have since been miniaturized and refined for handheld applications. Figure 3.3 shows the schematic comparing diode bars with single and multiple active regions.

A recent milestone was achieved by Liu et al. [19], who introduced a handheld embedded laser-diode optoacoustic system integrating multiple high-power LD arrays (808–940 nm) with compact driver electronics and real-time reconstruction hardware. Their prototype achieved three-dimensional angiographic imaging of human vasculature up to ~2 cm deep at >10 Hz frame rates, demonstrating

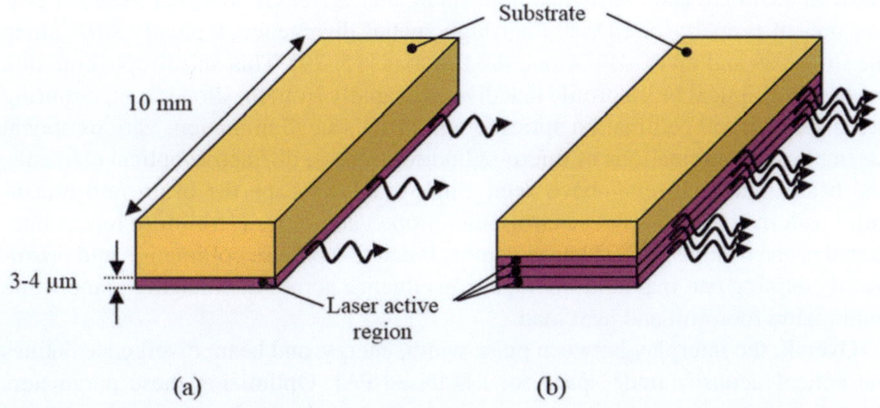

Fig. 3.3 Schematic diagram of **a** a standard diode bar and **b** a multiple active region diode bar. Reproduced with permission from [14]

that modern LD architectures can deliver clinically relevant performance while maintaining low size, weight, and power requirements.

These advances underscore how increasing optical output through stacking, thermal management, and synchronized multi-diode driving continues to bridge the gap between semiconductor and solid-state laser excitation, paving the way for fully portable, real-time LD-PAI systems.

3.3.4 Pulse Width and Divergence

Pulse duration plays a critical role in determining both the frequency content and amplitude of the generated photoacoustic signals. To satisfy the stress confinement condition, the optical pulse must be shorter than the characteristic thermal and stress relaxation times of the tissue. While Q-switched lasers typically produce 5–10 ns pulses, commercially available pulsed LDs exhibit pulse widths in the range of 30–200 ns, depending on the driver circuitry and diode structure [12, 16, 17]. Such longer pulses yield narrower acoustic bandwidths, which are nevertheless well matched to the detection range (5–10 MHz) of clinical ultrasound probes used for deep-tissue imaging [10, 12]. For superficial or microscopic applications where higher spatial resolution is required, shorter pulses remain advantageous to generate broader-band acoustic spectra.

A practical trade-off exists between pulse width and optical energy: reducing pulse duration generally decreases the total energy per pulse because of driver current limitations. Canal et al. [14] demonstrated that 40 ns pulses can deliver up to 1.7 mJ per pulse when efficiently driven at 10 kHz, while other systems operating at 100–150 ns pulse widths achieve moderate energy (hundreds of μJ) suitable for handheld or tomographic applications. Although pulse-width tunability is routine in LED sources, such control has not yet been widely realized for LDs because of the tight coupling between electrical and thermal parameters in their driver circuits [12]. Another critical optical property of LDs is their high spatial divergence, typically ~10° along the slow axis and up to ~40° along the fast axis [15, 29]. This anisotropic emission creates an elliptical beam profile that diverges rapidly from the diode facet, requiring beam-shaping and collimation optics for uniform tissue illumination. Various optical assemblies—combinations of micro-cylindrical lenses, diffractive optical elements, and beam homogenizers—have been employed to reshape the beam into rectangular patterns matching linear ultrasound probe geometries [15]. More recent integrated probe designs [19–21] have demonstrated that precise collimation and prism-based coupling can maintain homogeneous fluence across the imaging plane while minimizing footprint and heat load.

Overall, the interplay between pulse width, energy, and beam divergence defines the optical–acoustic trade space for LD-based PAI. Optimizing these parameters through compact beam-shaping optics and efficient driver design remains central to achieving both high-resolution and deep-tissue performance in practical LD-PAI systems.

3.3.5 *Summary*

Pulsed LDs combine high repetition rates (kHz range) and moderate pulse energies (up to mJ) with tunable pulse widths (30–200 ns), allowing fine control over stress confinement and penetration depth. Beam divergence and non-uniform illumination remain technical challenges, but beam-shaping optics and diode stacking approaches have substantially improved illumination homogeneity. Modern driver circuits, high-speed gating, and efficient thermal management now enable robust, long-term operation for both benchtop and handheld systems.

3.4 Historical and Technological Developments

3.4.1 *Early Feasibility and Proof-of-Concept Studies (2005–2006)*

LD excitation for photoacoustics emerged in the mid-2000s as a compact, electronically driven alternative to bulky, low-PRR solid-state lasers. Proof-of-concept demonstrations by Allen and Beard showed that high-peak-power pulsed LDs at 905 nm can generate robust PA signals and that pulse duration governs the acoustic bandwidth under stress-confinement limits [17, 29]. Soon after, the first in-vivo feasibility was reported by Kolkman et al. (2006), who imaged superficial vessels in a human volunteer using a 905 nm LD module ($\approx$112 ns, 5 kHz), establishing clinical plausibility albeit with limited scanning speed and millimeter-scale depth [27].

3.4.2 *Integration of LDs with Handheld Dual-Modality Probes (2014–2016)*

The next major inflection point was handheld probe integration. Within a European consortium, Daoudi et al. (2014) packaged an 805 nm LD, beam-shaping optics (FAC/SAC and DOE), thermal management, and a 7 MHz linear array into a single dual-modality LD/US probe, enabling real-time co-registered imaging of finger joints and vascular phantoms at clinical frame rates [15]. In parallel, driver and source engineering advanced—Quantel's compact, on-board LD driver and stacked diode arrays achieved ~40 ns pulses at up to 1.7 mJ and $\approx$10 kHz PRR [14], proving that high pulse energy and high PRR could coexist in a palm-sized module. This milestone formed the foundation for modern probe-level implementations and multiwavelength LD stacks that later supported functional imaging.

3.4.3 Advances in Bench-Top Handheld and Tomographic Systems (2015–2020)

Concurrently, benchtop PLD-PAT systems matured. Upputuri and Pramanik (2015) characterized a low-cost, high-speed 803 nm setup (~1.4 mJ, 7 kHz) capable of tomographic imaging every few seconds and achieving ~2 cm phantom depth [16]. Follow-up studies extended to small-animal brain and venous-sinus applications [24, 28], demonstrating that kilohertz-rate PRR allows extensive frame averaging without compromising temporal resolution. Same group also reported LD-based PAI with a clinical US system and demonstrated its feasibility in phantom and small animal imaging in 2016 [31]. Receive-side enhancements, such as deformation-compensated averaging, later enabled handheld LD/US platforms to image the human carotid artery at ~1–1.5 cm depth despite motion, further validating real-time clinical feasibility [15, 22–25].

3.4.4 Recent Developments: Embedded and High-Performance LD Systems (2023–2025)

In the latest generation of LD-based systems, two main technological directions are apparent. First, overdriven LD optoacoustic microscopy has pushed diode operation into high-current, low-duty regimes, achieving sub-micrometer resolution for microscopic imaging without sacrificing compactness [18]. Second, embedded handheld LD systems now integrate light sources, driver electronics, and data acquisition directly into a single portable housing. Liu et al. (2025) demonstrated such an embedded laser-diode-based 3D angiography platform, performing real-time volumetric imaging of deep vascular structures [19]. Building on previous developments, Mitra et al. (2024) employed Quantel laser diode stacks integrated with the AcousticX DAQ and probe to perform raster-scanned 3D imaging, achieving fine vascular resolution in preclinical models [20]. Similarly, Thomas et al. (2025) demonstrated a multi-stack Quantel LD implementation with the same AcousticX acquisition backend for real-time imaging of the carotid and femoral arteries, showing the system's potential for breast and peripheral vascular imaging [21]. Together, these works exemplify a decisive shift toward compact, clinically oriented LD-PAI devices capable of delivering high-frame-rate, multi-centimeter-depth imaging using commercial ultrasound hardware.

3.4.5 *Summary and Outlook*

Figure 3.4 shows the detail timeline of the technical developments in LD-PAI. Overall, it has evolved from early benchtop feasibility studies into real-time handheld and tomographic imaging systems validated in both preclinical and human pilot investigations. Over the past two decades, developments in compact diode stacks, integrated driver electronics, and combined ultrasound–photoacoustic probe designs have pushed LD-based systems closer to clinical practicality. Despite these advances, it is noteworthy that no commercially available LD-based photoacoustic system currently exists, largely due to the complexity of optical packaging, safety compliance for coherent sources, and limited market demand relative to LED-based systems. Nevertheless, the maturity of semiconductor laser technology and the increasing adoption of hybrid US–PA architectures suggest that LD-based illumination could soon bridge the gap between laboratory prototypes and deployable clinical tools. The field is now poised for widespread translational exploration, particularly in vascular, musculoskeletal, and breast imaging through integration with affordable ultrasound electronics, semiconductor-driven light modules, and standardized imaging platforms.

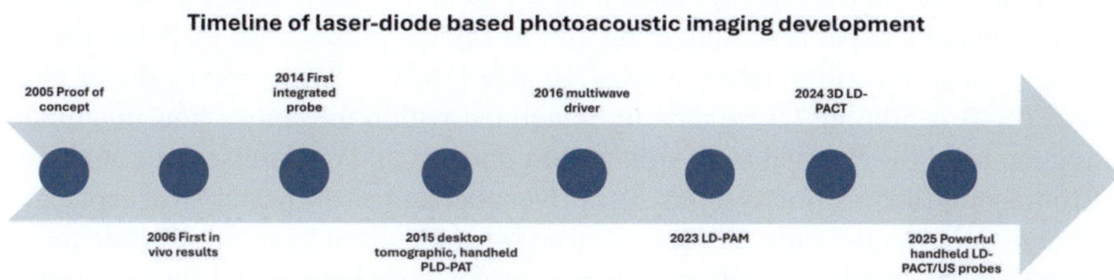

Fig. 3.4 Timeline of technological and translational milestones in laser-diode-based photoacoustic (LD-PA) imaging development. Key milestones in LD-PAI evolution from proof-of-concept to integrated handheld systems. Initial demonstrations of LD generated PA signals were reported by Allen and Beard [29] and followed by in vivo imaging by Kolkman et al. [27]. The first integrated LD/ultrasound probe was realized in the FULLPHASE project [15], and compact multiwavelength LD drivers were introduced by Canal et al. [14]. Concurrent and subsequent advances include benchtop LD-PAT systems (Upputuri and Pramanik [16]), handheld LD-PA system using clinical ultrasound system (Sivasubramanian and Pramanik [31]), overdriven LD-PAM (Seeger et al. [18]), and recent 3D and handheld LD-PACT/US platforms (Mitra et al. [20]; Thomas et al. [21]; Liu et al. [19])

3.5 System Architectures and Integration Efforts

3.5.1 Fully Integrated Handheld LD-PA/US Probe (FULLPHASE: Quantel + ESAOTE)

The first truly handheld LD-PAI probe was realized within the European FP7 FULL-PHASE project, which aimed to transition photoacoustics from benchtop prototypes to clinically compatible, dual-modality US and PA systems (Fig. 3.5a, b). The project brought together industry and academic partners including ESAOTE Europe, Quantel France, Slios France, and others, focusing on the development of a compact, multi-wavelength LD excitation unit integrated directly with a clinical linear-array US transducer for the first time [15]. The consortium's objective was to design a robust, portable, and laser-safe system suitable for clinical translation in vascular, rheumatologic, and oncologic imaging.

3.5.1.1 Optical Source and Driver

The illumination module incorporated a high-power 805 nm pulsed LD stack developed by Quantel, capable of delivering 130 ns optical pulses with ~0.5–0.6 mJ per pulse at a pulse-repetition rate (PRR) of 10 kHz [14, 15]. The compact driver electronics ($\approx$40 × 50 mm^2) were directly integrated within the source housing, eliminating the need for external high-voltage components and enabling synchronization with ultrasound acquisition systems. The driver operated on the principle of capacitor-discharge pulse modulation, designed for high-peak current generation while maintaining thermal stability at kHz-rate operation. Such co-located driving electronics marked a key engineering advance that allowed probe-level miniaturization.

3.5.1.2 Optical and Thermal Design

The handheld probe achieved efficient beam shaping through a combination of fast- and slow-axis collimation (FAC/SAC) lenses and diffractive optical elements (DOE) that homogenized the LD's inherently divergent beam. The final illumination field was a rectangular light sheet (~18 × 2.3 mm^2) that matched the active aperture of the 7 MHz linear-array transducer, ensuring uniform fluence across the imaging plane. To maintain optical and electronic stability during sustained high-PRR operation, the module incorporated an aluminum cooling rim and passive thermal management [15]. The entire assembly including LD stack, optics, cooling, and US transducer was enclosed in a single ergonomic casing, forming one of the first integrated LD/US probes (Fig. 3.5c).

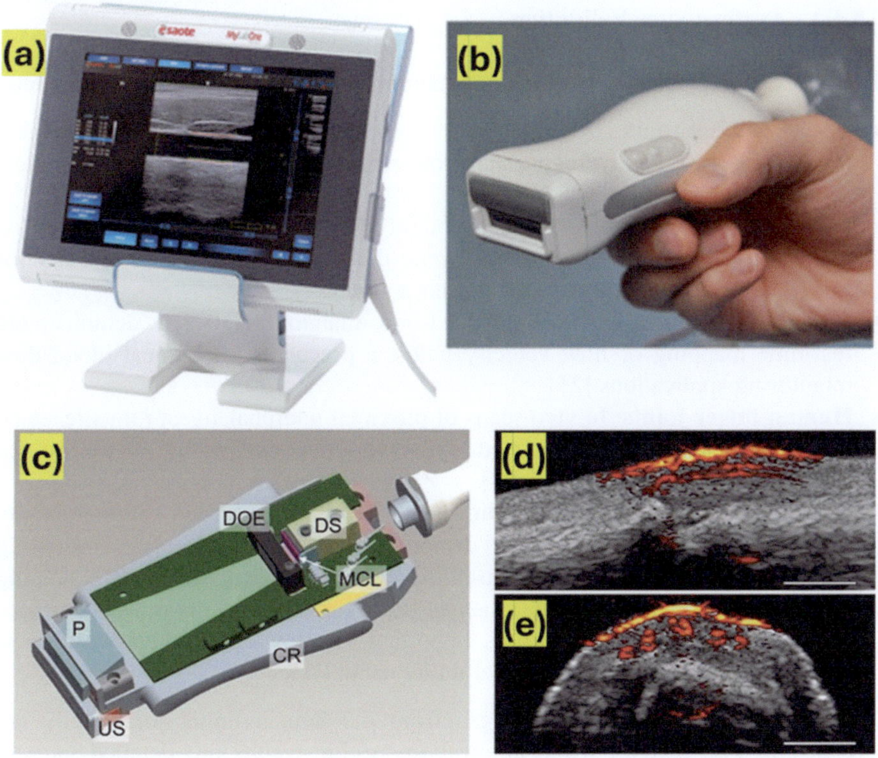

Fig. 3.5 Integrated laser-diode (LD)-based handheld photoacoustic and ultrasound (PA/US) imaging system. **a** Portable ultrasound scanner (MyLabOne, ESAOTE Europe) used as the acquisition platform. **b** Photograph of the dual-modality handheld probe integrating the LD module and linear-array transducer. **c** Schematic of the optical layout showing the deflecting prism (P), diffractive optical elements (DOE), diode stack (DS), micro-cylindrical lenses (MCL), and aluminum cooling rim (CR) incorporated within the probe housing. **d**, **e** Co-registered PA/US images of the human proximal interphalangeal joint acquired in sagittal and transverse planes, respectively. This prototype represented one of the earliest integrated LD–US systems capable of real-time handheld imaging. Adapted with permission from Daoudi et al. [15]

3.5.1.3 Data Acquisition and Real-Time Operation

The LD source was synchronized with ESAOTE's clinical-grade ultrasound acquisition unit, allowing simultaneous acquisition of B-mode ultrasound and photoacoustic frames. The 10 kHz PRR permitted substantial frame averaging (typically 500–1000 pulses) without loss of temporal resolution, resulting in real-time composite imaging at ~20 Hz. Such frame rates were sufficient to accommodate hand-induced motion, breathing artefacts, and dynamic blood flow, marking a significant milestone toward real-time PA imaging with semiconductor sources [15, 25].

3.5.1.4 Performance Validation

The system was evaluated in a range of phantom, ex vivo, and in vivo experiments:

- **Vascular phantoms**: Quantitative studies demonstrated SNR improvement through multi-frame averaging and flow sensitivity consistent with theoretical acoustic models [25].
- **Liver fibrosis (preclinical)**: Dual-modality PA/US imaging detected fibrosis-related changes in perfused rodent livers, illustrating the system's capacity for functional and structural tissue characterization [22].
- **Vascular dynamics**: Handheld pulsed-flow imaging at 10 kHz demonstrated real-time mapping of flow velocity fields, a precursor for clinical blood-flow monitoring applications [25].
- **Human finger joints**: In vivo scans of proximal interphalangeal joints revealed microvascular features associated with synovitis, confirming feasibility for rheumatologic diagnostics [15, 23].
- **Intraoperative clinical validation**: In a pilot study on carotid endarterectomy patients ($n = 16$), Müller et al. [32] demonstrated motion-compensated averaging that improved in-vivo PA signal-to-noise ratio by approximately 5 dB, enabling differentiation of hemorrhagic plaques from surrounding blood. This intraoperative feasibility study marks an important step toward non-invasive clinical assessment of carotid plaque vulnerability using PAI.

Representative system-level images from the human finger joint study are provided in Fig. 5.1d–e, while application-specific results are detailed later in Sect. 3.7.

3.5.1.5 Safety Considerations

As a coherent NIR source, the LD system adhered to ANSI MPE limits for skin exposure. The illumination geometry (rectangular spot) and PRR were selected to maintain per-pulse fluence below 1.3 mJ cm^{-2}, ensuring safe operation without exceeding permissible exposure during handheld scanning. Eye-protection measures remained mandatory [15].

3.5.1.6 Broader Impact and Legacy

The FULLPHASE handheld probe represented a technological inflection point demonstrating that laser-class pulse energy and ultrasound-compatible frame rates could be achieved from a diode source in a palm-sized format. It established the design principles: beam homogenization, on-board driving, and thermal stability that underpin all modern LD-based PA systems. The same architecture directly influenced later benchtop and portable systems, including the Quantel + AcousticX platforms described in Sect. 3.5.3.

Subsequent generations built upon this foundation: Upputuri and Pramanik developed a low-cost PLD-PAT system achieving ~2 cm imaging depth with ~1.4 mJ pulses at 7 kHz [16]; Rajendran et al. extended this to in vivo small-animal brain and venous-sinus imaging [24]; and Jaeger et al. implemented deformation-compensated averaging in FULLPHASE system to reduce motion clutter in handheld operation [25] and visualize carotid artery in real-time. More recently, Mitra et al. (2024) integrated Quantel LD stacks with the AcousticX DAQ for raster-scanned 3D microvascular imaging, and Thomas et al. (2025) used the same configuration for real-time carotid and femoral artery imaging, highlighting potential for carotid characterization and breast vascular applications [20, 21]. In parallel, Liu et al. (2025) presented a fully embedded handheld system with onboard light source, drivers, and DAQ achieving real-time 3D angiography [19].

Despite its strong technical foundation and successful in vivo demonstrations, the FULLPHASE handheld LD/US probe and its successor CVENT probe ultimately did not transition into a commercial product. This outcome can be partly attributed to inherent design compromises arising from the shared use of conventional ultrasound probes and acquisition electronics originally optimized for pulse-echo sonography rather than photoacoustic detection. In particular, standard clinical probes provide limited acoustic sensitivity and low front-end amplification, constraining the detection of weak PA signals at depth. Moreover, at the time, large ultrasound manufacturers such as ESAOTE remained cautious about extensive hardware modifications for a still-emerging imaging modality, limiting flexibility to implement PA-specific optimizations such as light pulse width and acoustic bandwidth matching or customized preamplifier designs.

Expert Perspective: PA-Specific Design as the Foundation for Affordable Systems

"Our experience in the FULLPHASE and CVENT projects showed that handheld photoacoustic imaging with integrated light sources is technically feasible, being the result of dedicated engineering efforts of industrial parties with expertise in photonics, high frequency high power electronics and ultrasound imaging. Handheld photoacoustic systems cannot simply be adapted from conventional ultrasound hardware. Front-ends designed for strong pulse-echo signals often lack the sensitivity and flexibility required for photoacoustic responses which are weak and polluted by clutter. Moving toward truly affordable and practical systems demands PA-specific design choices and sustained industrial commitment to tailored optics–acoustics integration and custom amplification. With such dedicated development, compact and accessible PA imaging can become a clinical reality."

—Prof. Wiendelt Steenbergen, Chair, Biomedical Photonic Imaging Group, University of Twente, Netherlands

Subsequent systems, such as the AcousticX platform, demonstrated how targeted engineering of ultrasound components could substantially enhance PA performance. The use of high-gain amplification (>100 dB) and specialized probes incorporating white acoustic lenses to suppress superficial reverberations enabled markedly improved SNR and deeper penetration. Notably, Thomas et al. [21] achieved superior imaging depth using Quantel laser diodes similar to those in the FULLPHASE prototype, underscoring that system-level integration, particularly probe acoustics and receiver electronics is as critical as the excitation source itself. These developments highlight that successful clinical translation of LD-based PAI hinges on co-designed ultrasound and optical subsystems optimized specifically for photoacoustic detection rather than retrofitted from conventional ultrasound hardware.

3.5.2 High Speed LD-Based Photoacoustic Imaging Systems

The next phase in LED-PAI development aimed at achieving real-time imaging speeds comparable to ultrasound, without compromising imaging depth or field of view. Upputuri and Pramanik (2015) first demonstrated the feasibility of tomographic imaging using a PLD source at 803 nm (Quantel, France) with a pulse energy of ~1.4 mJ, 7 kHz PRR, and ~130 ns pulse width [16]. Using a circular-scanning geometry and a single-element ultrasound detector, they obtained cross-sectional reconstructions of phantoms at depths up to 20 mm, showing that high-PRR LD excitation could deliver sufficient SNR for centimeter-scale imaging. This work established a baseline for high-speed, low-cost PLD-PAT instrumentation.

Building on this, Sivasubramanian and Pramanik (2016) reported a major step forward in imaging speed by integrating the same class of Quantel pulsed laser diodes (λ = 803 nm, 1.4 mJ, 7 kHz PRR) with a clinical ultrasound platform (E-CUBE 12R, Alpinion Medical Systems, South Korea) (Fig. 3.6) [31]. The system was synchronized through a custom trigger interface that allowed simultaneous acquisition of 128-channel photoacoustic data using a linear array transducer (L3-12, 128 elements, 5–12 MHz bandwidth). With optimized acquisition and buffering pipelines, they achieved photoacoustic frame rates up to 7000 frames per second, marking one of the fastest demonstrations of real-time PAI reported to date. Experimental validation on flow phantoms and ex-vivo tissue showed that even at kilohertz-scale acquisition rates, signal fidelity and image quality remained stable. The high PRR of the LD source allowed frame averaging and temporal compounding, improving SNR by ~9 dB without compromising temporal resolution. The imaging depth achieved was around 1.5–2 cm, comparable to laser-based systems, while the total system footprint remained compact and affordable.

This work proved that LD excitation could be seamlessly combined with existing clinical ultrasound hardware, requiring no modification to the transducer or DAQ electronics. The integration of a commercial scanner backend demonstrated the translational feasibility of LD-based photoacoustic imaging for dynamic physiological and flow imaging applications, where high frame rate and synchronization with

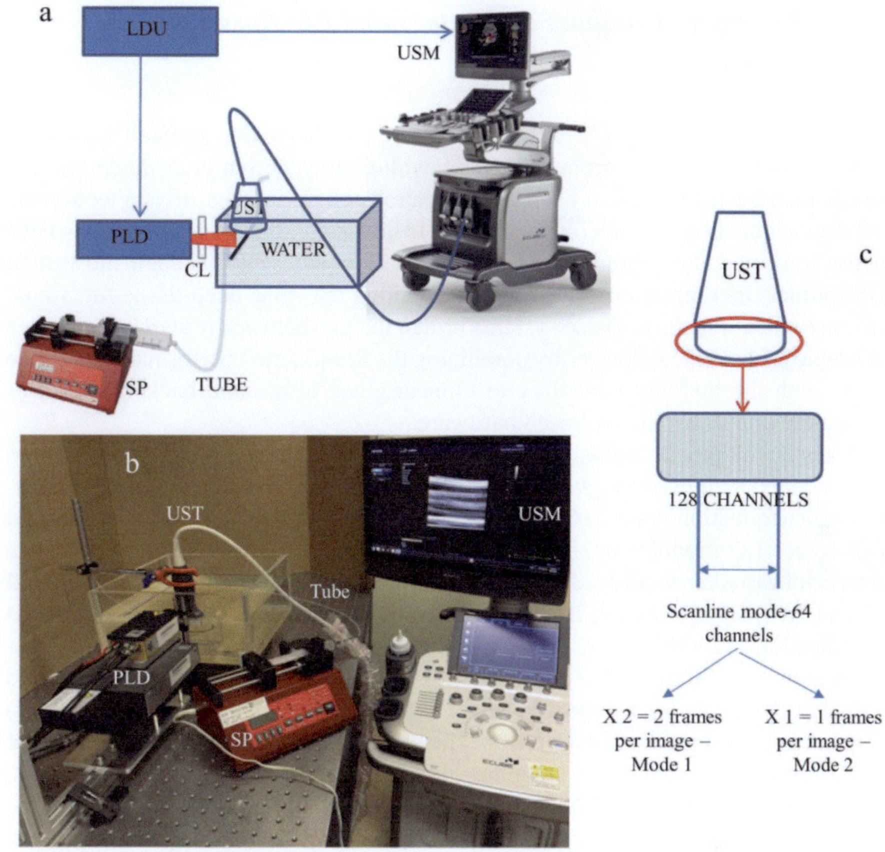

Fig. 3.6 **a** Schematic of the high-speed pulsed laser diode photoacoustic tomography (PLD-PAT) setup integrated with a clinical ultrasound system. PLD: pulsed laser diode; UST: ultrasound array transducer; CL: cylindrical lens; SP: syringe pump; LDU: laser driver unit; USM: ultrasound machine. **b** Photograph of the experimental setup showing the laser diode module, flow phantom, and E-CUBE 12R ultrasound platform (Alpinion Medical Systems). **c** Schematic of the L3-12 linear array transducer illustrating the 128-element configuration and dual acquisition modes (Mode 1: 64 + 64 channels; Mode 2: 128 channels). Reproduced with permission from [31]

ultrasound are essential. Overall, these advances mark an important milestone in LD-based PAI: demonstrating that affordable semiconductor sources, when coupled with high-speed ultrasound front-ends, can achieve imaging speeds rivaling those of ultrasound itself while maintaining clinically relevant imaging depths and field-of-view coverage.

3.5.3 Emerging Compact and Integrated LD-Based PAI Systems

Following the progression from benchtop and mechanically scanned LD-based systems to dual-modality probes, the next evolutionary step in laser diode photoacoustic imaging (LD-PAI) has been the integration of light sources, driver electronics, and data acquisition within compact, portable housings. Recent developments have focused on embedding high-power multi-stack LD modules into handheld ultrasound (US) probes or compact console systems, enabling real-time deep-tissue functional and anatomical imaging. These systems bridge the gap between research prototypes and deployable clinical devices by combining the illumination performance of diode stacks with the imaging versatility of clinical-grade ultrasound backends such as AcousticX or custom FPGA-based hardware.

A key development in this direction was reported by Thomas et al. (2025), who demonstrated a *multi-stack Quantel LD* implementation interfaced with the AcousticX data acquisition system (Cyberdyne Inc., Japan). The design employed stacked high-power LD modules operating in the near-infrared (NIR) range to illuminate superficial vasculature, while the AcousticX platform enabled real-time co-registered US and PA imaging through its 128-channel high-sensitivity receiver and >100 dB amplification chain (Fig. 3.7). Using this configuration, the authors achieved in vivo imaging of the carotid and femoral arteries, visualizing vascular morphology and perfusion dynamics with improved imaging depth and signal-to-noise ratio (SNR) compared to earlier LD-based probes. The study also highlighted the potential of such a configuration for breast imaging, where compactness and handheld operation facilitate superficial vascular and lesion characterization.

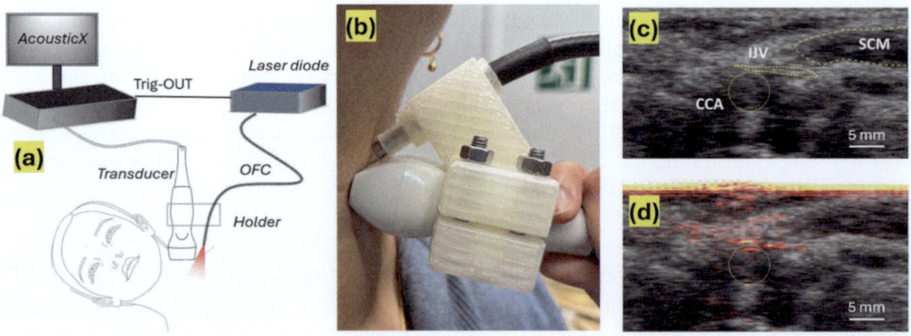

Fig. 3.7 Schematic and demonstration of a portable laser-diode–based photoacoustic and ultrasound system for carotid artery imaging. **a** Experimental setup showing the integration of multi-stack Quantel laser diodes with the AcousticX data acquisition system and ultrasound transducer. **b** Photograph of the probe configuration positioned on the neck during in-vivo imaging. **c** Co-registered B-mode ultrasound image of the common carotid artery (CCA), internal jugular vein (IJV), and sternocleidomastoid muscle (SCM). **d** Corresponding laser-diode–based photoacoustic image showing vascular contrast from the CCA region. CCA: common carotid artery; IJV: internal jugular vein; SCM: sternocleidomastoid muscle. Adapted with permission from authors [21]

Building on similar hardware foundations, Mitra et al. (2024) presented an advanced 3D LD-PA tomography (LD-PACT) platform using the same AcousticX DAQ system and Quantel multi-stack LD modules (Fig. 3.8). In their setup, a scanning configuration was employed to reconstruct volumetric images of tissue-mimicking phantoms and preclinical vascular networks. The combination of synchronized raster scanning and high-speed laser diode pulsing enabled high-resolution 3D multispectral visualization of vascular architectures at depths exceeding several millimeters. Their results demonstrated that combining a powerful LD source with a dedicated PA acquisition platform such as AcousticX can yield volumetric data at a fraction of the cost and size of conventional OPO-based systems. The work serves as an intermediate milestone between planar LD-PA imaging and fully embedded 3D tomography.

The latest milestone in this trajectory came from Liu et al. (2025), who reported the first embedded handheld LD-based 3D optoacoustic angiography system [19]. In this design, all key components—including multi-wavelength laser diodes, driver electronics, and data acquisition—were integrated within a single portable housing. By combining high-efficiency diode driving with FPGA-controlled synchronization and on-board reconstruction, the system achieved real-time volumetric (3D) imaging of deep vasculature, extending LD-based PAI to compact, fully self-contained architectures. The device successfully captured dynamic vascular networks with minimal latency and demonstrated a pathway toward low-power, wearable, or bedside

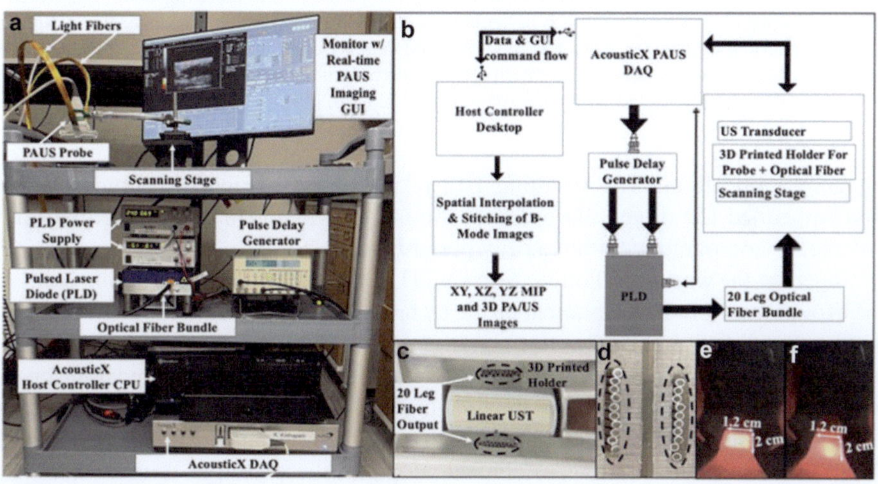

Fig. 3.8 Portable multiwavelength pulsed laser diode photoacoustic and ultrasound (PLD-PAUS) system. **a** Fully assembled PLD-PAUS setup on a mobile cart showing the AcousticX DAQ, host controller, 7 MHz linear ultrasound transducer, scanning stage, multiwavelength PLD source, and fiber bundle coupling. **b** Schematic diagram showing communication between the PLD, pulse delay generator, scanning stage, and AcousticX DAQ. **c**, **d** 3D-printed probe holder integrating a 20-leg optical fiber bundle arranged symmetrically around the transducer. **e**, **f** Optical illumination pattern showing rectangular focal spot formation at 14 mm and degradation at 17 mm from the transducer. Reprinted with permission from [20]

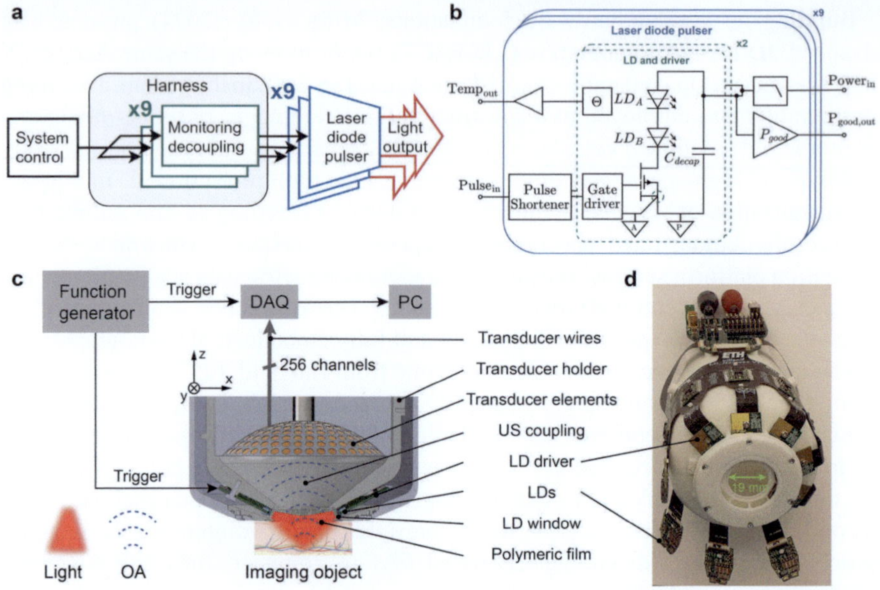

Fig. 3.9 Schematics of the handheld embedded laser-diode illumination optoacoustic system (HELIOS). **a** Overview of the modular LD light source consisting of monitoring decoupling and laser diode pulser modules. **b** Block diagram of the LD pulser (LDP) showing dual LD configuration and current control circuitry. **c** Schematic of the integrated optoacoustic system including the hemispherical transducer array, LD modules, and data acquisition setup (DAQ). **d** Photograph of the HELIOS prototype illustrating distributed LD drivers mounted around the imaging aperture (diameter: 19 mm). Reprinted with permission from Liu et al. [19]

imaging applications. The resulting HELIOS prototype (Fig. 3.9) represents the first embedded LD-driven 3D optoacoustic imaging platform capable of real-time volumetric angiography within a fully portable form factor.

Together, the works of Thomas, Mitra, and Liu represent a decisive turning point in the evolution of LD-based photoacoustic technology. The AcousticX-based systems emphasize clinical integration through high-sensitivity ultrasound backends, while embedded LD architectures focus on hardware miniaturization and standalone operation. Both directions underscore the growing maturity of semiconductor-based illumination and its seamless compatibility with modern ultrasound imaging ecosystems. As LD-based PAI moves toward clinical adoption, these designs demonstrate that careful system-level optimization encompassing light delivery, acoustic detection, and data handling is pivotal to achieving the performance and usability required for real-world diagnostic imaging. The examples presented here represent the most influential and technically distinctive LD-based PAI systems to date, illustrating the field's rapid evolution toward integrated, clinically viable architectures.

3.6 Advanced Variants of Laser-Diode-Based Photoacoustic Imaging

3.6.1 Laser-Diode-Based Photoacoustic Microscopy (LD-PAM)

While most photoacoustic microscopy (PAM) systems employ nanosecond-pulsed excitation to satisfy stress confinement and achieve high axial resolution, laser diodes uniquely offer the flexibility to operate in either pulsed or continuous-wave (CW) modulation modes. This capability has encouraged the development of laser-diode-based PAM (LD-PAM) platforms that balance performance, compactness, and affordability. PAM combines optical focusing with acoustic detection to achieve sub-cellular spatial resolution, enabling visualization of microvasculature, oxygenation, and cellular pigments with high sensitivity. Traditional PAM implementations rely on bulky, costly, Q-switched or fiber lasers that deliver sub-10 ns pulses with millijoule-level energies. These lasers, while powerful, limit portability and cost-effectiveness.

Recent developments in LD-based PAM bridge this gap by using compact, high-efficiency semiconductor diodes that produce nanosecond pulses with moderate energies (10–200 nJ) and high repetition rates (kHz range). Such sources enable compact microscopy systems suitable for preclinical and translational imaging. A recent review by Kalitsounakis et al. [33] summarized advances in low-cost PAM, identifying LDs as promising light sources for both time-domain (pulsed) and frequency-domain (CW-modulated) implementations. Time-domain LD-PAM employs short (<10 ns) pulses to generate broadband ultrasonic signals, while frequency-domain systems use intensity-modulated LDs or VCSELs at MHz frequencies to detect harmonic responses of absorbers. Compared to conventional lasers, LD-based PAM can reduce system cost by 20–40×, while maintaining micron-scale lateral resolution and imaging depths of 0.5–1 mm in tissue [18, 33–35].

Notable experimental demonstrations include Zeng et al. [34], who achieved label-free optical-resolution PAM (OR-PAM) of superficial vasculature using a compact visible LD source, and Erfanzadeh et al. [35], who introduced a laser-scanning LD-PAM system with miniaturized diode excitation and optical scanning mirrors. More recently, Seeger et al. [18] advanced the concept of overdriven continuous-wave laser-diode photoacoustic microscopy (CWLD-PAM), in which compact CW laser diodes were operated at high current bursts under low-duty-cycle "ALEX" (alternating-laser-excitation) mode to achieve multi-wavelength, sub-micrometer-resolution imaging of microvasculature and circulating melanoma cells in vivo. Remarkably, despite using CW sources, the system generated high-contrast images comparable to pulsed LD-PAM setups, confirming that carefully modulated CW diodes can emulate nanosecond pulse behavior for high-resolution microscopy applications (Fig. 3.10).

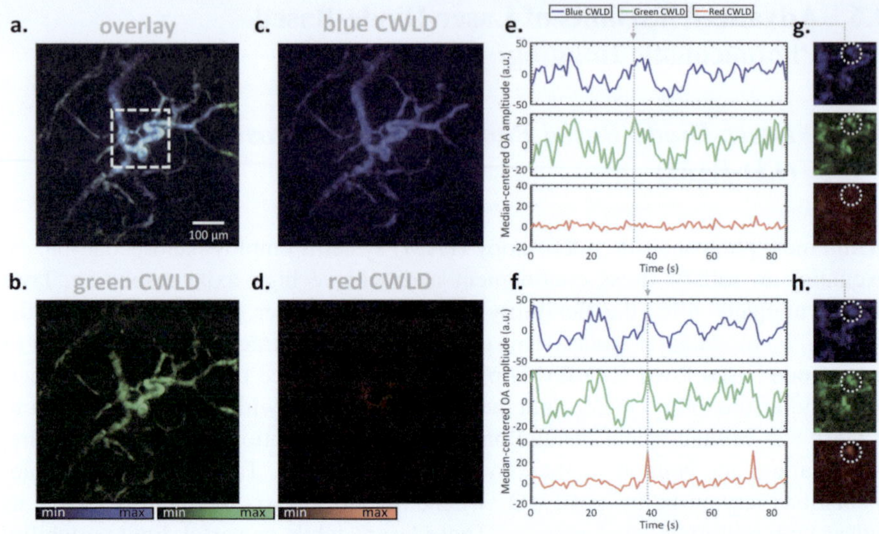

Fig. 3.10 Multi-wavelength photoacoustic microscopy of circulating melanocytes in vivo using overdriven continuous-wave laser diodes (CWLDs). Mouse ear vasculature was imaged under alternating excitation (ALEX) mode with blue, green, and red CWLDs. **a–d** Composite and separated multi-wavelength images showing the distribution of endogenous absorbers. **e**, **f** Temporal trajectories of optoacoustic (OA) signal intensities averaged within the regions of interest (ROIs) marked in **g**, **h**, recorded before and after injection of B16F10 melanoma cells. Reprinted with permission from [18]

Collectively, these works show that LD-based PAM can deliver high-contrast, real-time vascular imaging in compact and affordable configurations, an important step toward portable or even wearable photoacoustic microscopy platforms.

3.6.2 Vertical-Cavity Surface-Emitting Lasers for Photoacoustic Imaging

Vertical-cavity surface-emitting lasers (VCSELs) represent an emerging class of semiconductor light sources that combine many of the advantages of traditional LDs, such as coherence, high brightness, and nanosecond-scale pulsing—with the scalability and low-divergence characteristics of microfabricated emitters. Unlike edge-emitting LDs, VCSELs emit light perpendicular to the wafer surface, enabling two-dimensional array configurations that are inherently compact, thermally stable, and easily integrated with driving electronics. These features make VCSELs highly attractive for portable and handheld PAI systems, particularly in applications requiring uniform illumination and tight integration with ultrasound transducers [36, 37].

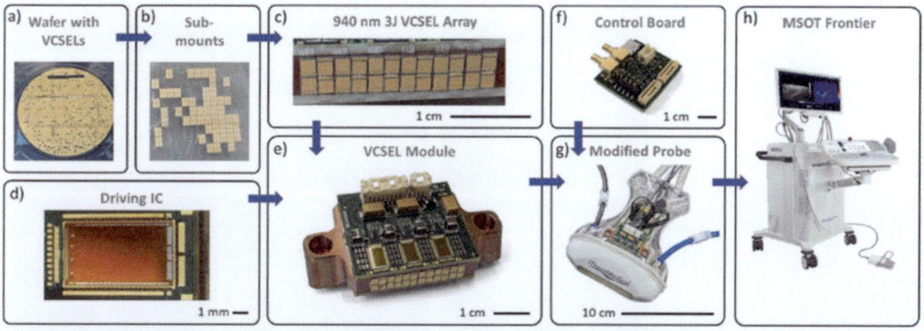

Fig. 3.11 Overview of different VCSEL-based handheld probe components (including the VCSEL module and the electronics) and clinical PAI system used for the experiments (MSOT Frontier, iThera Medical GmbH, Germany). **a** A wafer of triple-junction, 940-nm VCSELs is diced and the individual VCSELs are mounted onto **b** sub-mounts. The submounted VCSELs are assembled onto a **c** 2D array. **d** Driving ICs are affixed to driving circuit boards which are then **e** mounted onto either side of a copper block along with the VCSEL array. This VCSEL module is combined with a **f** custom FPGA control board into the **g** modified probe. Finally, this modified probe is connected with the **h** MSOT Frontier system. Reproduced with permission from [36]

Recent work by Eliason et al. (2025) and Konradl et al. (2025) demonstrated the first PAI systems based on 940 nm triple-junction VCSEL arrays integrated to Frontiers system from iThera Medical Gmbh, achieving robust photoacoustic excitation using compact, electronically driven modules (Fig. 3.11) [36, 37]. The arrays operated with pulse widths between 10 and 100 ns and a hardware pulse repetition rate (PRR) of 1 kHz, providing peak optical powers of approximately 3.5 kW and an impressive wall-plug efficiency of ~25%. Depending on pulse width, the per-pulse energy ranged from tens to a few hundred microjoules (for instance, ~35 μJ at 10 ns and ~350 μJ at 100 ns). These performance metrics are well suited for shallow to intermediate tissue imaging where both stress confinement and moderate optical penetration are required (Fig. 3.12).

In vivo feasibility was demonstrated in both studies. Human imaging of superficial vasculature and cutaneous structures was achieved by averaging multiple kHz-rate acquisitions, typically 1000 to 2000 frames per image, corresponding to effective imaging frame rates of 1 Hz and 0.5 Hz, respectively [36]. A comparative evaluation with laser-based PAI under identical acoustic detection and DAQ configurations demonstrated the feasibility of VCSEL illumination for superficial vascular imaging. The apparent low frame rate was limited by the need for frame averaging to boost signal-to-noise ratio (SNR), not by any intrinsic PRR limitation of the VCSEL source. These systems highlight the capability of VCSELs to deliver highly uniform, rectangular illumination with minimal optical alignment complexity, paving the way for practical, integrated, and potentially multi-channel clinical scanners. When compared with edge-emitting LDs, VCSEL arrays offer superior beam circularity, lower divergence, and simpler optomechanical packaging, attributes that greatly simplify their use in handheld ultrasound probes. Relative to LED arrays,

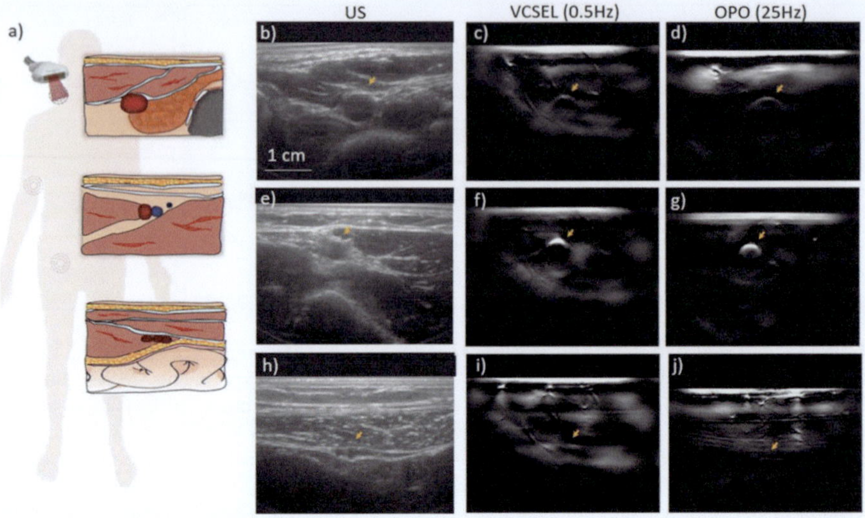

Fig. 3.12 VCSEL-based PA image quality in healthy volunteers. **a** Schematic of different anatomical spots of interest with different vessels of different sizes and located at different depths in the skin. The yellow arrows indicate the location of the vessels in the images. **b**, **e**, **h** Ultrasound reconstructed images; **c**, **f**, **i** photoacoustic VCSEL-based reconstructed images; **d**, **g**, **j** photoacoustic OPO-based reconstructed images. The anatomical locations scanned include the carotid artery (**b**–**d**); the brachial artery and vein (**e**–**g**); and the inferior epigastric vessels (**h**–**j**). Reproduced with permission from [36]

VCSELs provide substantially narrower spectral bandwidth and higher optical brightness, enabling higher SNR and improved light penetration. However, as coherent light sources, VCSELs still require laser safety measures for eye and skin protection, whereas LED-based systems maintain an advantage in inherent safety and open-environment use. Overall, these recent demonstrations confirm that kHz-rate VCSEL arrays can generate sub-100 ns optical pulses with hundreds of μJ per pulse in a handheld and energy-efficient form factor. While current frame rates are limited by averaging and SNR constraints, ongoing improvements in detector sensitivity, electronic synchronization, and optical packaging are expected to make VCSELs one of the most promising next-generation light sources for compact, high-performance PAI (Fig. 3.13).

Expanding beyond handheld systems, Gao et al. [38] demonstrated a wearable VCSEL-ultrasound patch capable of continuous, non-invasive vascular imaging and blood-volume monitoring in humans. The soft, skin-conformal device incorporated microscale VCSEL arrays and an ultrathin piezoelectric ultrasound transducer within an elastomeric substrate, enabling 3D imaging of subcutaneous vessels at multiple body sites, including the hand, wrist, forearm, and neck. The system captured real-time changes in vessel diameter during cuff inflation and deflation (Fig. 3.12), showcasing the potential of VCSEL-based illumination for wearable

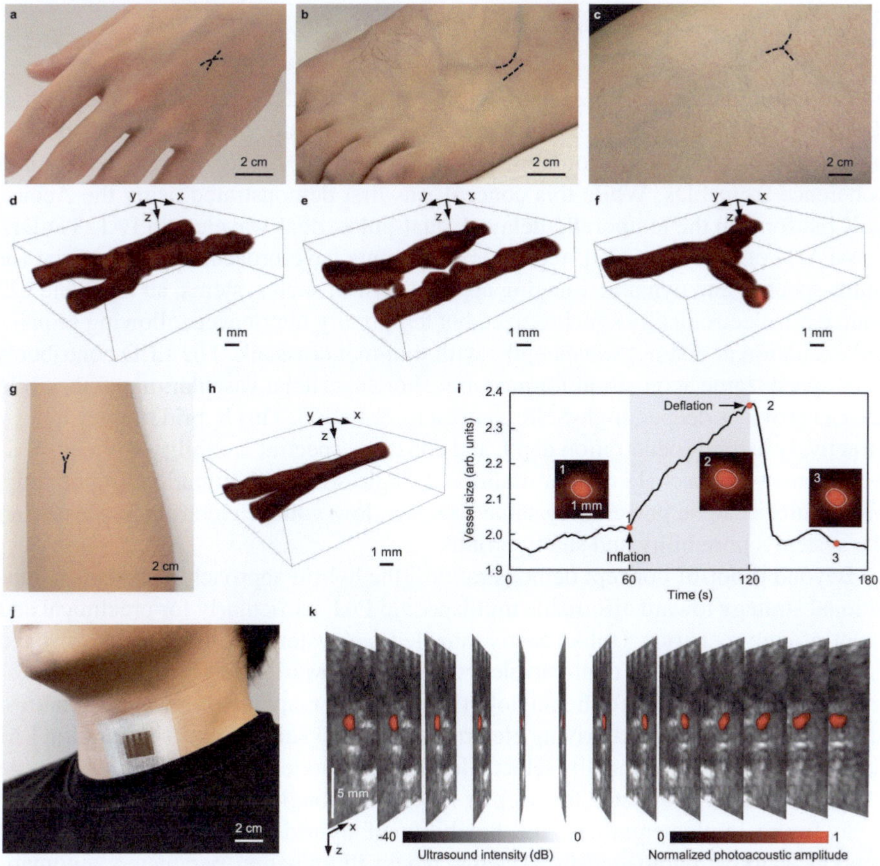

Fig. 3.13 Wearable VCSEL-based photoacoustic patch for real-time vascular imaging. **a–c**, **g**, **j** Photographs of measurement sites on the hand, wrist, forearm, and neck, respectively. **d–f**, **h** Reconstructed 3D photoacoustic images of subcutaneous blood vessels obtained using a soft, skin-conformal patch integrating VCSEL arrays and an ultrasound transducer. **i** Temporal variation in vessel diameter during cuff inflation and deflation, demonstrating real-time vascular monitoring. **k** Volumetric ultrasound (left) and photoacoustic (right) images acquired at the neck, revealing carotid vasculature. Reproduced with permission from [38]

photoacoustic sensing, continuous hemodynamic monitoring, and potentially for personalized cardiovascular diagnostics.

Together, these advancements position VCSELs as an intermediate bridge between high-energy LDs and broad-emission LEDs, combining the coherence and directionality of lasers with the modularity and safety advantages of semiconductor emitters. As array miniaturization, heat management, and driver integration continue to improve, VCSEL-based PAI may form the technological foundation for next-generation wearable or patch-type photoacoustic devices capable of real-time vascular and metabolic monitoring in daily life.

3.6.3 *Hybrid LED–LD Excitation Strategies*

Hybrid illumination schemes that combine LEDs and LDs represent an emerging direction in PAI, aiming to exploit the complementary advantages of both sources, high repetition rate and affordability from LEDs, and higher energy density and coherence from LDs. While this concept was first demonstrated using the AcousticX platform in the temporally delayed, dual-pulse, dual-wavelength PAI (TD-DP-DW-PAI) configuration [39], recent developments have broadened its potential for multi-spectral and dynamic imaging applications. In such systems, an LED and LD source are electronically synchronized but temporally interleaved, allowing sequential excitation at distinct wavelengths with minimal crosstalk. The LED contributes high-speed frame acquisition for anatomical or superficial vascular mapping, while the LD provides deeper, high-SNR contrast information. This hybrid timing strategy effectively extends penetration depth and enhances spectral flexibility without sacrificing frame rate, a challenge for traditional single-source illumination. Importantly, both sources can be powered by same compact, low-voltage electronics, preserving the system's portability and safety profile.

Beyond proof-of-concept demonstrations, the hybrid approach represents a transitional strategy toward affordable multispectral PAI, particularly for preclinical and point-of-care scenarios. Unlike costly tunable laser systems, combining narrowband LDs with broadband or dual-wavelength LEDs allows limited but practical spectral separation sufficient for hemoglobin oxygenation mapping or molecular contrast studies. Moreover, hybrid driving electronics are well suited to next-generation PAI devices that can dynamically select illumination modes, LED-only for real-time monitoring and LD-assisted for deep or functional imaging.

In future implementations, the flexibility of hybrid LED–LD excitation may support adaptive photoacoustic systems, where illumination parameters automatically adjust based on target depth, motion, or spectral feedback. Integration with intelligent driver modules and compact ultrasound front-ends, as seen in commercial platforms such as AcousticX, can enable seamless transitions between shallow and deep imaging modes. The TD-DP-DW-PAI framework [39] thus serves as a model for how multiple semiconductor-based light sources can be harmonized to balance cost, depth, and diagnostic performance.

3.7 Preclinical and Clinical Applications

Despite the technological maturity of laser diodes, the number of preclinical and clinical applications of LD-based PAI remains relatively limited compared with LED-based systems. This discrepancy primarily stems from the absence of commercially available LD-PAI platforms and the technical complexity of integrating coherent, safety-regulated laser-diode sources into compact imaging probes. Most reported systems have therefore emerged from dedicated research laboratories or academic–industry collaborations, such as the European FULLPHASE FP7 consortium, which pioneered the first integrated handheld LD/ultrasound probe. Nevertheless, these studies have provided strong evidence that LD illumination can enable both tomographic and handheld imaging of deep vasculature, soft tissues, and pathological conditions, bridging the gap between benchtop laser-based systems and emerging affordable clinical devices.

3.7.1 Preclinical Applications

3.7.1.1 Liver Fibrosis Assessment

One of the earliest preclinical demonstrations of in-vivo LD-based photoacoustic imaging was conducted by van den Berg et al. [22], who applied a hand-held dual-modality LD/ultrasound (US) probe to detect and stage liver fibrosis in rats. This work represented a milestone within the FULLPHASE consortium, integrating Quantel's pulsed laser-diode module (805 nm, ~1.5 mJ per pulse, ~7 kHz PRR) with an ESAOTE linear-array ultrasound transducer (7 MHz). The objective was to evaluate whether optical absorption differences associated with fibrotic remodeling and collagen accumulation could be non-invasively detected using a compact, diode-based platform (Fig. 3.14). In this study, fibrotic and healthy rat livers were imaged in vivo through the abdominal wall without the need for invasive optical coupling or water immersion. The photoacoustic contrast between the two tissue types correlated with histological staging, as confirmed by Masson-trichrome staining. Increased PA signal intensity and structural heterogeneity were observed in fibrotic livers, reflecting elevated hemoglobin concentration and altered vascular architecture. Furthermore, co-registered US images provided complementary anatomical context, allowing accurate differentiation of hepatic lobes and lesion boundaries. The experiment demonstrated that LD-based PAI can detect fibrotic alterations several millimeters beneath the surface, achieving sufficient penetration and signal-to-noise ratio (SNR) for organ-level functional imaging. Compared with systems driven by bulky OPO or Nd:YAG lasers, the LD-based setup offered advantages in portability, safety, and frame-rate flexibility, enabling real-time co-registered imaging in a small-animal laboratory setting. The success of this approach underscored the promise of LD illumination for quantitative molecular imaging of chronic liver disease. Its compatibility with conventional ultrasound architectures makes it an attractive candidate for future clinical translation, where combined hemodynamic and structural information could guide disease staging or treatment response monitoring.

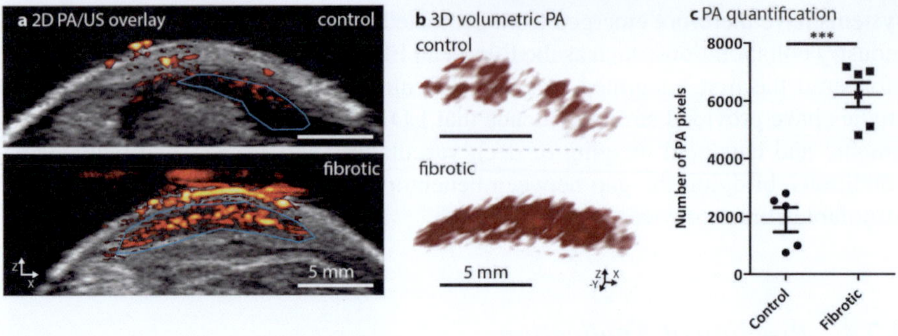

Fig. 3.14 Non-invasive photoacoustic (PA) imaging and quantitation in control and CCl_4-induced liver fibrosis mouse models. **a** Representative PA/US overlay scanned images of liver from healthy (control) mice (n = 5) and CCl_4-treated fibrotic mice (n = 6). The confined blue lines denote the region of interest (ROI) registered during PA imaging and quantitation for the PA images. **b** Cross-view representation of the 3-dimensional volumetric reconstruction of PA imaging in the control livers and fibrotic mouse livers based on the ROI as shown in overlay images (see online Visualization 1 and Visualization 2). **c** The graph depicts the number of PA pixels (calculated in ROI within scanned livers, detailed in methods) in control livers and 6 weeks fibrotic livers. Each symbol represents an individual mouse. Data represent mean ± SEM. *** $p < 0.001$ denotes statistical significance. Reproduced with permission from [22]

3.7.1.2 Functional and Dynamic Flow Imaging

Beyond structural visualization, LD-based PAI has also demonstrated the ability to quantify dynamic physiological processes such as blood flow and oxygenation, which are critical for vascular diagnostics and tissue viability studies. A seminal demonstration was reported by van den Berg et al. [25], who utilized FULLPHASE LD–US probe to perform *pulsed photoacoustic flow imaging* at multiple flow speeds in tissue-mimicking phantoms. The system incorporated an 805 nm pulsed laser diode coupled with a 7.5 MHz linear array ultrasound transducer integrated into a compact probe head. Using cross-correlation-based motion estimation of photoacoustic signals, they reconstructed 3D flow velocity maps with both magnitude and direction encoded in color, achieving accurate flow quantification up to 75 mm/s.

As shown in Fig. 3.15, the reconstructed flow profiles revealed a clear linear relationship between set and measured velocities, confirming the feasibility of quantitative flow imaging using LD excitation. This work established that the high pulse repetition rate (up to 10 kHz) achievable with LDs enables real-time acquisition of multiple frames per second, supporting functional studies of vascular dynamics. Although demonstrated in phantoms, the methodology has since inspired in vivo extensions for small-animal and superficial human imaging, reinforcing the potential of LD-based PAI as a low-cost, portable platform for dynamic hemodynamic assessment.

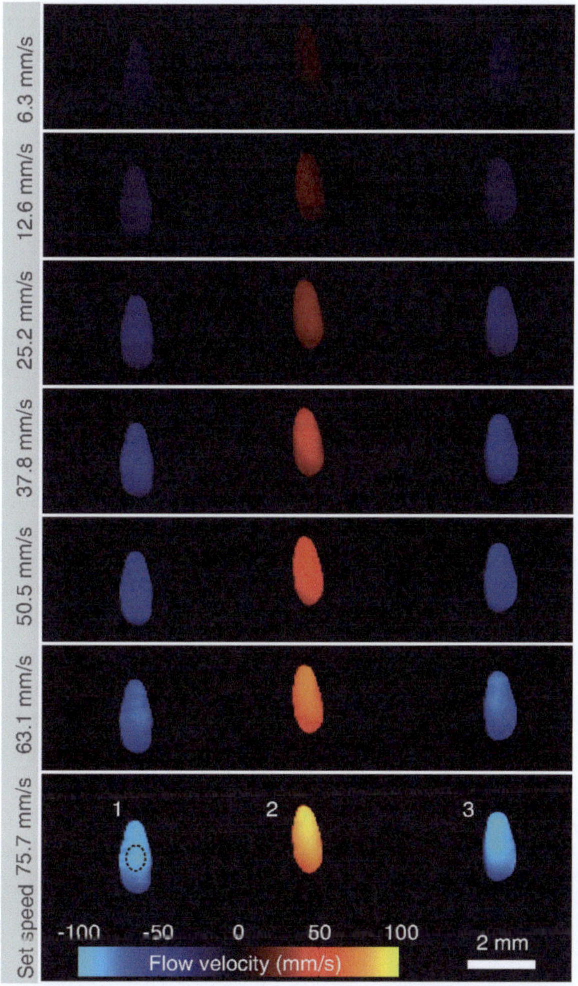

Fig. 3.15 Photoacoustic flow imaging with a handheld LD-based system. Flow velocity maps at varying set speeds from 6.3 to 75.7 mm/s obtained using cross-correlation analysis of photoacoustic signals. Blue and red indicate opposite flow directions. The study demonstrated accurate quantification of flow velocity using LD excitation at 805 nm with a handheld ultrasound–photoacoustic probe. Reprinted with permission from [40]

3.7.1.3 Brain and Intracranial Imaging

LD-PAI has also been applied to image cerebral vasculature and monitor intracranial physiological changes in small animals. Rajendran et al. [24] developed a second-generation high-speed PLD-PAT system operating at 803 nm with a pulse repetition rate of 7 kHz and pulse energy around 1.4 mJ. The system achieved cross-sectional imaging of the rat brain with a penetration depth sufficient to visualize the superior sagittal sinus (SSS) through the intact skull. By inducing cerebrospinal fluid (CSF) leakage to simulate intracranial hypotension, they observed significant distension of the venous sinus and variations in photoacoustic signal intensity, reflecting venous pressure changes.

These results demonstrated that LD-PAT can capture functional alterations in intracranial hemodynamics with high temporal resolution, using affordable and compact light sources suitable for small-animal neurovascular studies. The findings suggest potential applications in monitoring intracranial pressure (ICP) or cerebral blood volume in preclinical models, paving the way for translation into portable brain monitoring tools.

3.7.1.4 3D Functional and Multispectral Imaging

LD-PAI has matured to the point where compact multispectral systems can visualize functional hemodynamics in 3D in vivo. Mitra et al. [20] demonstrated one of the first dual-wavelength PLD-PAUS platforms capable of real-time small-animal imaging using Quantel diode stacks synchronized with the AcousticX data-acquisition unit. This system was applied to map oxygen saturation changes and microvascular dynamics in live rodents, underscoring the translational potential of LD-driven multispectral PAI. The study employed three diode wavelengths (760 nm, 808 + 940 nm, and 915 nm) to interrogate hemoglobin oxygenation states under controlled CO_2-induced vasodilation. The experimental configuration allowed simultaneous recording of ultrasound, photoacoustic, and oxygen-saturation (sO_2) maps across the same anatomical plane. As shown in Fig. 3.16, the 3D reconstructions reveal distinct vascular responses to hypercapnia: an overall increase in PA signal amplitude and spatial expansion of vessel cross-sections when CO_2 was introduced. Corresponding sO_2 maps (panels G–L) captured functional shifts between oxy- and deoxy-hemoglobin with sub-millimeter precision, demonstrating the feasibility of functional oxygen-metabolic imaging using entirely diode-based sources.

The authors also generated depth-encoded reconstructions and differential ΔsO_2 maps (panels P–Q), showing that changes were most pronounced in arteriolar regions at ~1 cm depth, an encouraging depth range for preclinical vascular and metabolic studies. Importantly, these results were achieved without high-energy solid-state lasers or OPOs, validating that stacked PLD arrays can deliver sufficient energy for dual-wavelength, real-time functional imaging. Such studies emphasize how LD-PAI can bridge the gap between low-cost imaging and high-information content functional mapping. Beyond metabolic monitoring, this approach may be extended to tumor hypoxia assessment, perfusion monitoring, and pharmacodynamic studies in small animals.

The HELIOS system, developed by Liu et al. [19], represents one of the most advanced embodiments of LD-based PAI to date. This embedded, handheld 3D angiography system integrates a compact multi-channel LD driver and diode stack directly with a hemispherical ultrasound array, achieving real-time volumetric imaging of vascular networks at centimeter depths. Using optimized overdriven nanosecond LD pulses and fully embedded acquisition electronics, HELIOS generated detailed angiographic maps of the human palm and fingers, resolving both superficial and subdermal capillary beds (Fig. 3.17). The vascular morphology and depth-resolved images were comparable in quality to those obtained using conventional

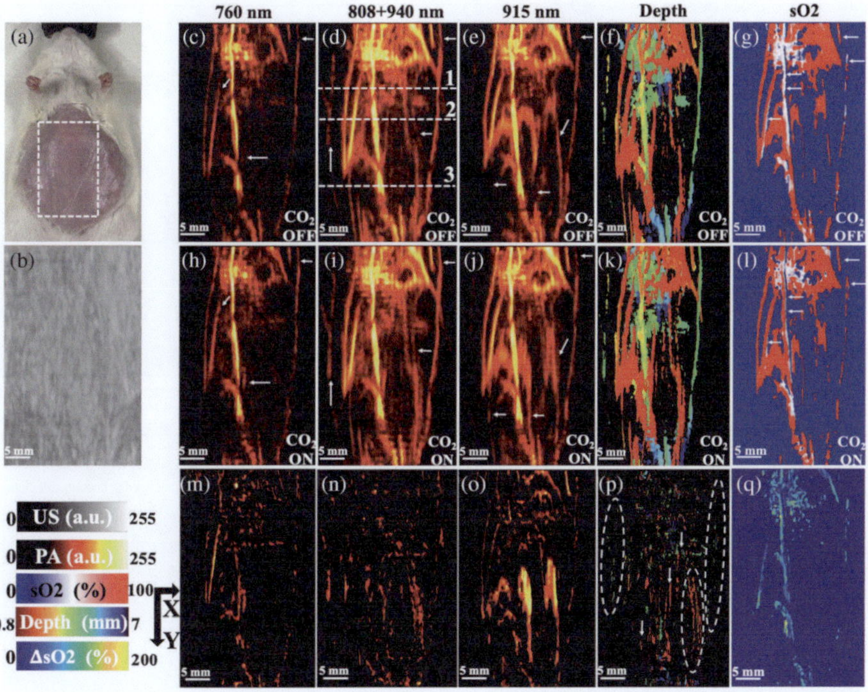

Fig. 3.16 Functional PA imaging of CO_2-induced hemodynamic modulation in a rat model using a dual-wavelength LD-PAUS system. **a**, **b** Imaging setup and co-registered ultrasound image. **c–e**, **h–j** PA intensity maps acquired at three excitation wavelengths with CO_2 off and on. **f**, **g**, **k**, **l** Depth- and oxygen-saturation-encoded reconstructions. **m–q** Differential PA and ΔsO_2 maps illustrating vascular response. Reprinted with permission from [20]

solid-state laser (SSL)-based systems, demonstrating the readiness of LD excitation for portable 3D functional imaging of microcirculation.

3.7.1.5 Translational Real Time Deep-Vasculature Imaging

Among the most recent advancements in LD-PAI is its translation toward deep-vascular and peripheral imaging in human subjects. Thomas et al. [21] demonstrated a real-time dual-modality LD-PA/US system that integrates Quantel multi-stack pulsed laser diodes with the AcousticX data acquisition and beamforming platform. The system operates at 850 nm, delivering nanosecond pulses with energy up to 1.2 mJ per pulse at 1 kHz PRR, coupled through a custom-designed optical holder that mounts directly onto a custom-made 5 MHz linear array ultrasound transducer.

In their study, the authors imaged the common carotid and femoral arteries (Fig. 3.18) in healthy volunteers, achieving penetration depths of approximately 15

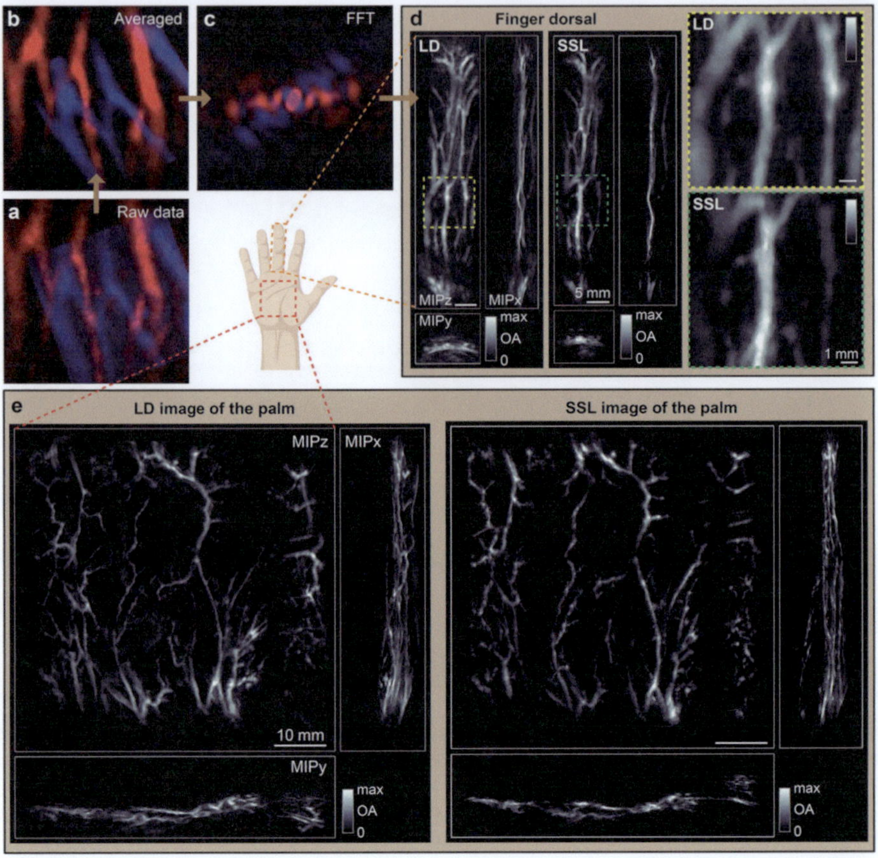

Fig. 3.17 High-resolution LD-based 3D optoacoustic imaging using the HELIOS handheld system. **a–c** Processing pipeline showing raw data, frame averaging, and Fourier-domain reconstruction. **d** Comparative finger vascular imaging between LD and SSL excitation, with detailed magnified views. **e** 3D maximum-intensity-projection (MIP) images of the palm acquired using LD and SSL illumination showing comparable microvascular structure and contrast. Adapted with permission from [19]

mm, sufficient to visualize both superficial and deeper vascular branches. Furthermore, the system demonstrated remarkable capability for deep-seated tissue imaging when applied to the breast. Leveraging photoacoustic vascular contrast, the system achieved PA imaging depth of up to 23 mm, which clearly delineated the anatomical structures and deep vascular distribution (Fig. 3.19). This 23 mm depth represents the highest reported imaging depth in LD-PAI to date, achieved in a highly scattering medium like the breast. The combined photoacoustic–ultrasound imaging enabled co-registered anatomical and functional views, with photoacoustic contrast clearly delineating hemoglobin-rich regions within the arterial lumen and breast vasculature.

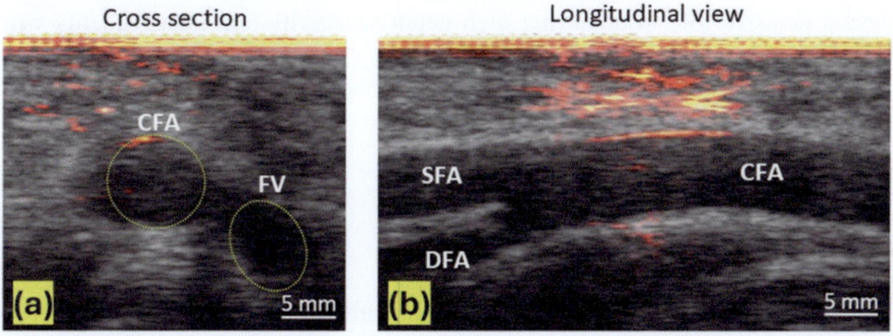

Fig. 3.18 Dual-modality laser-diode-based photoacoustic and ultrasound imaging of the human femoral artery. **a** Cross-sectional and **b** longitudinal co-registered photoacoustic (color overlay) and ultrasound (grayscale) images acquired using the portable AcousticX–Quantel LD-PAI system. The common femoral artery (CFA) and femoral vein (FV) are clearly visualized in **a**, while **b** shows longitudinal mapping of the CFA, superficial femoral artery (SFA), and deep femoral artery (DFA) with distinct vascular contrast up to ~15 mm depth. The strong photoacoustic signals delineate hemoglobin absorption within the arterial lumen, validating the system's deep-vascular imaging capability. Adapted with permission from authors [21]

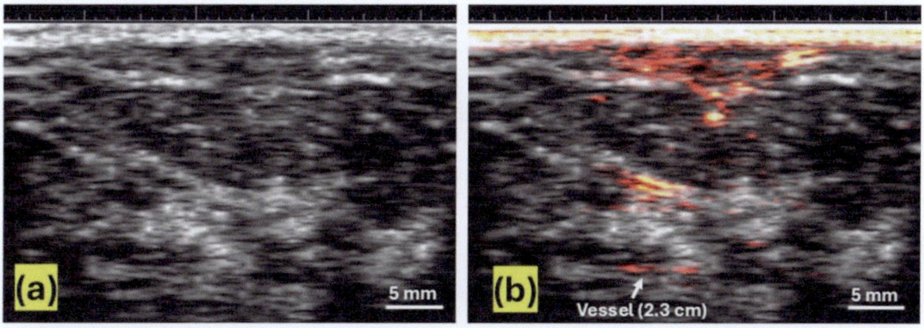

Fig. 3.19 LD-PAI of breast tissue demonstrating deep vascular penetration. **a** Grayscale ultrasound (US) image of the breast parenchyma. **b** Co-registered Photoacoustic (PA, red/yellow overlay) and US image, clearly visualizing a deep-seated vessel at a depth of 2.3 cm. This depth represents the maximum penetration depth for vascular imaging reported in highly scattering media using an LD-PAI system. Image courtesy of Dr. Francis Kalloor Joseph (Erasmus MC, Netherlands)

These results indicate that compact LD-based light sources, when combined with high-sensitivity photoacoustic-specific ultrasound probes and amplifiers, can offer imaging performance approaching that of conventional laser systems while retaining safety, affordability, and portability. The success of this demonstration underscores the translational potential of LD-PAI for applications such as vascular disease monitoring, carotid plaque characterization, and breast lesion assessment, particularly in resource-limited or point-of-care clinical settings. It also reinforces the notion that photoacoustic-specific probe and electronics design, as implemented in AcousticX, plays a decisive role in achieving high depth sensitivity and image fidelity with diode-based excitation sources.

3.7.2 Clinical Pilot Studies

3.7.2.1 Rheumatoid Arthritis Imaging

Inflammatory arthritis, such as rheumatoid arthritis (RA), involves synovial hyperemia and neoangiogenesis in early stages, making it a compelling target for PAI. In a clinical feasibility study, van den Berg et al. (2017) employed FULLPHASE LD-PA/ultrasound system to image synovitis in human finger joints (proximal interphalangeal joints) using 805 nm pulsed laser diodes combined with conventional ultrasound detection and compared the results with doppler imaging, which is the gold standard [41]. As illustrated in Fig. 3.20, co-registered PA and B-mode US images clearly reveal vascular signals in inflamed joints, which correlate with clinical signs of swelling and tenderness. The PA contrast in synovitic joints was significantly higher than in non-inflamed control joints, confirming the modality's sensitivity to increased hemoglobin concentration in microvasculature. This study provides robust evidence that LD-PAI can non-invasively detect hyperemic changes in human joint inflammation. This demonstration underscores the translational potential of LD-based PAI in rheumatology, offering a path toward portable, functional imaging of arthritis at the point of care.

3.7.2.2 Carotid Plaque Imaging

A key translational demonstration of LD-based PAI is the intraoperative study by Müller et al. (2021), conducted during carotid endarterectomy in human patients. Using a handheld PA/ultrasound probe based on the FULLPHASE design, they imaged excised carotid plaques in situ inside the surgical field [32]. Motion-compensated processing yielded an approximate 5 dB improvement in PA signal-to-noise ratio (SNR) compared to raw acquisitions, enabling visualization of diffuse PA contrast in the upper regions of plaques. Notably, the hemorrhagic regions within plaques exhibited distinct spectral and temporal signatures relative to the lumen blood pool, suggesting that LD-based PAI can help delineate plaque composition and vulnerability in a clinical setting. Figure 3.21 shows In vivo PA and US image of a human carotid artery with intraplaque hemorrhage. This intraoperative pilot shows that even without non-invasive access, LD-PAI can be integrated into surgical workflows, supporting future development of non-invasive carotid imaging devices.

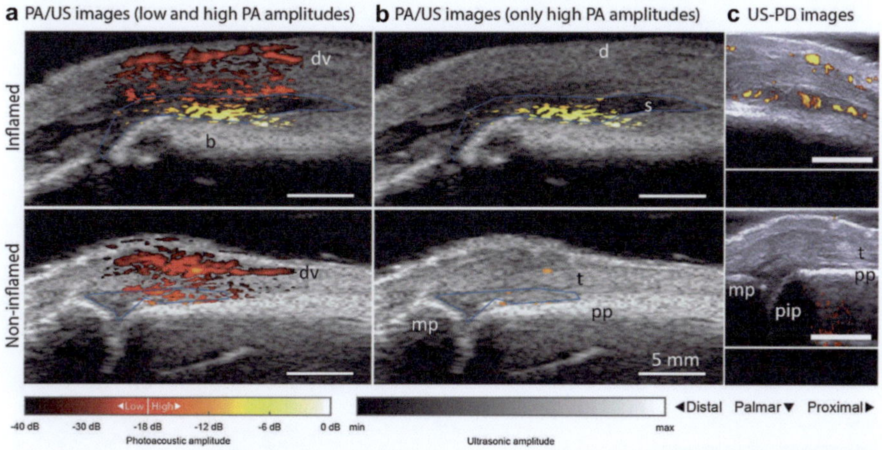

Fig. 3.20 PA/US and US/PD images of an inflamed (upper row) and non-inflamed contra-lateral joint (bottom row) of an RA patient. PA/US images in **a** show a difference in color between inflamed and non-inflamed corresponding to an increase in amplitude levels. When discarding low PA amplitudes in **b**, only features in the inflamed joint are visible. Corresponding US-PD images are shown in **c**. The blue line in the PA/US images indicates the ROI used for quantification of PA features in the synovial space. The 0 dB level is the maximum PA amplitude from the inflamed joint. d = dermis; dv = dorsal vein; pp = proximal phalanx; pip = proximal interphalangeal joint; mp = middle phalanx; s = synovium; t = extensor tendon. Reproduced with permission from [41]

3.7.3 Summary and Outlook

Table 3.2 shows preclinical and clinical applications of LD-PAI. Over the past decade, LD-PAI has evolved from proof-of-concept demonstrations into a versatile platform capable of addressing a range of preclinical and early clinical applications. Preclinical studies have showcased its potential in functional imaging, such as vascular oxygenation mapping, blood-flow monitoring, and deep-tissue tomography, while maintaining a compact and affordable footprint. The introduction of integrated systems such as the FULLPHASE handheld probe and Quantel–AcousticX configurations has enabled real-time dual-modality imaging of small-animal organs and human vasculature. On the translational side, clinical pilot investigations including rheumatoid arthritis assessment and intraoperative carotid plaque imaging have demonstrated that LD-PAI can deliver diagnostically relevant contrast in vascularized and inflammatory tissues using safe, portable light sources.

Despite these advances, no commercial LD-PAI system is currently available, largely due to limited industrial adoption, optical packaging complexity, and the need for optimized ultrasound front-ends tailored to photoacoustic detection. Nonetheless, the steady improvement in semiconductor laser efficiency, driver miniaturization, and integrated cooling is rapidly closing the gap between laboratory prototypes and deployable clinical devices. With its ability to balance performance, safety, and scalability, LD-PAI stands poised to become a bridge technology between LED-based

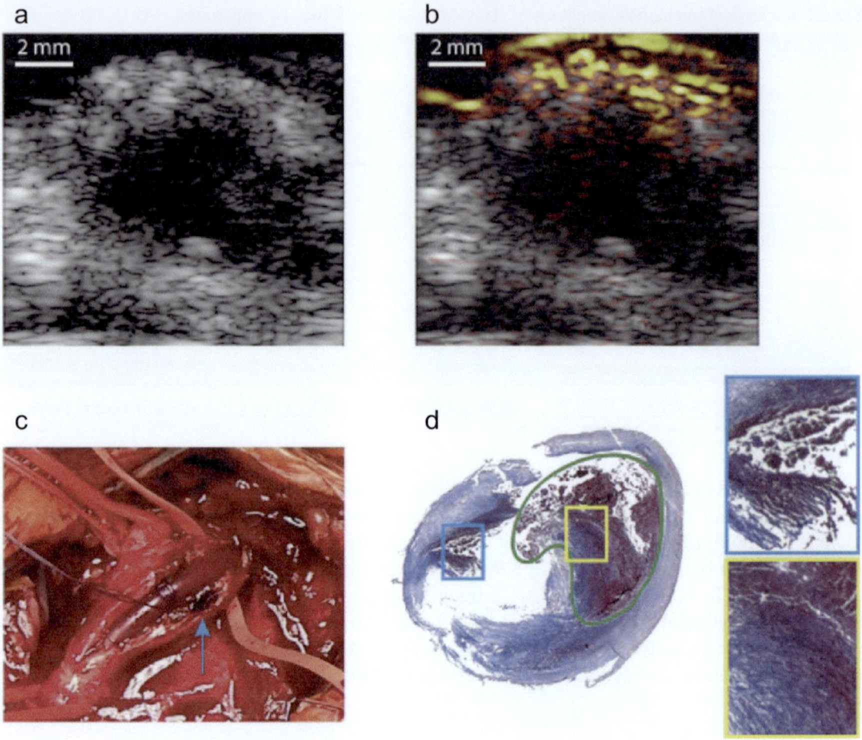

Fig. 3.21 In vivo PA and US image of a human carotid artery with intraplaque hemorrhage; **a** US image; **b** overlaid PA/US image (808 nm, dynamic range 23 dB); **c** photo of the carotid plaque during the CEA surgery; **d** Masson's trichrome staining of the artery. The area indicated in green is a lipid core filled with a large hemorrhage. The highlighted boxes show two regions of hemorrhages found in the plaque. Reproduced with permission from [32]

and solid-state laser-based platforms, particularly suited for vascular diagnostics, musculoskeletal imaging, and intraoperative guidance in point-of-care environments. Future developments in modular probe design, AI-enhanced reconstruction, and cost-efficient manufacturing could accelerate the pathway toward commercial translation and widespread clinical utility. Very recently, researchers from LumaWave Inc and Tulane University introduced a handheld, ultrasound-guided microvascular oxygenation assessment tool based on LD excitation [43], representing an important step toward clinically deployable LD-PAI systems and signaling the beginning of early commercial translation in this field.

Table 3.2 Preclinical and clinical applications of LD-PAI. Adapted with permission from [42]

Target	Application		Depth (mm)	Contrast agent	Wavelength (nm)
Vasculature	Detection of intraplaque haemorrhage in carotid artery [26]	Phantom and ex vivo studies	20	N/A	808
Vasculature	Dynamic imaging studies (for example in cardiovascular medicine) [31]		20	N/A	803
Red blood cells	Noin-invasive blood flow imaging [40]		7	N/A	805
Vasculature	Detection of liver fibrosis [22]	Animal in vivo	5	N/A	808
Cortical vasculature	Brain imaging [28]		5	ICG	803
Cerebro-spinal fluid volume level	Detection of venous sinus distension by measuring intra-cranial hypertension [24]		5	N/A	803
Vasculature	3D functional imaging [20]		7	N/A	760/808/940/915
Vasculature and perfusion	Vascular/dermal pathologies [23]	Healthy human in vivo	5	N/A	805
Vasculature	Carotid imaging [25]		15	N/A	808
Finger joints	Imaging of Rheumatoid arthritis [15]		5	N/A	808
Vasculature	Breast, Carotid, Femoral artery imaging [21]		23	N/A	850
Vasculature	VCSEL-based vascular imaging [36]		10	N/A	940
Vasculature	High resolution large-scale 3D vascular imaging [19]		8	N/A	915
Finger joints	Imaging of Rheumatoid arthritis [41]	Patient in vivo	5	N/A	808
Vasculature	Detection of intraplaque hemorrhage (intra-operative clinical study) [32]		6	N/A	808/940

3.8 Challenges, Optimization, and AI-Enhanced Reconstruction

LD-PAI has made remarkable progress in recent years, evolving from benchtop feasibility demonstrations to fully integrated handheld and tomographic prototypes. Yet, translating these systems into robust clinical tools requires overcoming several technical and operational challenges. Compared with conventional laser systems, LD sources deliver lower per-pulse optical fluence, exhibit larger beam divergence, and demand sophisticated synchronization and cooling strategies. This section outlines the key limitations that currently define LD-PAI performance, summarizes recent optimization strategies, and briefly highlights the emerging role of artificial-intelligence-based reconstruction and denoising algorithms.

3.8.1 Practical and Technical Challenges

Several factors currently constrain LD-PAI performance:

- **Optical Output and Illumination Uniformity**:
  Despite diode stacking, the optical energy per pulse (typically 0.5–2 mJ) remains significantly lower than that of OPO or Nd:YAG lasers. Limited optical fluence reduces imaging depth and signal-to-noise ratio (SNR) in deep tissues. Divergent beam profiles also require careful beam shaping using fast/slow-axis collimators or diffractive optical elements to maintain homogeneous illumination across the region of interest [14–17, 29].
- **Thermal Stability**:
  High-current pulsed operation leads to temperature-dependent wavelength drift (~0.25 nm $°C^{-1}$) and gradual output degradation. Efficient heat dissipation through aluminum cooling rims or thermoelectric modules remains essential to maintain spectral stability, as shown in Daoudi et al. [15].
- **Synchronization and Timing Jitter**:
  Achieving nanosecond-level synchronization between the laser-diode driver and the ultrasound data-acquisition system is critical for accurate delay-and-sum reconstruction. Even ± 10 ns timing jitter can introduce several percent variation in reconstructed amplitude [16, 24].
- **Limited Sensitivity of Conventional Ultrasound Electronics**:
  Many early LD-PAI prototypes reused commercial ultrasound (US) front-ends not optimized for photoacoustic signals. Low front-end gain (<60 dB) and frequency-specific filtering reduced sensitivity. AcousticX and other modern DAQs now employ >100 dB amplification with wide-band receivers and white acoustic lenses to suppress surface clutter [20, 21].

- **Regulatory and Eye-Safety Barriers**:
  Unlike LEDs and VCSELs, coherent LD sources often fall under Class 3B or Class 4 laser safety regulations, necessitating controlled environments and limiting straightforward point-of-care deployment.

3.8.2 *Optimization Strategies*

Multiple hardware and algorithmic refinements have been introduced to mitigate these constraints (Table 3.3):

- **Optical Design and Coupling Efficiency**:
  Compact diode stacks combined with micro-cylindrical lenses (MCLs) and diffractive beam-shaping optics can improve coupling uniformity by 30–40% [15]. Some groups employ reflective diffusers or angled dual-illumination geometry to minimize shadowing near tissue interfaces [20, 21].
- **Thermal and Electrical Management**:
  Advanced GaN MOSFET drivers with pulse-width control and feedback loops allow stable high-current operation up to 25 A without wavelength drift [19]. Multi-channel temperature sensors integrated into the probe ensure consistent illumination during extended scans.
- **Probe and Detection Optimization**:
  Dedicated US probes with optimized acoustic lenses and high dynamic-range amplifiers (>100 dB) have shown substantial SNR gains compared with reused diagnostic probes [20, 21]. Proper electrical shielding and impedance-matched cabling further reduce electromagnetic interference from high-current LD drivers.
- **Signal Averaging and Motion Compensation**:

Table 3.3 Optimization levers and resulting performance improvements in LD-based PAI

Optimization parameter	Typical implementation	Reported improvement
Beam-shaping optics	FAC/SAC + DOE combination [15]	30–40% higher fluence uniformity
Thermal management	Active cooling rim or TEC [19]	<0.05 nm °C^{-1} wavelength drift
Amplifier gain	AcousticX DAQ [20, 21]	+10–15 dB SNR vs. conventional US DAQ
Frame averaging (kHz PRR)	500–1000 frames [16, 25]	4–6 dB noise reduction
Motion compensation	Deformation-based registration [25, 32]	5 dB in-vivo SNR gain
Dual-wavelength operation	Multi-stack LD arrays [20]	Enables functional hemoglobin mapping

High-repetition-rate (kHz-range) LDs permit frame averaging to suppress noise. Jaeger et al. [25] introduced deformation-compensated averaging, improving in-vivo SNR by $\approx$5 dB during handheld imaging of carotid artery.

- **Spectral and Functional Optimization**:
  Multi-stack LD arrays enable time-multiplexed dual-wavelength operation for functional mapping of oxy- and deoxy-hemoglobin. Systems such as the PLD-PAUS platform [20] have demonstrated accurate spectral unmixing for dynamic flow and oxygenation studies.

3.8.3 AI-Assisted and Data-Driven Reconstruction

While LD hardware continues to mature, signal reconstruction and denoising are increasingly enhanced through machine learning. Conventional delay-and-sum or back-projection algorithms perform poorly in low-SNR and limited-view conditions typical of compact LD systems. Deep learning-based methods can overcome these constraints by learning the nonlinear mapping between noisy and artifact-free images [44].

Francis et al. [45] developed a generative-adversarial-network (GAN) framework for reconstructing high-quality PA images from bandlimited data, while Paul and Mallidi [46] implemented an optimized conditional GAN that enables real-time LED/LD-based imaging with up to 25$\times$ faster reconstruction speed. Convolutional and transformer-based neural networks are now being adapted for joint PA-US data fusion, noise suppression, and depth-dependent fluence compensation. Emerging *physics-informed neural networks* (PINNs) further integrate photoacoustic wave equations into network architectures, enabling data-efficient training while preserving quantitative accuracy. Together, these techniques promise to elevate low-fluence LD-PAI systems to clinical image quality standards without adding hardware complexity.

3.9 Summary and Outlook

LD-PAI has evolved from fundamental proof-of-concept demonstrations in the early 2000s to compact, handheld platforms capable of real-time in-vivo visualization of functional vasculature. The journey from benchtop prototypes to clinically oriented systems reflects the steady maturation of semiconductor laser technology and its adaptation for biomedical imaging. Compared to traditional Q-switched and OPO laser sources, LDs offer a distinctive balance of portability, high repetition rate, and moderate optical energy, placing them between LED-driven and high-power laser-driven systems in both cost and performance.

Over the past decade, LD-PAI has delivered convincing results across preclinical and early clinical studies, including microvascular imaging, functional hemodynamic mapping, joint inflammation detection, and intraoperative plaque characterization. Integration with commercial ultrasound platforms, such as the FULLPHASE and AcousticX systems, has demonstrated that coherent semiconductor light sources can be coupled effectively with standard acoustic hardware. These achievements confirm that LD-PAI is not merely an academic curiosity but a practical illumination approach capable of addressing clinically relevant problems where millimeter- to centimeter-scale penetration is sufficient.

Technological progress continues on several fronts. Stacked diode arrays and multi-wavelength LD modules have improved optical fluence and spectral versatility, while advances in driver electronics have enhanced stability, pulse control, and safety compliance. Optical packaging innovations including fast/slow-axis collimation, diffractive beam homogenization, and improved thermal management have yielded higher output uniformity and reduced wavelength drift. Concurrently, dedicated ultrasound front-ends with high-gain preamplifiers and custom acoustic lenses have closed the sensitivity gap that once limited handheld implementations. Together, these improvements have enabled LD-based imaging depths approaching 15–20 mm in vivo while preserving fine vascular contrast.

From a broader perspective, the semiconductor industry's trajectory strongly supports continued growth in LD-PAI capability. Ongoing miniaturization, efficient heat management, and high-power operation of gallium-based diodes will likely yield sub-mJ, multi-kilohertz pulse trains from sources no larger than a coin. Emerging VCSELS and hybrid LD–LED architectures further extend the design space by combining the spatial coherence and modulation speed of LDs with the safety and cost benefits of LEDs. These hybrid strategies could underpin compact multispectral and wearable imaging platforms, particularly for vascular monitoring, dermatology, and wound-care applications.

While no widely available commercial LD-PAI system currently exists, the foundation for clinical translation is now strong. The demonstrated compatibility with portable ultrasound DAQs, the development of real-time AI-assisted reconstruction algorithms, and the convergence of standardized safety guidelines all point toward near-term clinical feasibility. As semiconductor sources continue to shrink and diversify, it is foreseeable that dual-modality US–PA probes powered entirely by LD or hybrid LD–LED illumination will become part of routine diagnostic workflows.

In summary, LD-PAI occupies a unique middle ground, combining the efficiency and scalability of LED-based systems with the precision and depth capability of laser-driven platforms. With further improvements in source engineering, optical coupling, and intelligent reconstruction, LD-PAI is poised to serve as a cornerstone technology in the next generation of affordable, real-time, point-of-care PAI systems.

Key Takeaways

- **LD-PAI bridges the gap between LED and laser systems**, offering higher depth and resolution at a fraction of the size and cost of solid-state lasers.
- Technological maturity in semiconductor fabrication and driver electronics is enabling **compact, thermally stable, and spectrally tunable LD sources**.

- **Integration with ultrasound hardware and AI-based image reconstruction** supports clinically relevant, real-time imaging performance.
- Hybrid LD–LED and VCSEL architectures point toward **miniaturized, battery-operated, and wearable future systems**.
- The next phase of progress lies in standardization, safety certification, and commercial translation to bring **LD-PAI from laboratory prototypes to practical clinical devices**.

References

1. P. Beard, Biomedical photoacoustic imaging. Interface Focus **1**, 602–631 (2011)
2. L.V. Wang, J. Yao, A practical guide to photoacoustic tomography in the life sciences. Nat. Methods **13**, 627–638 (2016)
3. S. Manohar, D. Razansky, Photoacoustics: a historical review. Adv. Opt. Photon. **8**, 586–617 (2016)
4. M. Omar, J. Aguirre, V. Ntziachristos, Optoacoustic mesoscopy for biomedicine. Nat Biomed Eng, 3(5):354-370 (2019)
5. L.V. Wang, Multiscale photoacoustic microscopy and computed tomography. Nat. Photon. **3**, 503–509 (2009)
6. J.J. Niederhauser, M. Jaeger, R. Lemor, P. Weber, M. Frenz, Combined ultrasound and optoacoustic system for real-time high-contrast vascular imaging in vivo. IEEE Trans. Med. Imaging **24**, 436–440 (2005)
7. C. Lutzweiler, D. Razansky, Optoacoustic imaging and tomography: reconstruction approaches and outstanding challenges. Sensors **13**, 7345–7384 (2013)
8. I. Steinberg, D.M. Huland, O. Vermesh et al., Photoacoustic clinical imaging. Photoacoustics **14**, 77–98 (2019)
9. M. Erfanzadeh, Q. Zhu, Photoacoustic imaging with low-cost sources: a review. Photoacoustics **14**, 1–11 (2019)
10. H. Zhong, T. Duan, H. Lan, M. Zhou, F. Gao, Review of low-cost photoacoustic sensing and imaging based on laser diode and light-emitting diode. Sensors **18**, 2264 (2018)
11. Y. Zhu, T. Feng, Q. Cheng, X. Wang, S. Du, N. Sato, J. Yuan, M.K.A. Singh, Towards clinical translation of LED-based photoacoustic imaging: a review. Sensors **20**, 2484 (2020)
12. S.K. Kalva, M. Pramanik, *Photoacoustic Tomography with High-Energy Pulsed Laser Diodes. SPIE Spotlight Series Book (SL56)* (2020)
13. A. Hariri, J. Lemaster, J. Wang et al., The characterization of an economic and portable LED-based photoacoustic imaging system to facilitate molecular imaging. Photoacoustics **9**, 10–20 (2018)
14. C. Canal, G. Rousset, S. Passard et al., Portable multiwavelength laser diode source for handheld photoacoustic devices. Proc. SPIE **9887**, 98872B (2016)
15. K. Daoudi, P.J. van den Berg, O. Rabot et al., Handheld probe integrating laser diode and ultrasound transducer array for ultrasound/photoacoustic dual-modality imaging. Opt. Express **22**, 26365–26374 (2014)
16. P.K. Upputuri, M. Pramanik, Performance characterization of low-cost, high-speed, portable pulsed laser diode photoacoustic tomography (PLD-PAT) system. Biomed. Opt. Express **6**, 4118–4129 (2015)
17. T.J. Allen, P.C. Beard, Pulsed near-infrared laser diode excitation system for biomedical photoacoustic imaging. Opt. Lett. **31**, 3462–3464 (2006)
18. M. Seeger, A. Stylogiannis, L. Prade et al., Overdriven laser diode optoacoustic microscopy. Sci. Rep. **13**, 19542 (2023)

19. X. Liu, F. Villani, I. Esteban, X. Chang, B. Zhang, A. Cossettini, X.L. Deán-Ben, L. Benini, D. Razansky, Handheld embedded laser-diode illumination optoacoustic system for real-time 3D angiography of deep tissues. Laser Photon. Rev. e01006 (2025)
20. M. Mitra, A. Haworth, P. Gaddale, F. Badran, N. Lagno, C. Pameijer, F. Aziz, S.-R. Kothapalli,Multiwavelength laser diode based portable photoacoustic and ultrasound imaging system for point-of-care applications, J. Biophoton. **17**(7), e202400058 (2024)
21. A. Thomas, G. van Soest, K.J. Francis, Peripheral vascular photoacoustic imaging using a portable laser diode, in *European Conference on Biomedical Optics* (Optica Publishing Group, 2025), pp. S1D-4
22. P.J. van den Berg, R. Bansal, K. Daoudi, W. Steenbergen, J. Prakash, Preclinical detection of liver fibrosis using dual-modality photoacoustic/ultrasound system. Biomed. Opt. Express **7**, 5081–5091 (2016)
23. H.M. Heres, P.J. van den Berg, J. Jose et al., Visualization of vasculature using a hand-held photoacoustic probe: phantom and in vivo validation. J. Biomed. Opt. **22**(4), 041013 (2017)
24. P. Rajendran, P.K. Upputuri, S.K. Kalva et al., In vivo detection of venous sinus distension due to intracranial hypotension in small animal using pulsed-laser-diode photoacoustic tomography. J. Biophoton. e201960162 (2020)
25. M. Jaeger et al., Deformation-compensated averaging for deep-tissue LED and laser diode-based photoacoustic imaging integrated with handheld echo ultrasound, in *LED-Based Photoacoustic Imaging. Progress in Optical Science and Photonics*, vol. 7, ed. by M. Kuniyil Ajith Singh (Springer, Singapore, 2020)
26. M.U. Arabul, J.D. Biesheuvel, A.F.W. van der Steen, G. van Soest, Toward the detection of intraplaque hemorrhage in carotid artery lesions using photoacoustic imaging. J. Biomed. Opt. **22**(4), 041010 (2016)
27. R.G. Kolkman, E. Hondebrink, W. Steenbergen, T.G. van Leeuwen, In vivo photoacoustic imaging of blood vessels with a pulsed laser diode. Lasers Med. Sci. **21**, 134–139 (2006)
28. S.K. Kalva, P.K. Upputuri, M. Pramanik, High-speed, low-cost, pulsed-laser-diode-based second-generation desktop photoacoustic tomography system. Opt. Lett. **44**, 81–84 (2019)
29. T.J. Allen, P.C. Beard, A.P. Dhillon, B.T. Cox, S.R. Arridge, Generating photoacoustic signals using high-peak-power pulsed laser diodes (Proc. SPIE, 2005), p. 5697
30. Nissinen J., Kostamovaara J. A 4 A peak current and 2 ns pulse width CMOS laser diode driver for high measurement rate applications. In: Proc. 39th European Solid-State Circuits Conference (ESSCIRC), Bucharest, Romania. IEEE; 2013.
31. K. Sivasubramanian, M. Pramanik, High frame rate photoacoustic imaging at 7000 frames per second using clinical ultrasound system. Biomed. Opt. Express **7**, 312–323 (2016)
32. J.-W. Muller, R. van Hees, M. van Sambeek, P. Boutouyrie, M. Rutten, P. Brands, M. Wu, R. Lopata, Towards in vivo photoacoustic imaging of vulnerable plaques in the carotid artery. Biomed. Opt. Express **12**, 4207–4218 (2021)
33. P. Kalitsounakis, G. Zacharakis, G.J. Tserevelakis, Towards affordable biomedical imaging: Recent advances in low-cost, high-resolution optoacoustic microscopy. Journal of Microscopy, 298, 3–9 (2025)
34. L. Zeng, S. Wang, R. Zhang et al., Label-free optical-resolution photoacoustic microscopy of superficial microvasculature using a compact visible laser diode excitation. Opt. Express **23**, 31026–31033 (2015)
35. M. Erfanzadeh et al., Laser-scanning laser diode photoacoustic microscopy system. Photoacoustics **9**, 1–9 (2018)
36. B. Eliason, A. Longo, C. Dehner, L. Lilaj, S. Gronenborn, C. Klett, A. Weigl, D. Dörich, U. Malzahn, G. Zahnd, J. Konradl, P. Leisching, Performance characterization of a high-power, compact, low-cost, and high-efficiency VCSEL subsystem for photoacoustic imaging, in *Photons Plus Ultrasound: Imaging and Sensing 2025*, vol. 13319, (Proc. SPIE, 2025), p. 133190S
37. J. Konradl, S. Gronenborn, A. Weigl, C. Klett, D. Dörich, U. Malzahn, A. Longo, C. Dehner, B. Eliason, C. Reichl, G. Zahnd, P. Leisching, High-power compact VCSEL subsystem as game-changer for photoacoustic medical imaging and sensing, in *Vertical-Cavity Surface-Emitting Lasers XXIX*, vol. 13384, (Proc. SPIE, 2025), p. 133840G

38. X. Gao, X. Chen, H. Hu et al., A photoacoustic patch for three-dimensional imaging of hemoglobin and core temperature. Nat. Commun. **13**, 7757 (2022)
39. V. Periyasamy, A. Das, M. Pramanik, Temporally delayed, dual-pulse, dual-wavelength photoacoustic imaging (TD-DP-DW-PAI): advancing capabilities of a commercial light-emitting diode-based photoacoustic imaging system with a pulsed laser diode. Biomed. Opt. Express **16**(7), 2945–2958 (2025)
40. P.J. van den Berg, K. Daoudi, W. Steenbergen, Pulsed photoacoustic flow imaging with a handheld system. J. Biomed. Opt. **21**(2), 026004 (2016)
41. P.J. van den Berg, K. Daoudi, H.J. Bernelot Moens, W. Steenbergen, Feasibility of photoacoustic/ultrasound imaging of synovitis in finger joints using a point-of-care system. Photoacoustics **8**, 8–14 (2017)
42. M. Kuniyil Ajith Singh, W. Xia, Portable and affordable light source-based photoacoustic tomography. Sensors **20**(21), 6173 (2020)
43. V.D. Vincely, M.A. Dickerson, C.E. Hawkins, C.L. Bayer, Near-infrared-II pulsed laser diode array for spectral photoacoustic imaging of deep tissue oxygenation. Biomed. Opt. Express **16**, 4970–4978 (2025)
44. M. Kuniyil Ajith Singh, K. Sivasubramanian, N. Sato, F. Ichihashi, Y. Sankai, L. Xing, Deep learning-enhanced LED-based photoacoustic imaging, in *Photons Plus Ultrasound: Imaging and Sensing 2020*, vol. 11240 (Proc. SPIE, 2020), p. 1124038
45. K.J. Francis, A. Arora, P. Kancharla, M.K.A. Singh, W. Steenbergen, S.S. Channappayya, Generative adversarial network-based photoacoustic image reconstruction from bandlimited and limited-view data, in *Photons Plus Ultrasound: Imaging and Sensing*, vol. 11642 (Proc. SPIE, 2021), p. 1164235
46. A. Paul, S. Mallidi, Facilitating real-time LED-based photoacoustic imaging with DenP2P: an optimized conditional generative adversarial deep learning solution. AIP Adv. **15**(5), 055227 (2025)

Chapter 4
Affordable Detection and Data Acquisition for Photoacoustic Imaging

Learning Objectives

After reading this chapter, the reader will be able to:

- **Explain the fundamental principles of acoustic wave detection** in photoacoustic imaging and how detector parameters like bandwidth, sensitivity, and impedance affect image quality.
- **Differentiate between detector technologies**, including piezoelectric, polymer-based, micromachined, and optical methods, highlighting their respective advantages, limitations, and cost–performance trade-offs.
- **Describe the structure and operation of low-cost detector designs**, such as PVDF/PVDF-TrFE films, CMUTs, and PMUTs, and understand their fabrication and integration pathways.
- **Identify key design requirements for affordable data-acquisition systems**, including analog front-end electronics, synchronization, and compatibility with LED/laser diode excitation sources.
- **Evaluate emerging data acquisition architectures**, such as compact multi-channel and FPGA-based platforms, and recognize how open-source control software supports cost reduction and scalability.
- **Integrate the co-design of light source, detector, and DAQ subsystems** to conceptualize complete, low-cost photoacoustic imaging systems for preclinical, clinical, and point-of-care use.

M. Kuniyil Ajith Singh, *Photoacoustic Imaging Using Affordable Resources*, Progress in Optical Science and Photonics 40, https://doi.org/10.1007/978-981-95-7175-8_4

4.1 Principles of Acoustic Detection in Photoacoustic Imaging

Photoacoustic imaging (PAI) relies on detecting ultrasound waves generated by the rapid thermo-elastic expansion of optical absorbers following pulsed light excitation. The efficiency and fidelity of this acoustic detection stage directly determine the achievable resolution, sensitivity, and imaging depth [1, 2]. When a short optical pulse irradiates tissue, localized heating induces a transient pressure rise that propagates as an acoustic wave. The amplitude and frequency spectrum of this wave depend on the absorber size, optical fluence, and the tissue's thermo-mechanical properties. For microscopic absorbers, the signal extends into tens of megahertz, whereas for larger vascular structures it lies predominantly in the 1–5 MHz range. Therefore, detectors must possess sufficiently broad frequency response to capture the full information content of the photoacoustic signal [3].

Key performance parameters include bandwidth, sensitivity, noise-equivalent pressure (NEP), and impedance matching [2, 4]. A wide fractional bandwidth (typically 50–150% of the center frequency) ensures faithful reconstruction of fine spatial details, while high sensitivity enables detection of weak signals from deeper tissue regions. NEP defines the minimum detectable pressure corresponding to the detector's intrinsic noise floor; lower NEP values indicate higher sensitivity. Acoustic impedance matching between the transducer, coupling medium, and tissue minimizes reflection losses, and proper electrical impedance matching between the detector and front-end electronics optimizes power transfer and signal-to-noise ratio. Because photoacoustic signals are often weak on the order of sub-kilopascal pressure, careful acoustic coupling, low-noise amplification, and electromagnetic shielding are essential [5]. Depending on system geometry, detection may be performed using single-element transducers, linear or two-dimensional arrays, or optical sensors. Each configuration involves trade-offs between spatial coverage, acquisition speed, and cost [6].

Foundational analyses by Lutzweiler and Razansky (2013) emphasized the influence of detection geometry on image fidelity. As shown in their work, planar, cylindrical, and spherical detector arrangements govern the angular coverage of detected acoustic fields and thus affect spatial resolution and artifact suppression (Fig. 4.1) [2]. The photoacoustic wave equation typically solved through spherical Radon or time-reversal formulations formalizes the relationship between the initial pressure distribution and the detected signal. These theoretical foundations continue to guide modern affordable PAI platforms, which increasingly employ simplified planar or partial-view detection geometries optimized for cost and portability [1, 7].

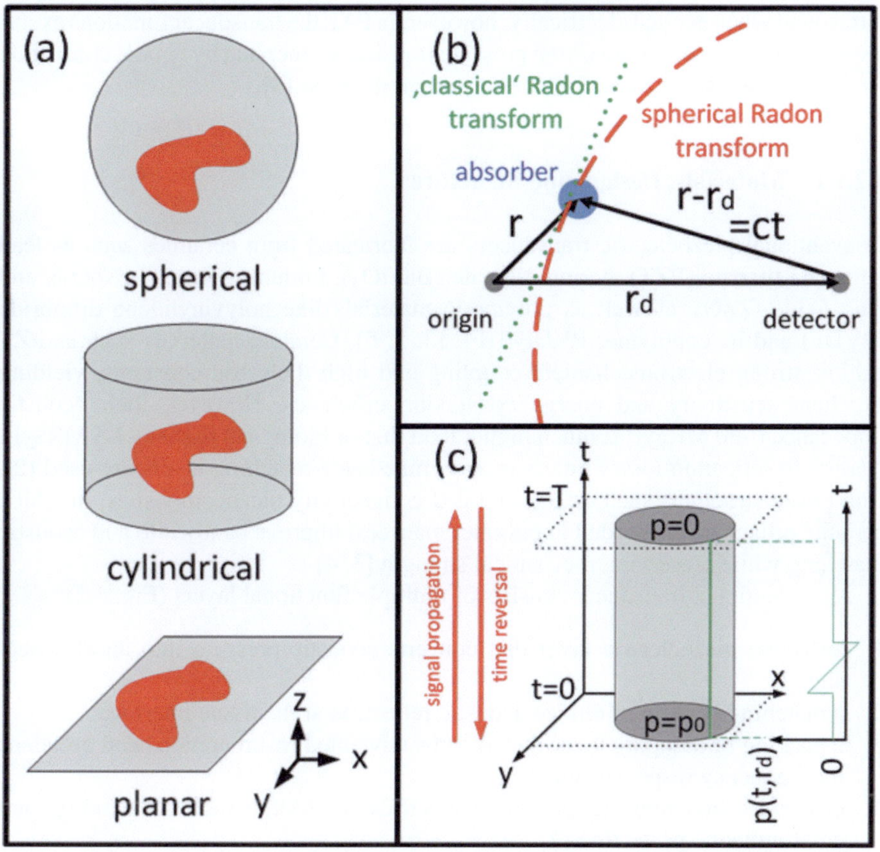

Fig. 4.1 Common photoacoustic detection geometries and signal propagation principles. **a** Spherical, cylindrical, and planar detector configurations used for tomographic acquisition; **b** Spherical Radon transform illustrating how the detector integrates pressure over constant time-of-flight surfaces; **c** Time-reversal reconstruction concept. Reproduced with permission from [2]

4.2 Photoacoustic Transducer Technologies

4.2.1 Conventional Piezoelectric Transducers

Piezoelectric transducers remain the most established and widely used detectors in PAI, owing to their mature fabrication technology, high signal conversion efficiency, and compatibility with commercial ultrasound systems [2, 3, 6]. These transducers operate on the direct piezoelectric effect, in which mechanical stress produced by impinging acoustic waves induces a proportional electrical potential across the electrodes of a piezoelectric element. The converse effect allows the same element to emit ultrasound when excited electrically; however, in PAI, the transducer functions exclusively in receive mode, converting pressure transients generated by optical absorption into electrical signals for amplification and digitization [3].

4.2.1.1 Materials, Design, and Structure

Conventional piezoelectric transducers are fabricated from ceramics such as lead zirconate titanate (PZT), barium titanate ($BaTiO_3$), lithium niobate ($LiNbO_3$), and zinc oxide (ZnO), as well as polymeric materials like polyvinylidene difluoride (PVDF) and its copolymer PVDF-TrFE [3, 5, 6]. Ceramic materials such as PZT exhibit strong electromechanical coupling and high dielectric constants, yielding excellent sensitivity and energy conversion efficiency. However, their acoustic impedance (~30 MRayl) is much higher than that of biological tissue (~1.5 MRayl), leading to reflection losses unless proper impedance-matching layers are used [2]. Composite structures such as 1–3 and 2–2 connectivity piezocomposites, in which ceramic pillars are embedded in polymer matrices, improve bandwidth and acoustic matching while preserving mechanical strength [3, 4].

A conventional transducer consists of multiple functional layers (Fig. 4.2):

1. an active piezoelectric layer that converts acoustic pressure into an electrical signal;
2. a matching layer to minimize acoustic reflection at the tissue interface;
3. a backing or damping layer that absorbs unwanted reverberations and broadens the frequency response; and
4. electrodes and housing providing electrical contact, structural stability, and environmental protection [2].

   Typical configurations achieve fractional bandwidths of 60–80% and sensitivities in the mV/kPa range [3]. The operating frequency is governed by element thickness, with thinner elements suited for high-frequency, shallow imaging and thicker ones for deeper imaging at lower frequencies.

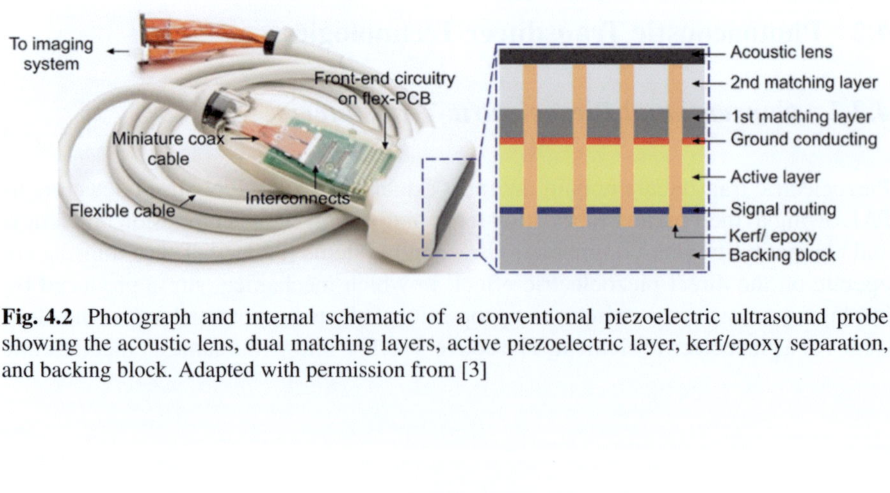

Fig. 4.2 Photograph and internal schematic of a conventional piezoelectric ultrasound probe showing the acoustic lens, dual matching layers, active piezoelectric layer, kerf/epoxy separation, and backing block. Adapted with permission from [3]

4.2.1.2 Performance, Challenges, and Recent Developments

The main performance metrics of piezoelectric detectors include center frequency, fractional bandwidth, sensitivity, and NEP. Commercial ultrasound probes exhibit excellent sensitivity but are expensive and mechanically rigid, limiting their adaptation to wearable or handheld PAI systems [1, 4]. High-frequency devices (>20 MHz) require tight fabrication tolerances, further increasing cost and complexity [3]. Recent work aims to modernize conventional piezoelectric transducers for affordable and miniaturized systems. As reviewed by Choi et al. [6], new designs employ flexible and optically transparent piezoelectric materials, including PVDF-TrFE films and transparent electrodes such as indium tin oxide (ITO) or conductive polymers. These approaches enable coaxial optical illumination and acoustic detection, simplifying alignment and promoting integration with compact LED- and LD-based PAI systems. Hybrid designs that merge piezoelectric and optical sensing principles further enhance bandwidth and sensitivity [5, 6].

Detector geometry strongly influences image fidelity. As shown by Lutzweiler and Razansky [2], planar, cylindrical, and hemispherical detector layouts define the angular aperture of detection and thus affect spatial resolution and artifact suppression. Recent advances in piezopolymer technology have achieved NEP values below 1 Pa with bandwidths extending from 1 to 30 MHz, representing a major step toward cost-effective, high-performance PAI detection [5].

4.2.2 Low-Cost Piezoelectric Polymer and Composite Detectors

The pursuit of affordable, wideband, and flexible photoacoustic detectors has driven substantial progress in polymer-based piezoelectric materials and composite structures. These approaches aim to maintain the high sensitivity of traditional ceramic elements while improving mechanical flexibility, acoustic impedance matching, and ease of large-area fabrication [3, 4]. Polymers and composites provide an attractive route for low-cost, portable PAI systems where conventional bulk ceramics are either too rigid or too expensive.

4.2.2.1 Polymer-Based Detectors

Among polymeric materials, PVDF and its copolymer PVDF-TrFE are the most widely employed piezoelectric films in PAI. They offer several intrinsic advantages:

- Low acoustic impedance (3–5 MRayl), closely matching that of soft tissue;
- Broad bandwidth due to lower mechanical Q factor and efficient acoustic damping;
- Mechanical flexibility, enabling conformal transducer geometries; and

- Ease of large-area film fabrication, making them suitable for scalable, low-cost production [4, 5].

However, their lower piezoelectric coefficients ($d_{33} \approx -20$ to -30 pC/N) compared to ceramics (PZT ≈ -200 pC/N) result in reduced sensitivity. To counter this, multilayer PVDF film stacks and charge-optimized electrode configurations have been developed, increasing output voltage without compromising flexibility [4]. Recent studies have demonstrated that ultrawideband PVDF-TrFE detectors can achieve noise-equivalent pressures below 1 Pa and bandwidths up to 30 MHz, bridging the performance gap between polymer and ceramic sensors. Their cost-effectiveness, environmental stability, and biocompatibility also make them ideal for wearable and implantable photoacoustic devices. In addition, flexible electrodes made of graphene or silver nanowire composites enable transparent, flexible detectors for integration with coaxial optical paths [6].

Recent experimental studies have demonstrated that affordable piezopolymer detectors can now achieve sub-pascal sensitivity without complex electronics. Kurnikov et al. (2024) systematically analyzed PVDF and PVDF-TrFE films of various thicknesses (20–120 μm) and reported noise-equivalent pressures as low as 0.3 Pa with a −6 dB bandwidth extending to 30 MHz [5]. The thicker (120 μm) PVDF-TrFE layer yielded higher voltage sensitivity ($\approx$15.5 $\mu V\ Pa^{-1}$) and lower noise, confirming that material composition and film thickness strongly influence acoustic response. Simulation of angiographic imaging in that study produced <3% reconstruction error relative to an ideal array, validating their imaging fidelity for wideband photoacoustic applications (Fig. 4.3). These detectors, fabricated using standard spin-coating and polarization steps, rely on inexpensive materials and simple readout circuitry, making them highly suitable for low-fluence LED- or LD-based PAI systems where affordability and flexibility outweigh extreme SNR demands.

4.2.2.2 Piezocomposite Detectors

Piezocomposites combine the sensitivity of ceramics with the acoustic matching and damping of polymers. Typical 2–2 and 1–3 connectivity patterns are shown in Fig. 4.4a, where ceramic elements are embedded within a polymer matrix to reduce impedance to 10–15 MRayl and broaden bandwidth [3, 4].

The dominant fabrication technique is the dice-and-fill (DF) method, illustrated in Fig. 4.4b [3]. In this process, a bulk piezoceramic plate is sequentially diced along orthogonal directions, the kerfs are filled with polymer, the block is reversed, and secondary dicing and filling steps create isolated ceramic pillars in a polymer matrix. The structure is then metallized to form electrodes. This scalable process yields uniform 1–3 composites with predictable mechanical and acoustic properties. Alternative fabrication strategies such as laser micromachining, injection molding, or 3-D printing offer finer pillar pitch, lower cost, and design freedom for curved or flexible arrays. Composites produced through these methods show wideband responses (60–100%) and sensitivity improvements of 2–3× over pure polymer films [5, 6].

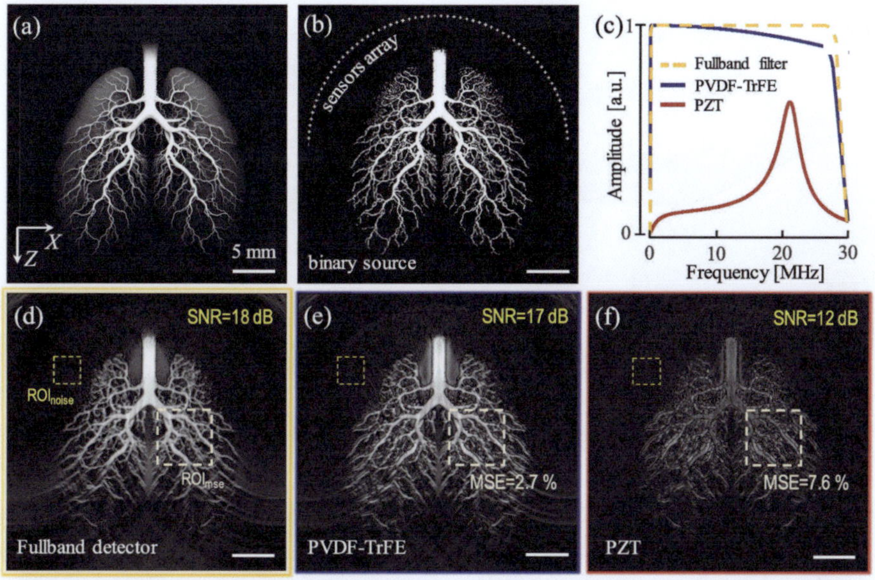

Fig. 4.3 Numerical simulation of photoacoustic angiographic imaging using detector arrays with different spectral sensitivities. **a** Digital vascular phantom and **b** its binarized version used for simulation; **c** Fourier filters representing the frequency responses of three detector types—full-band ideal, PVDF-TrFE piezopolymer, and PZT ceramic; **d–f** corresponding simulated angiographic reconstructions illustrating that wideband polymer detectors (PVDF-TrFE) provide image fidelity comparable to full-band detection while conventional PZT arrays lose fine vascular detail. Reprinted with permission from [5]

4.2.2.3 Emerging Trends and Applications

Recent innovations extend polymer and composite detection beyond traditional PZT arrays toward flexible, transparent, and printed detectors. Additive-manufactured PVDF-TrFE films using electrospinning and inkjet printing have achieved excellent homogeneity and reproducibility, allowing roll-to-roll fabrication of low-cost transducer sheets [6]. Transparent polymer-based detectors integrated with optical fibers or waveguides support coaxial illumination, simplifying optical alignment for portable systems. Furthermore, hybrid composites incorporating nanomaterials (e.g., CNTs, graphene) show enhanced charge mobility and improved piezoelectric response, pointing to next-generation flexible detectors for wearable PAI.

In summary, polymer and composite piezoelectric materials are redefining the affordability, versatility, and ergonomics of photoacoustic detectors. They enable wideband, conformal, and cost-effective acoustic sensing, marking a critical step toward accessible clinical translation of photoacoustic technology.

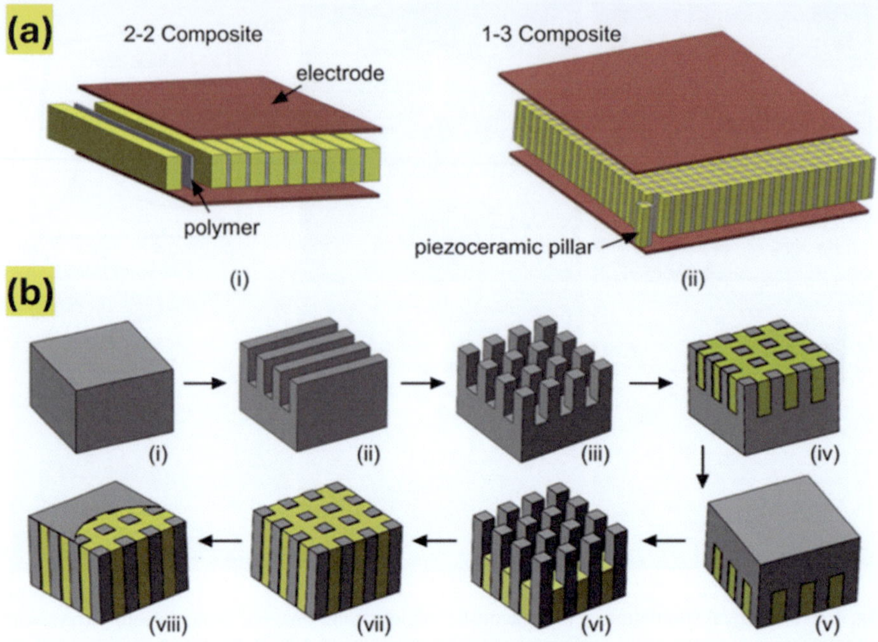

Fig. 4.4 **a** Representative 2–2 and 1–3 piezocomposite architectures showing ceramic elements embedded in a polymer matrix for improved impedance matching and bandwidth. **b** Process flow of the conventional dice-and-fill (DF) fabrication method for 1–3 piezocomposite transducers, including: (i) piezoceramic material, (ii) dice in x direction, (iii) dice in y direction, (iv) epoxy filling, (v) reverse, (vi) back-side dicing in x and y directions, (vii) second epoxy filling, and (viii) deposition of conductive layer. Adapted with permission from [3]

4.2.3 Micromachined Ultrasonic Transducers (CMUTs and PMUTs)

Micromachined ultrasonic transducers (MUTs) represent one of the most significant innovations in photoacoustic detector technology, offering miniaturization, array scalability, and low-cost batch fabrication that traditional piezoelectric transducers cannot achieve.

Two primary MUT categories—capacitive MUTs (CMUTs) and piezoelectric MUTs (PMUTs)—differ in their transduction mechanisms but share the benefits of MEMS-compatible microfabrication, enabling fine-pitch 2D arrays, flexible integration with ASIC/CMOS electronics, and potential wafer-scale production [3, 6, 8–11]. Beyond miniaturization, these devices are driving a transition toward wearable and portable photoacoustic imaging PAI platforms, particularly when paired with low-cost excitation sources such as LDs, LEDs, and VCSELs [6, 11].

4.2.3.1 Capacitive Micromachined Ultrasonic Transducers (CMUTs)

A CMUT consists of a conductive membrane suspended above a cavity, forming a variable capacitor between the membrane and the substrate electrode [8, 9]. When a DC bias voltage is applied, the membrane is attracted toward the substrate, creating an equilibrium gap. Acoustic pressure—such as that produced by photoacoustic waves—causes small deflections of the membrane, modulating the capacitance and generating an AC output signal.

CMUTs exhibit several major advantages:

- Broadband frequency response (up to 100% fractional bandwidth);
- Low acoustic impedance, which matches tissue more naturally than bulk PZT;
- Planar geometry, enabling high-density 1D/2D arrays fabricated lithographically;
- Compatibility with silicon micromachining, allowing monolithic integration with CMOS electronics [8–10].

However, CMUTs also face challenges:

- Dielectric charging and reliability issues under high DC bias;
- Pull-in instability, where excessive bias causes membrane collapse;
- Vacuum cavity sealing requirements; and
- Nonlinear electromechanical coupling, which complicates calibration [8, 9].

Recent strategies such as collapse-mode operation, high-k dielectrics (Al_2O_3, HfO_2), charge-neutralized operation, and fully encapsulated CMUT-on-CMOS architectures have significantly improved robustness and sensitivity [8–11]. When operated in receive-only mode for PAI, CMUT arrays achieve sub-millipascal NEP levels, rivaling piezoceramic arrays while offering a thinner, flexible form factor [10, 11].

4.2.3.2 Piezoelectric Micromachined Ultrasonic Transducers (PMUTs)

PMUTs rely on the direct piezoelectric effect rather than capacitive modulation. Each PMUT cell consists of a piezoelectric thin film (e.g., AlN, Sc-AlN, PZT, or PVDF-TrFE) deposited on a flexural membrane. Acoustic pressure induces mechanical bending of the membrane, producing voltage via the piezoelectric film [3, 6, 10].

Key advantages of PMUTs include:

- Low drive voltage and no need for DC bias;
- CMOS process compatibility, particularly for AlN and Sc-AlN films;
- Good mechanical robustness and small footprint;
- Ease of integration into flexible or curved geometries; and
- Potential optical transparency when using PVDF-TrFE or ITO electrodes [3, 6, 10, 11].

However, PMUTs typically operate in resonant bending modes, limiting bandwidth relative to CMUTs. Their sensitivity is determined by film thickness, residual

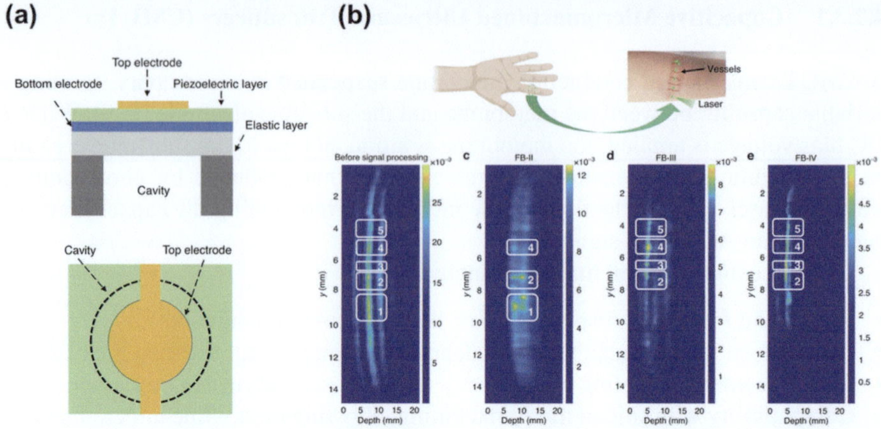

Fig. 4.5 **a** Cross-sectional view (top) and top view (bottom) schematic of the structure of PMUT, **b** PAI of a human hand using PMUT. Comparison of the PA images between their fundamental modes. FB frequency band. Reprinted with permission from [6]

stress, and electrode geometry. Advanced designs now incorporate multi-mode resonance, composite membranes, and thin-film piezopolymers, achieving >50% bandwidth and <2 Pa NEP [10, 11]. Emerging transparent PMUT arrays fabricated on glass or flexible polymer substrates are particularly promising for coaxial LED/VCSEL excitation, where direct optical coupling.

In a recent review by Choi et al. (2025), authors discussed the photoacoustic detection performance of MEMS-fabricated PMUT arrays [6]. Although the study focused on receive-mode operation, the piezoelectric membrane design inherently supports dual-mode capability, enabling both ultrasound transmission and reception in future implementations. The PMUTs, composed of thin piezoelectric membranes suspended over etched cavities, operate at low voltage and are fully CMOS-compatible—key attributes for compact and affordable PAI probes. As shown in Fig. 4.5, optimized frequency-band processing substantially enhanced the contrast and depth resolution of in-vivo vascular images, highlighting the potential of PMUT arrays for portable, low-cost photoacoustic imaging [6, 11].

The fundamental structural differences between conventional piezoelectric transducers, CMUTs, and PMUTs are illustrated in Fig. 4.6, which highlights the cavity-based architecture of CMUTs and the flexural piezoelectric membrane configuration of PMUTs [12].

4.2.3.3 Integration with Affordable Light Sources

The miniaturization of MUTs naturally aligns with the trend toward compact optical excitation. LEDs, edge-emitting LDs, and VCSELs have been successfully integrated with MEMS arrays to create fully solid-state PAI modules.

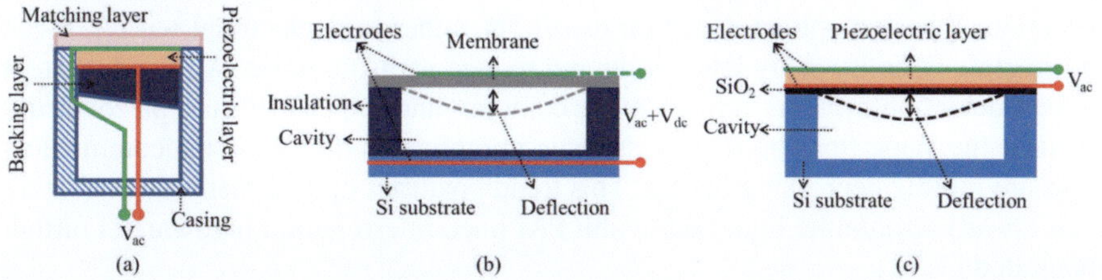

Fig. 4.6 **a** Conventional bulk piezoelectric transducer with matching and backing layers. **b** Capacitive micromachined ultrasonic transducer (CMUT) showing a suspended conductive membrane over a vacuum cavity. **c** Piezoelectric micromachined ultrasonic transducer (PMUT) employing a piezoelectric thin film on a flexible silicon membrane. Reprinted with permission from [12]

- LEDs offer low-cost, wide-area illumination but low pulse energy.
- LDs provide higher intensity and nanosecond pulse modulation.
- VCSELs, the newest entrants, combine the coherence of LDs with the form factor and 2D scalability of LEDs. Their low divergence and uniform beam profile make them ideal for handheld or wearable PAI systems [6, 11].

Similar to the VCSEL–PZT modules described in Chap. 3, the planar geometry VCSELs offers natural compatibility with MEMS-based acoustic detectors. Extending VCSEL illumination to micromachined ultrasonic transducers (CMUTs and PMUTs) could enable monolithic, wafer-level photoacoustic integration. While such compact implementations remain under development, recent analyses identify this as a promising path toward fully solid-state, low-power PAI systems [11, 12].

4.2.3.4 Outlook

MUT technology bridges the gap between traditional piezoelectric and emerging optical detection, combining wide bandwidth, integration potential, and affordability.

CMUTs offer unmatched broadband performance for tomographic systems, while PMUTs are ideally suited for wearable, transparent, or flexible probes. When co-designed with compact illumination (LD/VCSEL) and miniature data acquisition electronics, MUT arrays form the foundation of next-generation portable and affordable PAI systems.

4.2.4 Optical and Hybrid Detectors

Optical ultrasound detection represents an important technical alternative to piezoelectric receivers in PAI. These detectors transduce ultrasound-induced pressure into changes in optical phase or intensity, offering very broad bandwidths, immunity to electromagnetic interference, and transparency that permits coaxial excitation and detection. Because optical detection uses light rather than electrical wiring, it can in principle simplify array integration and reduce parasitic noise. Although current optical detection systems remain more complex and expensive than piezoelectric sensors, the same fiber-optic and photonic components that drive telecommunications are rapidly decreasing in cost. This trend positions optical methods as a future path toward affordable, high-bandwidth PAI once interrogation hardware is further integrated [3, 11].

4.2.4.1 Principles of Optical Detection

Acoustic pressure alters the optical path length of a cavity or waveguide through small refractive-index and thickness variations. When a continuous-wave laser interrogates the cavity, these perturbations produce phase or amplitude modulation of the reflected light, which can be detected by a photodiode. The most widely adopted geometry is the Fabry–Pérot interferometer (FPI)—a thin polymer film bounded by partially reflective mirrors. Common spacer materials such as Parylene C, SU-8, and PDMS can be deposited with low-cost spin-coating processes. The FPI design provides high sensitivity without electrical pre-amplifiers and achieves sub-nanometer displacement resolution [11]. At present, optical detection is not generally used in low-budget systems, but its reliance on inexpensive continuous-wave diode lasers (850 nm or 1550 nm) and standard fibers suggests strong potential for cost reduction as compact interrogators mature.

4.2.4.2 Fabry–Pérot Interferometric Sensors

Polymer-film Fabry–Pérot sensors are now considered a benchmark for optical ultrasound detection in PAI. A transparent polymer spacer, a few micrometers thick, defines the optical cavity between dielectric mirrors deposited on a glass or polymer substrate. Continuous-wave interrogation near a cavity resonance converts pressure-induced thickness changes into measurable intensity modulation. Beard (2011) reported noise-equivalent pressures below 5 mPa $Hz^{-1/2}$ for such polymer FPIs [11]. While the fabrication processes are straightforward and low in material cost, the required interrogation modules like precision tunable lasers and photodetectors currently dominate total system expense. Nonetheless, progress in miniature diode-laser interrogators and monolithic photonic circuits indicates that similar performance may become achievable with significantly reduced cost and footprint in the near future. Figure 4.7 shows the schematic showing scanner read-out architecture and the latest Fabry–Perot ultrasound sensor (FPUS) from the group of Beard [13].

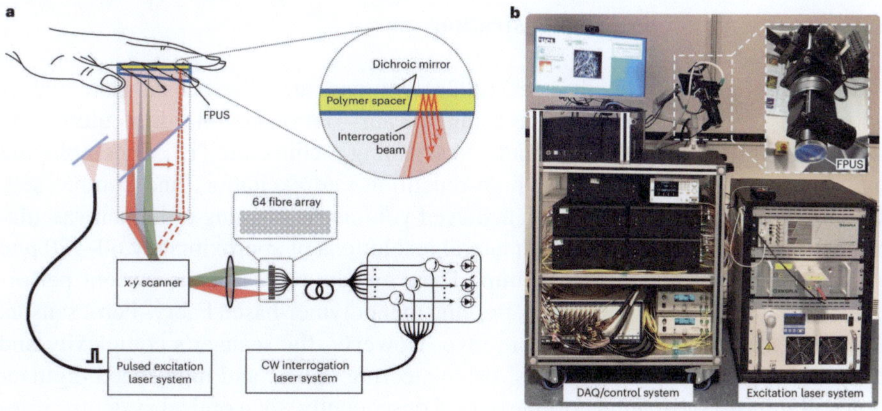

Fig. 4.7 **a** Schematic showing scanner read-out architecture and FP ultrasound sensor (FPUS) structure. **b** Photograph of scanner showing cart-based acquisition system, imaging head (inset) and OPO excitation laser system. Adapted with permission from [13]

4.2.4.3 Fiber-Optic and Waveguide Detectors

Fiber-optic detection extends interferometric sensing into flexible and alignment-free geometries. In fiber-tip FPI or fiber Bragg grating (FBG) sensors, acoustic pressure modulates either cavity length or grating period, producing a wavelength-dependent intensity change in the reflected signal [11]. These detectors already benefit from the mature telecommunications supply chain: standard single-mode fibers, low-cost diode lasers, and photodiodes are readily available and reliable.

Although the interrogators needed for high-sensitivity PAI still cost more than piezoelectric amplifiers, prototype systems demonstrate that compact, single-fiber detectors can combine optical excitation and detection within one fiber—reducing cabling complexity and paving the way for inexpensive, integrated probes once interrogation electronics are simplified.

4.2.4.4 Hybrid Optical–Piezoelectric Detectors

Hybrid detectors merge piezoelectric transduction with optical readout, aiming to combine the robustness of ceramic or polymer sensors with the low-noise advantages of optical interrogation. Configurations include PVDF-TrFE films forming one mirror of an optical cavity and micromachined ultrasonic transducers (MUTs) coupled to external FPIs for optical monitoring [10, 12]. Optical readout removes the need for high-impedance amplifiers and complex shielding, simplifying circuit design. Although not yet lower-cost in practice, such hybrid approaches show how optical interrogation could eventually reduce system complexity as scalable photonic components become available.

4.2.4.5 Recent Clinical Demonstration

A recent advance by Huynh et al. (2024) demonstrated the clinical potential of all-optical photoacoustic imaging using a Fabry–Pérot polymer-film ultrasound sensor integrated into a fast, parallelized scanning architecture [13]. By employing multibeam optical interrogation, high-repetition-rate excitation, and compressed-sensing reconstruction, the system achieved volumetric imaging of human vasculature within seconds (Fig. 4.8), with lateral resolutions of approximately 60–120 μm and penetration depths up to 15 mm. These results represent the current performance frontier of optical detection, proving that polymer-based Fabry–Pérot sensors can deliver clinical-grade images in vivo. However, the scanner's complexity and reliance on precision laser sources, beam-steering optics, and high-speed digitization still impose substantial cost barriers. Consequently, such optical systems exemplify technological feasibility rather than affordability at present. Bridging this gap requires parallel progress in simplified optical interrogation modules, low-power laser diodes, and cost-optimized data-acquisition electronics. The latter forms the next crucial component of affordable photoacoustic imaging, discussed in Sect. 4.6, which addresses data acquisition and signal conditioning strategies for cost-effective system implementation.

4.2.4.6 Comparison and Outlook

Optical and hybrid detection techniques already match or exceed the acoustic bandwidth of conventional piezoelectric sensors while providing transparency and immunity to electromagnetic interference. Their present limitations lie mainly in the price and stability of interrogation lasers and in temperature-induced drift. With ongoing miniaturization of laser diodes, development of passive multiplexing schemes, and polymer-based cavity fabrication, optical detection is moving steadily toward cost-efficient integration.

At this stage, optical detectors should be regarded as complementary to piezoelectric technologies within affordable PAI, offering performance benefits and design flexibility rather than immediate cost savings. Continued research into integrated photonic circuits and wafer-level polymer FPIs is expected to close the cost gap and enable future low-cost, all-optical photoacoustic modules suitable for portable or point-of-care applications.

4.2.5 Global Comparison and Outlook on Detector Technologies

The broad range of detector technologies described in previous sections, from conventional piezoceramic and polymer films to micromachined and all-optical

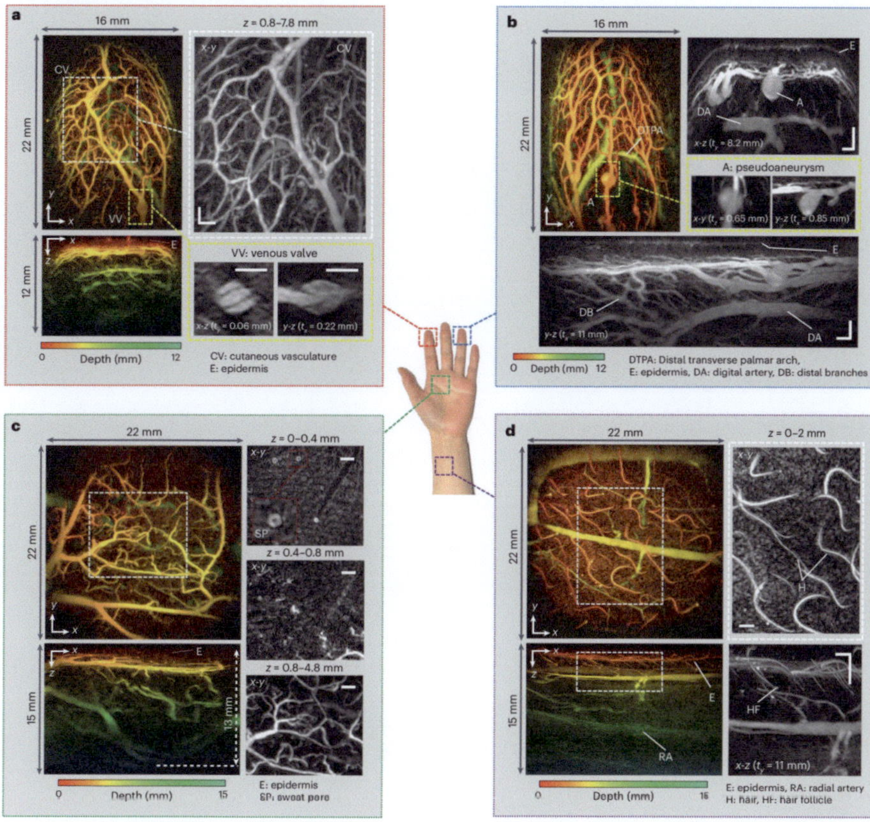

Fig. 4.8 In-vivo volumetric photoacoustic images acquired using a fast all-optical Fabry–Pérot photoacoustic tomography scanner: **a** index fingertip showing cutaneous vasculature and venous valve; **b** ring fingertip revealing dermal microvascular anatomy and digital artery; **c** palm region displaying vascular structures to a depth of $\sim$13 mm; **d** wrist region showing the radial artery and hair follicles. Imaging wavelength = 850 nm; lateral sampling = 108 μm; total scan time $\approx$5 s. Adapted with permission from [13]

sensors, illustrates the diversity of approaches currently pursued in PAI. Each technology represents a unique balance between sensitivity, bandwidth, fabrication complexity, and cost. To aid in system selection, the key characteristics of representative detector classes are summarized in Table 4.1, enabling a direct comparison of their suitability for affordable and clinically translatable PAI.

The data in Table 4.1 emphasize the fundamental trade-offs between performance, cost, and functionality. A general inverse relationship exists between cost and bandwidth, polymer and piezoceramic detectors remain economical but relatively narrowband, whereas optical interferometric systems deliver exceptional bandwidth at higher expense. Piezoceramic composites and PMUTs occupy the most balanced

Table 4.1 Comparative characteristics of photoacoustic ultrasound detectors and their suitability for dual-mode operation

Detector type	Sensitivity (NEP)	Bandwidth (−6 dB) (MHz)	US transmission capability	Cost level	Integration complexity	Key comments
PVDF/PVDF-TrFE film	0.3–1.2 Pa (@ 1–30 MHz, Kurnikov 2024)	1–30	Moderate (piezoelectric)	Low	Simple	Flexible polymer film; sub-pascal NEP with inexpensive materials and standard electronics; ideal for affordable and educational PAI
Piezoceramic composite (1–3 or 2–2)	1–10 mPa $Hz^{-1/2}$	1–10	High	Medium	Moderate	Strong electromechanical coupling ($k_t > 0.5$); established for dual-mode clinical probes
CMUT	5–50 mPa $Hz^{-1/2}$	5–30	High	Medium–high	High (bias required)	Dual-mode operation; CMOS-compatible; higher fabrication cost
PMUT	1–10 mPa $Hz^{-1/2}$	1–10	Potential (bidirectional design)	Medium	Medium	MEMS-based; demonstrated PA detection; low-voltage drive; compact array potential

(continued)

Table 4.1 (continued)

Detector type	Sensitivity (NEP)	Bandwidth (−6 dB) (MHz)	US transmission capability	Cost level	Integration complexity	Key comments
Optical (FPI)	100–200 mPa $Hz^{-1/2}$ (Huynh 2024)	10–25	Receive only	High (current)	Moderate–high	All-optical, broad bandwidth; moderate sensitivity; system cost dominated by interrogation optics
Hybrid optical–piezoelectric	5–20 mPa $Hz^{-1/2}$	5–20	Potential (design-dependent)	Medium–high	High	Combines optical readout and piezoelectric TX; promising but early-stage

Data synthesized from [3–5, 10–13]. *Note* NEP values for PVDF/TrFE detectors are reported in Pa over the stated bandwidth rather than normalized per $\sqrt{Hz}$

region of this design space, combining sufficient sensitivity and bidirectional operation with manageable fabrication complexity. In contrast, optical and hybrid detectors provide superior frequency response and immunity to electromagnetic interference but depend on costly interrogation optics and precise alignment.

The choice of detector should therefore align with the system's primary design objective.

Cost-driven implementations, for instance, portable or educational PAI systems, benefit from PVDF or simple piezoceramic receivers, which are inexpensive and easy to integrate.

Performance-driven or dual-mode clinical systems justify the added complexity of CMUT, PMUT, or hybrid architectures, which enable both ultrasound transmission and reception.

The comparative relationships among detector sensitivity, bandwidth, cost, and ultrasound transmission capability are further visualized in Fig. 4.9. This bubble-plot representation highlights the trade-off between performance and affordability across the major detector classes. Polymer-based PVDF/PVDF-TrFE detectors occupy the low-cost, moderate-sensitivity region, while piezoceramic composites and MEMS transducers (CMUT, PMUT) achieve higher performance at greater fabrication complexity. Optical and hybrid detectors extend bandwidth but remain limited by receive-only operation and interrogation cost, emphasizing the need for integrated, low-power solutions in next-generation portable PAI systems.

Selecting an appropriate detector for PAI requires balancing sensitivity, bandwidth, manufacturability, and functional versatility. For systems based on low-fluence LED or LD sources, detector sensitivity is the primary factor determining image quality. At the same time, dual-mode capability, transmission and reception remains crucial for clinical translation, where co-registered ultrasound and photoacoustic images are expected. Piezoceramic composites currently provide the most practical compromise between cost, robustness, and transmit efficiency. PVDF-based piezopolymers remain the lowest-cost option for portable or instructional systems, offering sub-pascal sensitivity with simple readout electronics. CMUTs and PMUTs introduce wafer-scale integration and compactness, while optical Fabry–Pérot detectors achieve unmatched bandwidth and sensitivity at higher cost. Emerging hybrid optical–piezoelectric approaches may ultimately combine optical readout with acoustic actuation, bridging high-performance and affordable operation in a single device class. From an affordability perspective, piezopolymer and simple piezoceramic detectors presently offer the best balance of performance and cost. Continued progress in MEMS and photonic integration is expected to yield hybrid and micromachined detectors that achieve both high sensitivity and dual-mode capability at lower cost, paving the way toward clinically deployable, low-cost photoacoustic systems within a compact footprint.

The following section examines how data-acquisition hardware and signal-conditioning electronics influence the overall sensitivity, dynamic range, and cost of photoacoustic systems. Particular attention is given to simplified architectures based on single-element detectors, multiplexed arrays, and compact digitizers, which are increasingly used to realize low-cost, portable, and educational photoacoustic

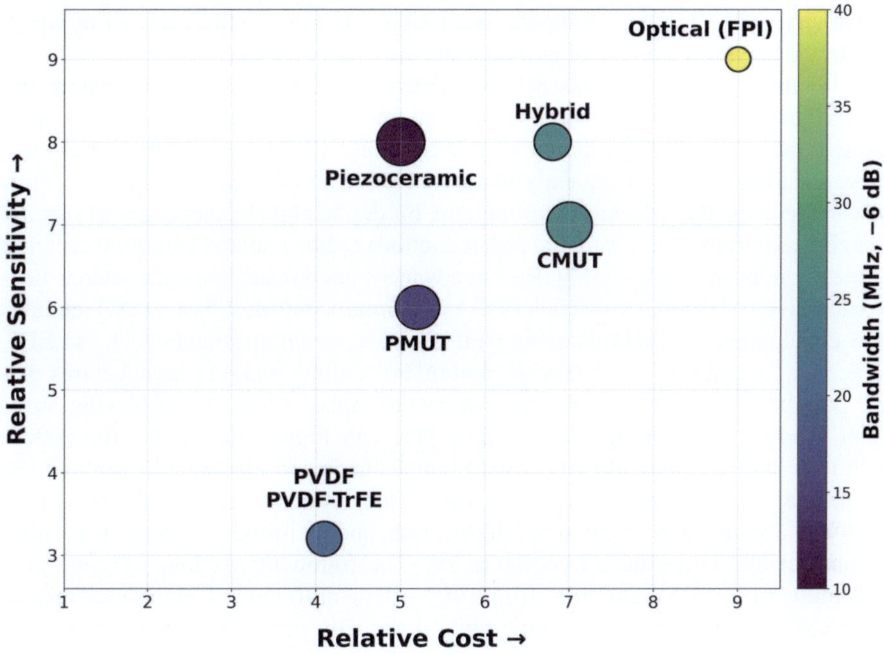

Fig. 4.9 Detector design landscape for affordable photoacoustic imaging. Bubble position shows the trade-off between relative cost (x-axis) and relative sensitivity (y-axis). Color encodes detector bandwidth (MHz, −6 dB), and bubble size reflects ultrasound transmission capability (larger = stronger TX; receive-only for optical FPI). Polymer piezopolymers (PVDF/PVDF-TrFE) occupy the low-cost, moderate-bandwidth region; piezoceramic composites, CMUTs, and PMUTs provide higher bidirectional performance at increased complexity; optical Fabry–Pérot detectors deliver the widest bandwidth but are receive-only with higher interrogation cost; hybrid detectors bridge optical and piezoelectric approaches, promising balanced performance for future portable PAI systems

imaging platforms. These approaches illustrate how thoughtful system-level design can extend affordability without compromising diagnostic performance.

4.3 Affordable Data Acquisition and System Integration

4.3.1 Introduction

The data-acquisition (DAQ) subsystem forms the bridge between the photoacoustic detector and image-reconstruction stage, directly influencing both diagnostic performance and overall system cost. Regardless of detector type—piezoelectric, optical, or hybrid—the analog photoacoustic signal must be amplified, filtered, digitized, and synchronized with the excitation source to reconstruct accurate spatial maps of absorbed optical energy. Consequently, DAQ design dictates key parameters such as sensitivity, dynamic range, temporal resolution, and power efficiency, all of which determine a system's practical usability and scalability [2, 11].

Affordable PAI systems must achieve a delicate balance between performance and cost. High-end laboratory instruments often employ multi-channel digitizers operating at hundreds of mega samples per second and 12–16-bit resolution; however, these specifications are frequently unnecessary for portable, educational, or point-of-care devices. By tailoring sampling rate, bit depth, and channel count to specific imaging requirements, substantial cost reductions can be realized without sacrificing clinically relevant fidelity [1, 7]. Recent advances in compact and open-source electronics have accelerated this trend. FPGA- and microcontroller-based DAQ modules now enable precise synchronization with low-cost excitation sources such as LEDs, LDs, and VCSELs [14–19]. These combinations allow imaging architectures that are portable, battery-operated, and low-maintenance, while still achieving sufficient bandwidth for tissue-level imaging [15, 16]. Figure 4.10 shows the system architecture of a handheld LED-based photoacoustic and ultrasound imaging platform, adapted from Xia et al. [21]. The design integrates LED drivers, analog front-end amplification, high-speed digitization, and real-time reconstruction within a compact and cost-efficient configuration. The following sections examine how streamlined DAQ configurations from single-element to multichannel implementations can maintain imaging fidelity while drastically reducing system complexity and price.

4.3.2 Single-Element and Minimalist PAI Architectures

Single-element PAI systems embody the most economical realization of the technique, prioritizing simplicity over speed. In contrast to array-based implementations that require dozens to hundreds of parallel acquisition channels, each with dedicated amplifiers, filters, digitizers, and synchronization lines, a single-element configuration reduces the entire front-end to one detection channel. This immediately eliminates the need for high-cost multichannel analog front ends (AFEs), high-bandwidth data buses, and large-scale storage and reconstruction hardware. Each acquisition channel in a conventional array typically contributes not only the cost of its transducer element but also its corresponding analog and digital circuitry. For example, a 64-element array requires 64 low-noise preamplifiers, 64 anti-aliasing filters, and a multichannel analog-to-digital converter (ADC) with simultaneous sampling capability. In contrast, a single-element detector can use a single low-noise amplifier (LNA) and a cost-efficient high-resolution ADC operating at modest sampling rates. The simplified signal chain significantly reduces power consumption, printed circuit board (PCB) complexity, and electromagnetic interference management requirements.

Another major cost factor in parallel acquisition systems is data throughput. A 128-channel system sampling at 40 mega-samples per second (MS/s) and 12-bit depth produces nearly 80 MB/s of raw data per frame, demanding expensive high-speed interfaces and large memory buffers. Single-element scanning systems avoid

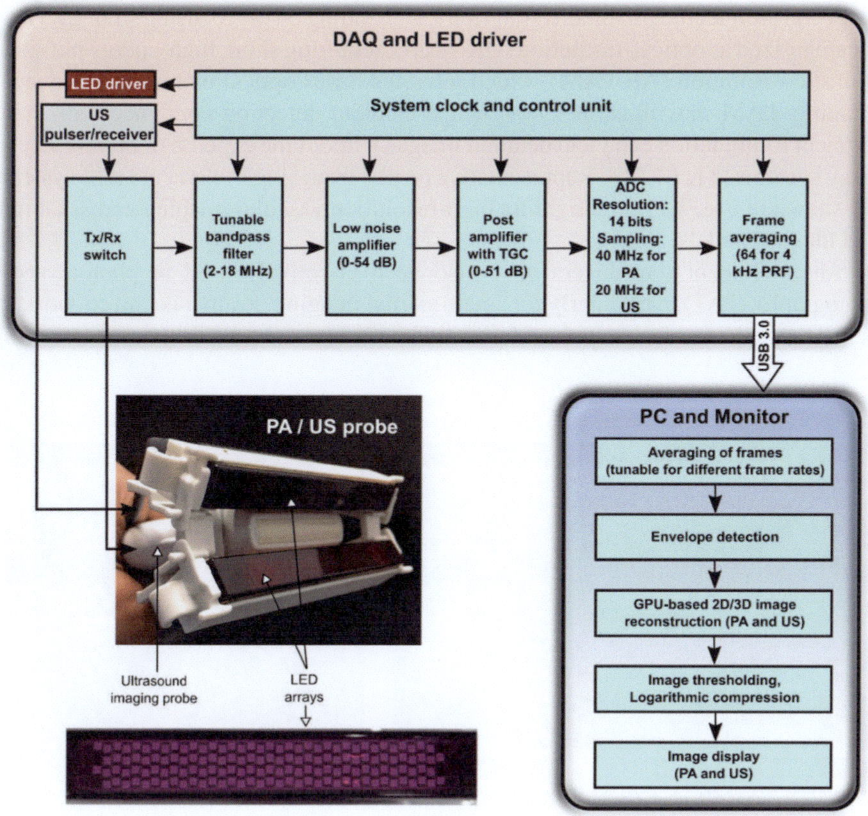

Fig. 4.10 Schematic of the handheld LED-based PA/US imaging system showing LED excitation, DAQ front-end, analog filtering/amplification, ADC/USB interface, and real-time processing. Adapted with permission from [21]

this overhead entirely, acquiring and storing only one time-domain signal at a time, which can be averaged in real-time to improve signal-to-noise ratio (SNR) using inexpensive microcontrollers or USB-based oscilloscopes [7]. The net result is a platform where every subsystem from transducer to digitizer is minimized in count and cost, yielding hardware budgets 10–50 times lower than array-based counterparts. These intrinsic efficiencies form the engineering basis for affordable, portable PAI suitable for low-resource or educational environments [1, 7, 15].

Single-element piezoelectric transducers are widely used in photoacoustic microscopy (PAM) for high-resolution imaging of small and superficial regions. In PAM, a focused pulsed laser beam is raster-scanned across the sample while a single detector records photoacoustic signals at each position. Recent efforts have explored replacing bulky pulsed lasers with compact LDs or VCSELs, particularly in acoustic-resolution PAM, where slightly longer pulse widths are acceptable. These affordable light sources enhance portability and reduce system cost, while the single-channel detection architecture maintains simplicity. Depending on the focusing strategy, PAM is categorized as optical-resolution (OR-PAM), requiring short, high-energy pulses or acoustic-resolution (AR-PAM), which tolerates lower optical intensities for deeper imaging. PAM thus illustrates how single-element detection combined with cost-efficient illumination can yield detailed images without the expense or complexity of array systems [11, 14, 20]. Representative results from single-element PAM systems are shown in Fig. 4.11, highlighting high-resolution vascular imaging across animal and human models.

Single-element transducers have also been effectively used in photoacoustic tomography (PAT), particularly for small-animal imaging. Pramanik and co-workers demonstrated a series of pulsed-laser-diode (PLD)-based PAT systems, showing

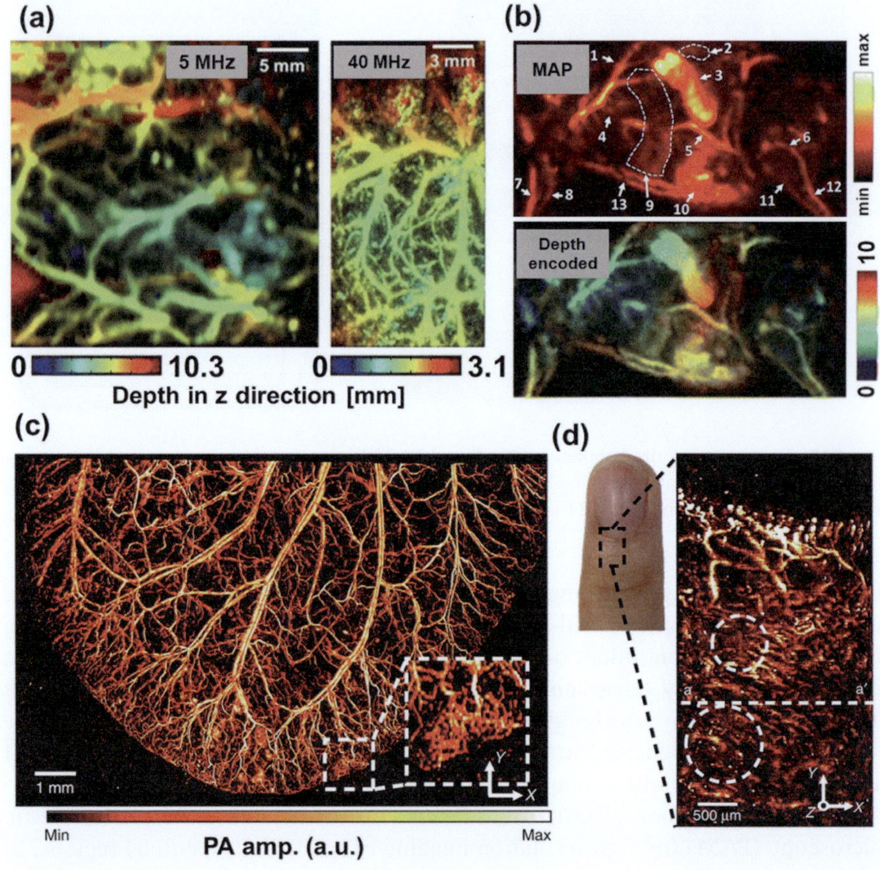

Fig. 4.11 Representative photoacoustic images acquired using single-element transducers. **a** Depth-encoded vasculature of a rat back imaged with 5 and 40 MHz detectors. **b** Whole-body sagittal photoacoustic image of a mouse. **c** Maximum amplitude projection (MAP) of a mouse ear showing microvasculature. **d** MAP image of the human fingertip illustrating superficial vascular networks. Reprinted with permission from [6]

that single-element configurations can achieve high-resolution tomographic imaging with very low hardware cost. Their early design by Upputuri and Pramanik (2015) achieved three-dimensional imaging of tissue phantoms and small animals with sub-200 μm resolution using a portable PLD source [22]. Subsequent desktop systems improved scanning speed and penetration depth through multi-diode excitation and optimized geometry [23]. Later, in vivo studies visualized venous sinus distension in a small-animal model of intracranial hypotension, highlighting the potential of compact PAT systems for preclinical functional imaging [24]. Together, these studies confirm that single-element PLD-PAT provides a practical route toward affordable, high-performance tomographic imaging.

In addition to laser-diode PLD-PAT systems, Pramanik's group recently demonstrated a single-element LED-PACT system using pulsed high-power LEDs as the illumination source [25]. By carefully pulsing the LEDs and synchronizing with a single broadband transducer, they achieved tomographic imaging with substantial cost savings. They provided quantitative estimates of system cost and component trade-offs, highlighting that LED-PACT substantially lowers optical source cost (vs. pulsed lasers), avoids large array hardware, and maintains clinically useful contrast in small-animal imaging. This example underscores that single-element tomography architectures with LED illumination are not only technically viable but also strongly aligned with affordability goals (Fig. 4.12).

Beyond cost-driven PLD-PAT and LED-PAT developments, innovative concepts from Lihong V. Wang's laboratory have extended single-element detection into high-performance tomography through photoacoustic topography through an ergodic relay (PATER) and its advanced version, photoacoustic computed tomography through an ergodic relay (PACTER) [26, 27]. In these systems, rather than mechanically scanning the detector or using arrays, acoustic waves from the entire object are redirected through an *ergodic relay*, a specially shaped acoustic cavity that scrambles spatial information into unique time-domain signatures. A single broadband transducer placed at the relay output records these temporally encoded signals, which are computationally decoded to reconstruct the original 3D photoacoustic field. The PATER concept was first demonstrated for small samples, showing full-field imaging using only one detector [26]. The later PACTER system extended this approach to in vivo human imaging, producing detailed vascular maps of a human hand with sub-millimeter resolution [27]. By converting spatial encoding into temporal diversity, these ergodic-relay systems eliminate the need for mechanical scanning or array detectors while maintaining high imaging speed and fidelity. In the PATER/PACTER system, the ergodic relay acts as a complex acoustic cavity that maps spatial information from the entire imaging volume into unique temporal delay patterns at the single detector, enabling full-field photoacoustic acquisition without mechanical scanning. Representative results from PATER are shown in Fig. 4.13, demonstrating wide-field functional imaging of a mouse brain through an intact skull. The system enabled real-time monitoring of oxygen saturation (sO_2) changes during alternating normoxic and hypoxic conditions, achieving imaging rates of 50 frames per second. The approach

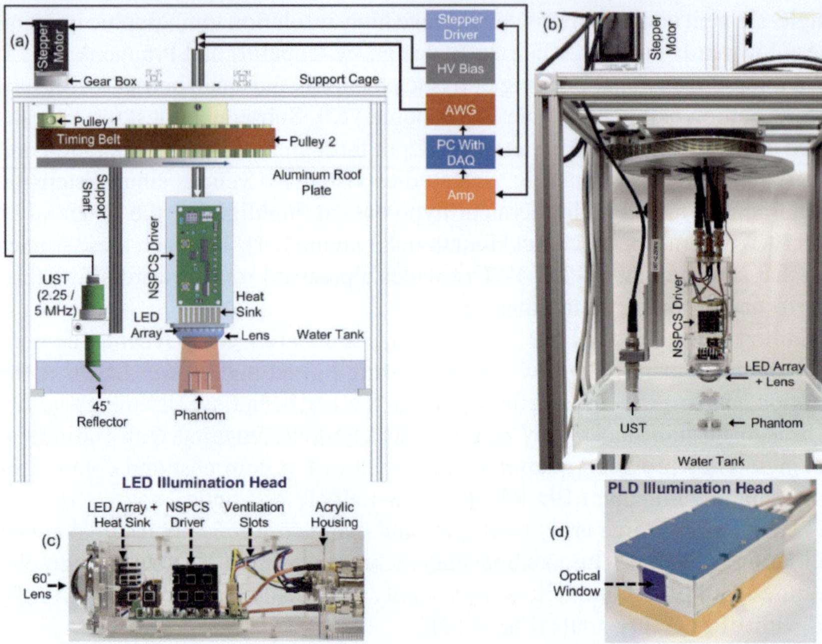

Fig. 4.12 **a** Schematic of the LED array-based PACT (LED-PACT) with single element UST. **b** Photograph of the LED-PACT system. **c** Photograph of the custom-built pulsed LED array illumination head. **d** Photograph of a commercial PLD illumination head used in the same system for comparison experiments. Reprinted with permission from [25]

represents a paradigm shift in single-element tomography, combining the affordability of single-channel acquisition with the volumetric capability of array-based systems.

Overall, single-element photoacoustic systems span a wide performance range from low-cost PLD-PAT designs for small-animal imaging to advanced ergodic-relay–based tomography capable of human in vivo studies. The simplicity of single-channel detection minimizes hardware cost, while innovations such as multi-diode excitation and temporal encoding extend functionality toward volumetric and functional imaging. These developments collectively demonstrate that affordable yet high-performance photoacoustic imaging is achievable through intelligent system architecture rather than expensive components [22–27].

4.3.3 Multi Channel Detection and Frontend Design

While single-element systems demonstrate the ultimate hardware simplicity, their sequential acquisition inherently limits imaging speed. Multichannel detection architectures overcome this constraint by capturing signals from many detector elements

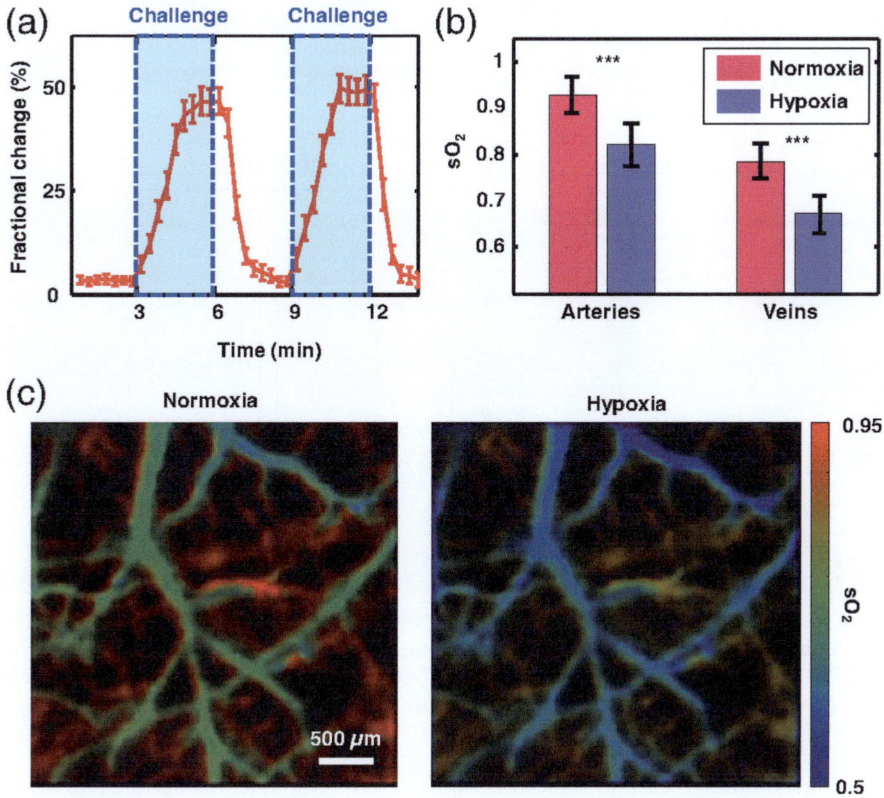

Fig. 4.13 Functional imaging using PATER. **a** Fractional change in photoacoustic signal during two oxygen challenge cycles in a mouse brain. **b** Quantified blood oxygen saturation (sO_2) in arteries and veins under normoxia and hypoxia. **c** Depth-resolved sO_2 maps before and after hypoxic challenge, imaged through an intact skull. Reprinted with permission from [26]

in parallel, enabling real-time or volumetric PAT. However, cost scales almost linearly with channel count because each channel requires its own AFE, comprising a low-noise preamplifier, anti-aliasing filter, and ADC. In conventional clinical ultrasound systems, these electronics dominate total system cost, often exceeding the transducer price itself [1].

To address this, recent studies have explored cost-optimized multichannel DAQ systems using integrated ultrasound AFEs and programmable logic platforms. Commercial multichannel ICs such as Texas Instruments AFE5832 and Analog Devices AD9670 provide up to 32 parallel receive paths with programmable gain and time-gain compensation, reducing board area and component cost (Fig. 4.14). In a 2025 implementation, Zafar et al. demonstrated a 30-channel PACT system built from low-cost evaluation modules and open-source firmware, achieving simultaneous acquisition with effective 12-bit resolution at a total hardware cost below USD 10,000 [7]. Their design highlighted that affordable DAQ performance can

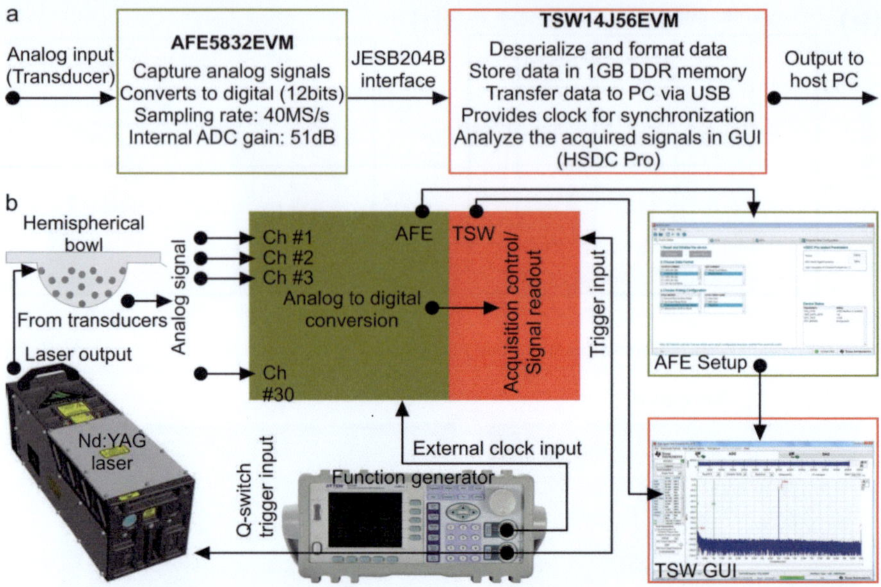

Fig. 4.14 Affordable multichannel DAQ for PACT. Block diagram of the TI AFE5832EVM (32-channel, 12-bit, 40 MS/s with TGC) feeding the TSW14J56EVM FPGA capture board via JESD204B, with synchronized laser and clock triggers; connections shown for a 30-element hemispherical PACT. Reprinted with permission from [7]

be achieved through careful trade-offs between sampling rate, resolution, and noise performance.

Beyond hardware selection, data throughput and synchronization are major challenges in parallel acquisition. A 128-channel system sampling at 40 MS/s produces several hundred megabytes of data per frame, requiring high-speed interfaces and large buffers. FPGA-based architectures provide deterministic timing and real-time control of excitation, amplification, and digitization, as demonstrated by Wu et al. (2018) in an intravascular photoacoustic–ultrasound (IVPA/US) imaging backend [18]. Such FPGA controllers allow precise triggering of both light source and detector arrays, reducing jitter and improving frame-to-frame stability while maintaining compact, portable form factors [7, 18].

Further cost reduction in multichannel systems is achieved through channel optimization and signal conditioning strategies that minimize the number of active acquisition paths without significant loss in image fidelity. The simplest approach is analog time-division multiplexing, in which groups of detector elements share the same amplifier and ADC through high-speed electronic switches or multiplexers. Although sequential channel switching introduces minor latency and may reduce frame rate, it substantially lowers hardware count and power consumption, an effective solution for portable PACT and LED-based systems [1, 28].

Advanced implementations employ frequency-division or coded excitation to distinguish signals from multiple detectors operating simultaneously, allowing partial parallelization even with limited digitizer channels [29, 30]. Meanwhile, compressed sensing and sparse-sampling reconstruction methods are increasingly applied to recover tomographic images from undersampled data, further reducing acquisition hardware and storage demands [31–33]. The AFE remains critical to overall sensitivity: wideband polymer and PVDF-based detectors typically require higher preamplifier gain and careful impedance matching to maintain sub-pascal noise performance [4, 5]. As demonstrated in recent PVDF and piezopolymer array work, properly tuned low-noise amplifiers can preserve dynamic range using inexpensive op-amps and minimal shielding [4, 5].

4.3.3.1 Amplification and Gain Management in Hybrid PA/US Systems

In PAI driven by low-fluence sources such as LEDs, laser diodes, or VCSELs, SNR depends critically on low-noise, multi-stage amplification. Since the generated acoustic pressures are several orders of magnitude weaker than those encountered in pulse-echo ultrasound, clinically usable SNRs typically require total gains approaching 90–100 dB, achieved through cascaded preamplifier and programmable gain stages [14, 15]. In contrast, conventional ultrasound systems apply only ~40–50 dB of gain, sufficient for high-pressure echoes but inadequate for PA signals. Modern hybrid platforms such as AcousticX therefore employ fast electronic gain switching, toggling between high-gain PA and low-gain US modes at kilohertz rates synchronized with excitation pulses [34]. This prevents saturation during ultrasound reception while maintaining maximum sensitivity for photoacoustic detection. Efficient gain management thus represents a key enabler of affordable, dual-modality PAI systems, ensuring that reduced light fluence does not compromise clinical imaging performance.

Through a combination of hardware multiplexing, intelligent sampling, and optimized signal conditioning, recent photoacoustic systems have achieved nearly an order-of-magnitude reduction in data-acquisition cost. These advances now enable large-area or multiview imaging using compact, battery-powered platforms, bridging the gap between laboratory prototypes and practical portable devices [1, 4, 5, 7, 28–33].

The comparative summary in Table 4.2 underscores that data-acquisition architecture is the core determinant of both cost and functionality in photoacoustic imaging systems. While single-element PLD- and LED-based configurations remain the most economical solutions, their limited frame rate and lack of true ultrasound capability confine them primarily to preclinical or educational use. By contrast, multichannel and embedded FPGA-based systems strike a compelling balance between affordability, speed, and multimodal compatibility, enabling simultaneous PA and US acquisition with real-time reconstruction. Emerging compressed and edge-AI architectures further reduce data volume and power demands, pointing toward compact, battery-powered designs suitable for bedside or point-of-care deployment. Ultimately, the pathway to affordable clinical PAI will depend on integrating cost-efficient excitation (LEDs, LDs, or VCSELs) with optimized front-end amplification, agile digitization, and intelligent acquisition strategies, transforming laboratory prototypes into practical diagnostic tools.

Table 4.2 Simplified comparison of data-acquisition architectures for affordable PAI

System type	Approx cost (USD)	Acquisition mode	Integration complexity	Ultrasound capability	Key features/ takeaway
Single-element (PLD-PAT) [22–24]	<3,000	Sequential scanning	Very low	Limited	Simplest low-cost design for PA microscopy/ tomography and preclinical use
Single-element (LED-PACT) [25]	≈4,000	Sequential scanning	Low–medium	Limited	Validated affordable 3D PA tomography using LED illumination
Multichannel (parallel) [7, 17, 18]	5,000–10,000	Real-time parallel	Medium	Native PA + US capability	Scalable array acquisition; basis for clinical dual-mode systems
Compressed/ multiplexed [28–33]	3,000–7,000	Time/ Freqmultiplexed	Medium–high	Possible but depends on reconstruction strategy	Reduces digitizer count; compact and power efficient
Embedded/ Edge-AI [17, 18, 33]	<5,000	On-device reconstruction	Medium	Supported in hybrid portable systems	Portable, real-time operation without external workstation
Research-Grade (Verasonics, AcousticX) [1, 15]	>70,000	Fully parallel (128–256 ch)	High	Benchmark clinical PA + US integration	Gold-standard performance; reference for affordable designs

* Cost of the system may vary based on design complexity

4.3.4 *Digitization, Processing, and Integration*

4.3.4.1 Real-Time Processing: FPGA, GPU, and Hybrid Approaches

To achieve real-time imaging, many systems offload reconstruction tasks to FPGA or GPU accelerators rather than relying on a host CPU. For example, Zafar et al. (2025) demonstrated FPGA-based real-time beamforming and acquisition control for photoacoustic computed tomography, enabling portable, multi-channel operation at significantly reduced cost [7]. Similarly, Wu et al. (2018) implemented an FPGA backend for intravascular photoacoustic and ultrasound imaging, achieving frame rates sufficient for real-time display [18]. On the GPU side, real-time volumetric PA/US imaging using GPU-accelerated reconstruction has already been reported, achieving several-fold faster processing than CPU methods while maintaining diagnostic image quality [35]. In another example, Gonzalez et al. (2020) demonstrated real-time short-lag spatial coherence (SLSC) beamforming on GPU hardware, improving photoacoustic contrast under low optical fluence conditions [36]. More recently, Zheng et al. (2024) introduced an FPGA-based model-based (s-Wave) reconstruction architecture, reporting acceleration factors of 26–270× over CPU implementations and >2700× improvements in energy efficiency, an important step toward palm-size photoacoustic tomography systems [37]. Such FPGA and GPU-based accelerations represent a convergence between low-cost acquisition and high-performance processing, vital for affordable portable photoacoustic platforms.

4.3.4.2 Integration: Synchronization, Control, and Embedded Systems

Beyond raw processing, system integration requires precise orchestration of excitation timing, gain switching, and data management. FPGAs typically serve as the central synchronization hub, generating triggers for LEDs or laser diodes, switching preamplifier gains between PA and US modes, and coordinating ADC sampling and data routing [7, 17, 34]. In dual-modality systems, tight synchronization between ultrasound transmission and photoacoustic reception is essential to prevent crosstalk or saturation. Increasingly, embedded SoC platforms (e.g., NVIDIA Jetson, ARM-based boards) are incorporated downstream of FPGA modules for real-time image reconstruction, AI-based denoising, or display. These compact embedded processors drastically reduce system size and power consumption while maintaining sufficient computational power for real-time rendering [17, 18].

Overall, an efficient digitization and processing chain combining low-noise amplification, moderate-speed ADCs, FPGA-based beamforming, and embedded GPU post-processing forms the backbone of affordable, high-throughput photoacoustic imaging. The main design trade-offs remain balancing throughput, compute power, and cost, with a growing emphasis on edge-compute architectures that eliminate dependence on bulky workstations.

4.3.5 System Integration and Synchronization

A well-integrated system architecture is fundamental to achieving cost-effective, high-performance PAI. In affordable implementations, careful orchestration between the excitation source, amplification stages, digitizers, and embedded processors determines not only image quality but also system reliability and portability.

4.3.5.1 Timing and Triggering Control

Accurate synchronization ensures that each optical pulse coincides with the acquisition window of the acoustic detector. In most systems, an FPGA acts as the master controller, generating precisely timed triggers for light sources (LED, LD, or VCSEL), controlling amplifier gain states, and initiating analog-to-digital conversion. In dual-mode PA/US systems, precise temporal coordination is particularly critical to prevent acoustic overlap and to switch gains appropriately between modalities. For example, Yoo et al. (2023) demonstrated a switchable preamplifier that rapidly toggles between high-gain and low-gain modes to accommodate PA and US signals within the same acquisition cycle [34]. Similarly, Zafar et al. (2025) implemented FPGA-based synchronization in a multichannel PAI system, achieving sub-microsecond timing precision with minimal hardware cost [7]. Such strategies maintain high signal-to-noise ratios (SNR) even under low-fluence conditions typical of affordable light sources.

4.3.5.2 Embedded Systems and SoC Integration

The recent shift toward embedded processors and system-on-chip (SoC) architectures enables portable and battery-operable PAI devices. Low-power computing platforms such as NVIDIA Jetson, ARM-based SoCs, or Raspberry Pi Compute Modules can perform data buffering, pre-processing, and even neural-network-based denoising on-board. For instance, Lan et al. (2021) demonstrated a deep-learning-assisted real-time PAT system that required only a single data-acquisition channel, effectively reducing system size and cost [33]. Combining such embedded control units with FPGA front-ends yields compact, self-contained systems capable of real-time image reconstruction and display without dependence on external computers.

4.3.5.3 Power and Thermal Management

Miniaturized and handheld systems must also address power efficiency and thermal stability. LED-based PAI units typically consume 10–30 W of electrical power, while nanosecond-pulsed laser-diode or VCSEL arrays can exceed 50 W during peak operation [14–16, 25]. Efficient heat dissipation via heat-sinks, active cooling fans, or thermoelectric elements is vital to maintain consistent pulse energy and transducer sensitivity. Compact power-distribution boards with isolated voltage domains (for digital, analog, and optical sections) minimize noise coupling and protect sensitive analog front-ends from switching interference [7, 17].

4.3.5.4 Communication Interfaces and Data Transfer

Affordable systems rely on low-cost, high-bandwidth communication links for transferring digitized data. USB 3.0 remains the most widely adopted interface, supporting sustained transfer rates above 300 MB/s at minimal hardware complexity. Higher-end configurations employ PCI Express or Gigabit Ethernet to accommodate real-time volumetric imaging or array-based acquisition. Some recent prototypes couple FPGA DMA streaming directly to embedded GPUs, enabling on-board reconstruction without external PCs, an essential step toward fully integrated, bedside-ready PAI systems [35–37].

A representative implementation is the AcousticX platform, which achieves real-time dual-modality (LED-PA/US) imaging using a pulse repetition rate (PRR) of 4 kHz. To balance data bandwidth with USB 3.0 transfer limits, 64 consecutive pulses are averaged on-board, resulting in an effective frame rate of 62.5 fps. For a 4 cm imaging depth, with 40 MS/s sampling for PA and 20 MS/s for US, the corresponding data throughput (~33 MB/s) comfortably fits within USB 3.0 capacity. This hardware averaging reduces data size by nearly two orders of magnitude compared with raw acquisition while preserving signal fidelity. The differing sampling rates also reflect bandwidth optimization: PA signals are wideband and benefit from higher sampling, whereas US echoes are narrower and adequately captured at 20 MS/s. This dual-rate, gain-switching strategy exemplifies how affordable systems maintain diagnostic SNR while operating within cost-limited DAQ and communication constraints.

4.3.5.5 Future Directions

Integration trends point toward monolithic SoC-based PAI engines, where FPGA logic, multi-channel ADCs, and embedded GPU or ARM cores coexist on a single board. Such platforms, exemplified by Xilinx Zynq UltraScale+ and Jetson Orin Nano, can handle precise timing, real-time reconstruction, and AI-based noise suppression within compact power envelopes. Adaptive gain control, intelligent triggering, and deep-learning-based signal enhancement will likely define the next generation of cost-efficient, intelligent photoacoustic imagers. By merging efficient hardware design with software intelligence, these systems will close the gap between laboratory prototypes and affordable clinical translation.

4.3.6 Emerging Trends and Outlook in Affordable PAI Systems

Recent progress in photoacoustic system design reveals a clear trend toward fully integrated, cost-optimized architectures that combine compact light sources, sensitive detectors, and intelligent electronics on a single board. The emergence of system-on-chip (SoC) and heterogeneous computing platforms such as Xilinx Zynq UltraScale+ and Jetson Orin Nano enables the integration of timing control, data acquisition, and reconstruction within a palm-size enclosure. By co-locating FPGA logic for real-time triggering with embedded GPUs for image reconstruction, these devices drastically reduce power and footprint while supporting advanced algorithms such as adaptive beamforming and compressed sensing [7, 17, 33]. Further integration with micromachined detectors including CMUT and PMUT arrays [10–12], and miniature diode or VCSEL light sources [14–16, 25] positions handheld, probe-style PAI as a practical clinical tool. Artificial-intelligence–assisted reconstruction is another rapidly expanding frontier. Deep learning models now provide image enhancement, noise suppression, and even direct inversion from limited or sparse data [29, 32, 33]. Such networks can be executed directly on embedded GPUs or low-power ARM NPUs, reducing dependence on external computers and accelerating real-time imaging. In addition, adaptive sampling and compressed-sensing schemes further alleviate data-bandwidth constraints by intelligently selecting or predicting signal components most critical to image fidelity [28–31]. Together, these developments lower hardware complexity while maintaining diagnostic image quality, an essential trade-off for affordable systems.

The next evolution will likely involve connected and cloud-supported imaging. Edge devices equipped with 5 G or Wi-Fi 6 interfaces can stream reconstructed volumes to remote servers for tele-diagnostics and collaborative clinical review. Such connectivity aligns PAI with emerging tele-ultrasound workflows, where low-cost imaging heads communicate with centralized AI servers for interpretation and storage. Parallel advances in cybersecurity and data compression are required to preserve patient privacy and maintain frame-rate continuity. Cost remains the defining constraint for translation. Adoption of commercial off-the-shelf (COTS) electronics, open-source FPGA firmware, and 3-D-printed acoustic housings has already reduced component prices by an order of magnitude [1, 20]. Manufacturers increasingly target modular designs that allow the same base hardware to support multiple excitation sources or detector types, extending lifetime and scalability. Collaborative academic–industry efforts, exemplified by Zafar et al. [7] and Darband et al. [17], demonstrate how simplified electronics and parallel software development can deliver clinically usable, low-cost prototypes.

Looking ahead, the convergence of affordable light sources, polymer or MEMS-based detectors, and embedded intelligent electronics is poised to redefine photoacoustic imaging. By prioritizing manufacturability and integration rather than sheer laser power, the next generation of systems will offer portable, maintenance-light, and cost-sustainable solutions for vascular, dermatologic, and interventional imaging.

Affordable PAI thus represents not a compromise in performance but an optimization transforming photoacoustics from laboratory innovation to everyday diagnostic reality.

4.4 Comparative Analysis of Affordable Photoacoustic Imaging Architectures

The diversity of recent PAI prototypes highlights that there is no single architecture that simultaneously maximizes sensitivity, speed, and affordability. Instead, each implementation represents a balance among light source cost, detector performance, data-acquisition complexity, and system integration level. A clear understanding of these trade-offs allows designers to select the most suitable configuration for a given clinical or preclinical target.

4.4.1 Performance–Cost Landscape

At the light-source level, LEDs and pulsed laser diodes (PLDs) dominate affordable systems due to their low cost, high repetition rate, and compactness [14–17, 22–25]. LED arrays provide broad illumination and sub-mJ pulse energies, sufficient for superficial vasculature or dermatologic imaging, whereas PLDs and VCSELs offer higher fluence and beam quality for tomography or deeper tissue studies. Compared with nanosecond solid-state lasers, these diodes reduce source cost by nearly two orders of magnitude while supporting portable operation.

On the detection side, polymer (PVDF or PVDF-TrFE) films remain the most economical option, providing sub-pascal sensitivity and wide bandwidth for simple setups [4, 5]. More advanced CMUT and PMUT arrays introduce improved bandwidth and integration with CMOS electronics but at higher fabrication cost [8–10, 12]. Optical Fabry–Pérot interferometers (FPI) now match piezoelectric bandwidth but remain costly due to laser interrogation optics [13]. Thus, from an affordability standpoint, polymer films and PMUTs represent the most realistic routes toward low-cost, flexible detectors.

Data-acquisition and processing architectures strongly influence overall expense. Single-element or low-channel designs drastically simplify front-end electronics and enable the use of compact FPGAs and microcontrollers for acquisition [22–27]. Conversely, multichannel array systems require high-speed ADCs and GPU acceleration but deliver 3-D imaging at video rate [35, 36]. Emerging hybrid FPGA–GPU SoCs now close this gap by providing real-time reconstruction within compact power envelopes [37].

4.4.2 Optimization Strategies for Affordability

System-level cost can be minimized through component reuse, modular electronics, and adaptive sampling. Low-cost multiplexing boards allow one high-quality amplifier chain to serve several channels sequentially, while compressed-sensing and AI-based beamforming reduce required sampling rates without losing spatial detail [28–33]. Open-source firmware and COTS components, combined with 3-D-printed enclosures, further lower manufacturing barriers [1, 7, 17, 20]. Such modular strategies also facilitate future upgrades—for example, swapping an LED array for a VCSEL source or integrating a new MEMS detector without redesigning the full DAQ chain.

4.4.3 Remaining Challenges and Opportunities

Despite rapid progress, affordable PAI faces several persistent challenges. Maintaining high sensitivity under laser-safety fluence limits requires continued innovation in detector materials and amplification design. Real-time 3-D reconstruction at limited bandwidth remains a computational bottleneck, though modern SoCs are narrowing the gap. Thermal management, electromagnetic compatibility, and mechanical durability are critical for portable and handheld use. Finally, large-scale translation will depend on standardized calibration procedures, robust safety certification, and demonstrated clinical benefit.

4.4.4 Outlook

Overall, the comparative landscape illustrated in Fig. 4.15 shows how affordable PAI technologies are converging toward more integrated and practical system designs. The figure highlights the key subsystems (light sources, detectors, data acquisition, and processing) and emphasizes the tradeoffs and engineering choices that shape low cost implementations. By combining LEDs or other economical light sources, polymer or MEMS-based detectors, FPGA-based acquisition hardware, and embedded processors capable of AI-assisted reconstruction, developers are increasingly able to assemble compact and modular PAI systems. These technological trends support the broader transition of PAI from research prototypes toward reproducible, scalable platforms suitable for real-world clinical environments. As such systems mature, they enable low-cost PAI modules for applications such as dermatology, vascular assessment, and bedside monitoring, helping move optical–ultrasound hybrid imaging closer to routine clinical deployment.

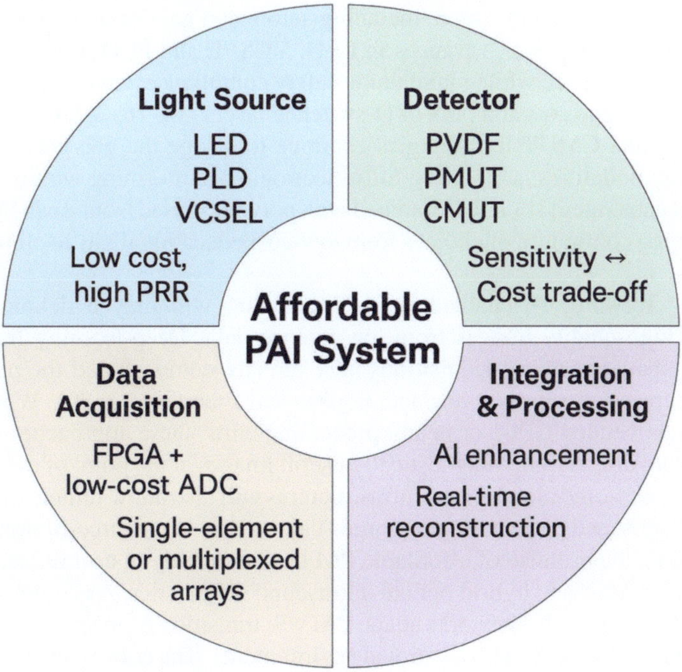

Fig. 4.15 Schematic overview of the affordable photoacoustic imaging (PAI) design landscape. Light-source, detector, data-acquisition, and processing subsystems are balanced to optimize cost, sensitivity, and system integration. Advances in embedded electronics and AI-assisted reconstruction now enable clinically relevant imaging performance using affordable, modular architectures

4.5 Future Perspectives and Translation Outlook

The trajectory of PAI has clearly shifted from laboratory-scale systems toward compact, integrated, and affordable platforms capable of clinical deployment. The rapid evolution of low-cost components ranging from LED and laser-diode sources to polymer and micromachined detectors has significantly lowered entry barriers for biomedical researchers and engineers [1, 5, 7, 10, 14–17, 22–25]. Emerging prototypes now integrate FPGA-controlled light excitation, programmable gain amplification, and embedded GPU processing onto single printed circuit boards, supporting real-time operation with minimal external hardware. These advances, coupled with improvements in 3D-printed acoustic housings and modular electronics, are establishing a foundation for reproducible and maintainable low-cost imaging systems [1, 7, 17, 20].

The next wave of innovation will focus on monolithic integration and clinical scalability. Compact system-on-chip (SoC) devices such as Zynq UltraScale+ and Jetson Orin Nano already combine FPGA logic and AI-capable cores, forming a self-contained processing unit for handheld probes. These platforms are poised to support dual-modality (PA/US) operation, including automated gain switching and adaptive frame-rate control [7, 34]. Advances in LED, VCSEL and PLD arrays will further enhance optical fluence while maintaining safety compliance, enabling deeper tissue imaging without the cost and bulk of Q-switched lasers [14–16, 25]. On the detector side, PMUT and CMUT technologies continue to bridge the gap between optical and acoustic modalities, supporting fully electronic beamforming within flexible or transparent substrates [10, 12]. Combined with polymer-based front-ends [4, 5], these devices offer a continuum of choices from low-cost educational kits to clinical-grade prototypes.

Looking forward, AI-enabled signal processing will play a defining role in extracting high-quality images from low-fluence data. Deep-learning frameworks trained on sparse or noisy acquisitions have already demonstrated the potential to restore contrast and suppress artifacts in near-real time [29, 32, 33]. When implemented on embedded GPUs or neural processing units, these approaches can transform raw data directly into diagnostically useful images at the point of care. Integration with wireless and cloud-based infrastructures will also allow remote monitoring, tele-diagnostics, and data sharing for large-scale studies in resource-limited settings.

Ultimately, the promise of affordable PAI lies not merely in cost reduction but in democratizing access to hybrid optical–ultrasound diagnostics. As compact systems mature and regulatory frameworks adapt, PAI will transition from specialized laboratories to routine clinical and educational environments. The convergence of scalable fabrication, embedded intelligence, and low-cost optoelectronic integration heralds an era in which photoacoustics becomes both accessible and impactful, realizing its long-envisioned potential as a safe, affordable, and versatile imaging modality.

Key Takeaways

- **Affordability and sensitivity are not mutually exclusive**. Polymer (PVDF) and micromachined (PMUT/CMUT) detectors, when paired with optimized amplification, can achieve sub-pascal sensitivity at a fraction of traditional system cost.
- **Optical detection technologies are approaching clinical performance**, but require cost reductions in interrogation lasers and optical alignment to compete with piezoelectric counterparts.
- **Single-element and low-channel architectures**, especially when combined with FPGA-based averaging and intelligent sampling offer an effective route to low-cost, high-SNR imaging suitable for preclinical and educational use.
- **Data-acquisition efficiency is central to affordability**. Real-time systems like AcousticX demonstrate that onboard averaging and dual-rate sampling can preserve image quality while operating within USB 3.0 data limits.
- **Embedded processing and AI integration** are transforming compact systems into self-sufficient imaging platforms. FPGA–SoC hybrids and on-device neural reconstruction now enable real-time performance without external computers.
- **System modularity and open-source electronics** accelerate innovation and accessibility, allowing researchers to reconfigure light sources, detectors, and DAQ boards without redesigning entire systems.

- **Future progress depends on manufacturability, calibration standards, and clinical validation**, the essential bridge between low-cost prototypes and certified medical devices.

References

1. A. Fatima, R. Manwar, K. Avanaki, Review of cost-reduction methods in photoacoustic computed tomography. Photoacoustics **15**, 100137 (2019)
2. C. Lutzweiler, D. Razansky, Optoacoustic imaging and tomography: reconstruction approaches and outstanding challenges in image performance and quantification. Sensors **13**(6), 7345–7384 (2013)
3. R. Manwar, K. Kratkiewicz, K. Avanaki, Overview of ultrasound detection technologies for photoacoustic imaging. Micromachines **11**(7), 692 (2020)
4. P. Subochev, M. Prudnikov, V. Vorobyev, A. Postnikova, I. Turchin, Wideband linear detector arrays for optoacoustic imaging based on polyvinylidene difluoride films. J. Biomed. Opt. **23**(9), 091408 (2018)
5. A. Kurnikov, A. Sanin, X.L. Dean Ben, D. Razansky, P. Subochev, Ultrawideband sub-pascal sensitivity piezopolymer detectors. Ultrasonics **141**, 107349 (2024)
6. S. Choi, J. Kim, H. Jeon, C. Kim, E.-Y. Park, Advancements in photoacoustic detection techniques for biomedical imaging. npj Acoust. **1**(1), 1–18 (2025)
7. M. Zafar, R. Manwar, S.M. Ranjbaran, K. Avanaki, Utility of low-cost multichannel data acquisition system for photoacoustic computed tomography. Photonics **12**(4), 385 (2025)
8. B.T. Khuri-Yakub, Ö. Oralkan, Capacitive micromachined ultrasonic transducers for medical imaging and therapy. J. Micromech. Microeng. **21**(5), 054004 (2011)
9. Ö. Oralkan et al., Capacitive micromachined ultrasonic transducers: next-generation arrays for acoustic imaging? IEEE Trans. UFFC **49**(11), 1596–1610 (2002)
10. Y. He, H. Wan, X. Jiang, C. Peng, Piezoelectric micromachined ultrasound transducer technology: recent advances and applications. Biosensors **13**(1), 55 (2023)
11. P. Beard, Biomedical photoacoustic imaging: signal generation and detection. Interface Focus **1**(4), 602–631 (2011)
12. Y. Qiu, J.V. Gigliotti, M. Wallace, F. Griggio, C.E.M. Demore, S. Cochran, S. Trolier-McKinstry, Piezoelectric micromachined ultrasound transducer (PMUT) arrays for integrated sensing, actuation and imaging. Sensors **15**(4), 8020–8041 (2015)
13. N.T. Huynh, E. Zhang, O. Francies et al., A fast all-optical 3D photoacoustic scanner for clinical vascular imaging. Nat. Biomed. Eng. **9**, 638–655 (2025)
14. Y. Zhu, G. Xu, J. Yuan et al., Light-emitting-diode-based photoacoustic imaging and potential clinical applications. Sci. Rep. **8**, 9885 (2018)
15. Y. Zhu, T. Feng, Q. Cheng, X. Wang, S. Du, N. Sato, J. Yuan, M. Kuniyil Ajith Singh, Towards clinical translation of LED-based photoacoustic imaging: a review. Sensors **20**(9), 2484 (2020)
16. X. Liu, S.K. Kalva, B. Lafci, D. Nozdriukhin, X.L. Deán-Ben, D. Razansky, Full-view LED-based optoacoustic tomography. Photoacoustics 31, 100521 (2023)
17. M.A. Darband et al., Implementation of a cost-effective, accurate photoacoustic imaging system based on high-power LED illumination and FPGA-based circuitry. Arch. Acoust. **50**(1), 37–48 (2025)
18. X. Wu, B. Wang, Y. Sun et al., An FPGA-based backend system for intravascular photoacoustic and ultrasound imaging. Sensors **18**(2), 332 (2018)
19. H. Kye, Y. Song, T. Ninjbadgar, C. Kim, J. Kim, Whole-body photoacoustic imaging techniques for preclinical small animal studies. Sensors (Basel) **22**(14): 5130 (2022)
20. H. Zhong, T. Duan, H. Lan, M. Zhou, F. Gao, Review of low-cost photoacoustic sensing and imaging based on laser diode and light-emitting diode. Sensors **18**(7), 2264 (2018)

21. W. Xia, M. Kuniyil Ajith Singh, E. Maneas, N. Sato, Y. Shigeta, T. Agano, S. Ourselin, S.J. West, A.E. Desjardins, Handheld real-time LED-based photoacoustic and ultrasound imaging system for accurate visualization of clinical metal needles and superficial vasculature to guide minimally invasive procedures. Sensors **18**(5), 1394 (2018)
22. P.K. Upputuri, M. Pramanik, Performance characterization of low-cost, high-speed, portable pulsed laser diode photoacoustic tomography (PLD-PAT) system. Biomed. Opt. Express **6**, 4118–4129 (2015)
23. S.K. Kalva, P.K. Upputuri, M. Pramanik, High-speed, low-cost, pulsed-laser-diode-based second-generation desktop photoacoustic tomography system. Opt. Lett. **44**, 81–84 (2019)
24. P. Rajendran, P.K. Upputuri, S.K. Kalva et al., In vivo detection of venous sinus distension due to intracranial hypotension in small animal using pulsed-laser-diode photoacoustic tomography. J. Biophoton. e201960162 (2020)
25. A. Das, M. Pramanik, Performance characterization of light-emitting diode array-based photoacoustic computed tomography (LED-PACT). Biomed. Opt. Express **16**, 3632–3647 (2025)
26. Y. Li, L. Li, L. Zhu, J. Shi, K. Maslov, L.V. Wang, Photoacoustic topography through an ergodic relay for functional imaging and biometric application in vivo. J. Biomed. Opt. **25**(7), 070501 (2020)
27. Y. Zhang, P. Hu, L. Li et al., Ultrafast longitudinal imaging of haemodynamics via single-shot volumetric photoacoustic tomography with a single-element detector. Nat. Biomed. Eng **8**, 712–725 (2024)
28. D. Jiang et al., Low-cost optoacoustic tomography system with programmable acoustic delay-line (time-sharing multiplexing) (2020). arXiv preprint arXiv:2012.03943
29. Y. Guo et al., Dual-compressed photoacoustic single-pixel imaging. Nat. Sci. Rev. **10**(1), nwac058 (2023)
30. T. Zhao, E. Zhang, P.C. Beard, W. Xia, MECOPE: Multifocal excitation compressive-sensing photoacoustic endomicroscopy through a multimode fibre. Photoacoustics 44, 100733 (2025)
31. M. Haltmeier et al., Design, implementation and analysis of a compressed sensing photoacoustic projection imaging system. J. Biomed. Opt. **29**(S1), S11529 (2024)
32. B. Yan et al., Compressed single-shot 3D photoacoustic imaging with a 3D encoder. Phys. Rev. Appl. **20**(2), 024010 (2023)
33. H. Lan et al., Deep learning enabled real-time photoacoustic tomography via single data acquisition channel. Photoacoustics **23**, 100283 (2021)
34. J. Yoo, D. Oh, C. Kim, H.H. Kim, J.-Y. Um, Switchable preamplifier for dual modal photoacoustic and ultrasound imaging. Biomed. Opt. Express **14**, 89–105 (2023)
35. Y. Wang, C. Li, Comprehensive framework of GPU-accelerated image reconstruction for photoacoustic computed tomography. J. Biomed. Opt. **29**(6), 066006 (2024)
36. E.A. Gonzalez, M.A.L. Bell, GPU implementation of photoacoustic short-lag spatial coherence imaging for improved image-guided interventions. J. Biomed. Opt. 25(7):077002 (2020)
37. X. Zheng et al., Hardware architecture design of model-based image reconstruction towards palm-size photoacoustic tomography (2024). arXiv preprint, arXiv:2408.06049

Chapter 5
Image Quality Enhancement in Affordable Photoacoustic Imaging

Learning Objectives

After reading this chapter, the reader will be able to:

- **Understand the sources and characteristics of image degradation** in affordable PAI systems, including low signal-to-noise ratio, limited bandwidth, and reflection-induced artifacts.
- **Explain the trade-off between computational and hardware cost**, and how it shapes design choices in low-cost photoacoustic imaging platforms.
- **Describe fast and efficient reconstruction techniques** such as optimized delay-and-sum, Fourier-domain, and compressed-sensing methods suitable for real-time or embedded applications.
- **Apply denoising and artifact suppression strategies**, including adaptive filtering, wavelet transforms, and reflection correction, to improve image quality under low-fluence conditions.
- **Recognize the contribution of AI-driven methods** (e.g., CNNs, U-Nets, and transfer learning) in achieving super-resolution and robust reconstruction from sparse or noisy data.
- **Evaluate algorithmic performance** using quantitative metrics such as signal-to-noise ratio (SNR), contrast-to-noise ratio (CNR), and reconstruction latency.
- **Assess the balance between algorithmic complexity, processing speed, and energy efficiency** in computational designs for affordable, portable photoacoustic systems.

M. Kuniyil Ajith Singh, *Photoacoustic Imaging Using Affordable Resources*, Progress in Optical Science and Photonics 40, https://doi.org/10.1007/978-981-95-7175-8_5

5.1 The Role of Computational Enhancement in Low-Cost Photoacoustic Imaging

Affordable photoacoustic imaging (PAI) systems built on light-emitting diodes (LEDs), laser diodes (LDs), or vertical-cavity surface-emitting lasers (VCSELs) provide compact and economical alternatives to class-IV OPO or Nd:YAG laser platforms. These sources substantially reduce system cost, footprint, and power consumption, yet their lower optical fluence and simplified detection/DAQ architectures lead to weaker signals and greater susceptibility to sampling and bandwidth limitations [1–4]. Representative implementations include LED-based PAI for bench-to-bedside translation [1, 5], LD-driven microscopy and tomography [2, 6, 7], and emerging VCSEL patches integrating source and detector arrays on flexible substrates.

Such constraints make software performance as critical as hardware capability. In affordable systems, image fidelity depends on computational strategies—reconstruction, denoising, and artifact suppression—rather than solely on excitation energy. This reflects a broader trend in medical imaging: algorithmic intelligence compensating for hardware simplification, as seen in low-dose CT or portable ultrasound. The computational cost ↔ hardware cost trade-off now favors computation, thanks to accessible embedded GPUs, FPGAs, and SoC accelerators that deliver real-time processing at minimal incremental cost [8–12].

Three interdependent pillars define computational enhancement in affordable PAI:

- **Fast Reconstruction**. Efficient back-projection or delay-and-sum (DAS) methods remain dominant for their simplicity and parallel scalability [8, 13]. GPU- and FPGA-accelerated frameworks achieve near-real-time reconstruction even for 2-D/3-D geometries, while iterative and model-based algorithms improve limited-view or sparse-data performance [9].
- **Denoising**. Low-fluence excitation introduces broadband, often non-Gaussian noise from electronic and thermal sources. Practical strategies combine frame compounding, adaptive temporal–spectral filtering, and lightweight priors that preserve vascular continuity [1, 5].
- **Artifact Suppression**. Compact geometries increase vulnerability to limited bandwidth, under-sampling, and reflection-type artifacts. Bandwidth compensation, synthetic-aperture and multi-view reconstructions, and data-consistency corrections mitigate these effects [9, 13].

A related enhancement is photoacoustic–ultrasound (PA/US) fusion, which leverages the shared detection chain of most affordable platforms. Real-time co-registration provides anatomical context for PA contrast, enhancing interpretability for applications such as lymph-node mapping, needle guidance, and superficial tumor assessment [14–19].

In summary, computational intelligence transforms affordability from a constraint into an opportunity. Optimized algorithms implemented on low-power parallel hardware restore much of the performance otherwise lost to hardware simplification.

The following Sect. 5.2 examines the physical origins of image degradation in such systems, forming the basis for the computational approaches detailed throughout this chapter.

5.2 Image Degradation in Affordable Photoacoustic Imaging Systems

Affordable PAI systems achieve compactness and reduced cost by simplifying optical, acoustic, and electronic components. However, these design optimizations inherently limit excitation fluence, detection bandwidth, and sampling density. The resulting physical and system-level constraints introduce characteristic degradation mechanisms that lower image fidelity [20]. Understanding these effects is essential to design computational strategies that restore image quality without raising hardware cost.

5.2.1 Low Optical Fluence and Stochastic Noise

LED–, LD–, and VCSEL-based light sources typically emit pulse energies in the microjoule range—several orders of magnitude below those of conventional nanosecond lasers [1–4, 21]. The corresponding photoacoustic amplitudes are therefore weak and often approach the noise floor of the electronics, producing a low signal-to-noise ratio (SNR). In addition, minor fluctuations in pulse energy and spatial illumination yield stochastic variations between frames, leading to grainy backgrounds and disrupted vascular continuity. While frame averaging and adaptive filtering can partially suppress these fluctuations, the fundamental limitation arises from the restricted photon budget inherent to affordable sources [5, 6, 21].

5.2.2 Reflection Artifacts in Low-Energy Handheld Illumination

In handheld and portable configurations, optical illumination is typically delivered obliquely from emitters positioned just below the ultrasound probe. This geometry improves optical coupling but also generates a strong superficial photoacoustic signal at the skin surface. The skin acts as a secondary acoustic emitter whose waves propagate deeper into tissue, reflect from impedance mismatches such as fascia, vessel walls, tendon, or bone, and return to the detector at delayed times. These reflections overlap with genuine photoacoustic signals, producing ring-like or diffuse ghost features that can mimic anatomical structures and reduce contrast and depth

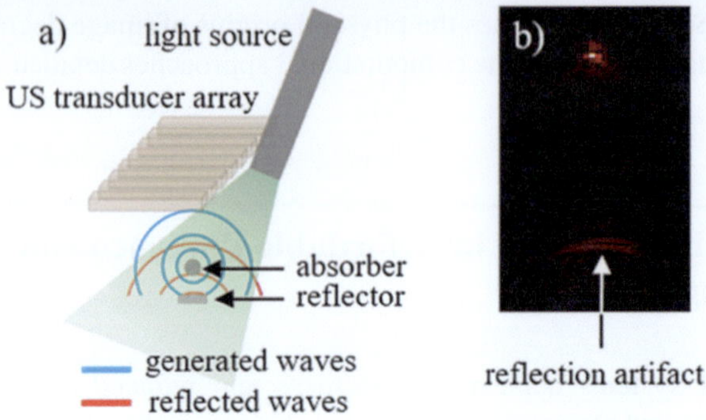

Fig. 5.1 Reflection artifacts in photoacoustic imaging (PAI). **a** Illustration of reflection artifact generation, where a strong photoacoustic signal from the superficial layer produces secondary echoes from deeper acoustic interfaces. **b** Representative PAI image showing a false signal appearing beneath the true absorber, mimicking deeper vascular structures—a common artifact in handheld, low-fluence clinical imaging. Reprinted with permission from [28]

fidelity [22–31]. The problem is accentuated in LED- and LD-based systems owing to broad illumination profiles, shallow penetration, and variable contact conditions. As illustrated in Fig. 5.1, these in-plane reflection artifacts constitute one of the most prominent sources of misinterpretation in low-fluence PAI.

Their physical mechanisms and suppression strategies including PAFUSion, deformation-compensated averaging, spectral correction, and AI-based reflection reduction have been extensively reported in the literature [22–31] and are discussed in detail in Sect. 5.4.4.

5.2.3 *Limited Bandwidth and Axial Blurring*

Miniaturized, low-cost transducers or MEMS detectors generally possess narrower acoustic bandwidths than laboratory-grade sensors. Reduced bandwidth attenuates high-frequency components, yielding axially elongated absorbers and blurred boundaries [13, 20]. Similarly, the longer optical pulse durations typical of low-cost drivers broaden the acoustic spectrum and degrade resolution. Frequency-domain compensation and deconvolution filters can recover some high-frequency detail, but bandwidth limits remain a key determinant of image sharpness in compact probes.

5.2.4 *Under-Sampling and Sparse Apertures*

Affordable systems often use fewer transducer elements, limited scanning ranges, or simplified acquisition electronics. Insufficient spatial sampling produces aliasing and streaking artifacts, particularly when conventional delay-and-sum reconstruction is applied [8, 9]. Sparse-view or compressed-sensing approaches, iterative model-based methods, and learned priors have shown promise in reconstructing accurate images from under-sampled data, although they increase computational demand [11].

5.2.5 Cost–Performance Causality

Each degradation mechanism directly reflects a hardware compromise made for affordability:

low optical fluence → weak signals;

limited bandwidth → axial blur;

fewer elements → under-sampling;

and shallow illumination → reflection artifacts.

In affordable PAI, image degradation is therefore a predictable consequence of cost-driven design trade-offs. The practical challenge lies in balancing minimal hardware with computational compensation capable of recovering lost information. Table 5.1 summarizes the principal degradation mechanisms and representative computational strategies suited for affordable PAI platforms.

5.3 Efficient and Real-Time Reconstruction Methods

Reconstruction forms the computational backbone of PAI. In affordable LED-, LD-, or VCSEL-based systems, where signals are weaker and sampling more limited, reconstruction efficiency directly determines image fidelity and clinical usability.

Traditional time-domain methods such as delay-and-sum (DAS) remain the default for low-cost devices, but frequency-domain and adaptive algorithms now offer better trade-offs between image quality, computation, and power consumption. This section reviews analytic, Fourier-domain, and model-based strategies, emphasizing the cost–computation–quality balance that defines affordable PAI.

5.3.1 Overview and Motivation

Affordable PAI platforms must deliver interpretable images under strict limits of cost, memory, and power. Real-time feedback is essential for handheld or point-of-care use, requiring methods that are both computationally lightweight and hardware-efficient.

Table 5.1 Image degradation mechanisms and computational remedies in affordable PAI

Degradation/artifact	Primary cause in low-cost systems	Key computational remedies	Notes on affordability/feasibility
Low fluence/noise	LED/LD/VCSEL micro-Joule pulses; weak signals	Frame averaging, adaptive filters, wavelet denoising, CNN-based denoisers	Software only; runs on embedded GPU/SoC
Reflection artifacts	Strong skin signal reflecting from subsurface inhomogeneities	PAFUSion, deformation-compensated averaging, LOVIT, model-based subtraction	Critical for handheld probes; low added cost
Limited view/Resolution anisotropy	Linear or partial aperture arrays	Multi-view synthesis, compressed sensing, model-based or AI reconstruction	Enhances sparse probes; real-time on GPU
Limited bandwidth	Narrowband, low-cost transducers; long laser pulses	Deconvolution, frequency compensation, spectral equalization	Light computation; compatible with FPGA/GPU
Under-sampling/Sparse aperture	Fewer channels or reduced frame rate	Compressed sensing, TV regularization, learned priors	Suited for portable or embedded PAI
Acoustic heterogeneity/SOS errors	Uncorrected tissue speed variations	Joint SOS–image reconstruction, segmentation-guided correction	Moderate compute cost; improves quantitation
Fluence non-uniformity	Oblique illumination, skin scattering	Model-based fluence correction, iterative estimation, spectroscopic unmixing	Computation-only; applicable to LED/VCSEL
Electronic noise/baseline drift	Compact electronics, minimal shielding	Digital filtering, adaptive thresholding	Simple, low-power implementation

Recent GPU- and FPGA-accelerated frameworks such as GAPAT [8] and learned iterative solvers [9, 10] have demonstrated that high-throughput reconstruction can now be achieved on commodity or embedded hardware. The design objective is maximum image quality per dollar-per-watt of hardware expenditure.

5.3.2 Delay-and-Sum, Fourier-Domain, and Adaptive Beamforming

Detailed mathematical formulations of the reconstruction algorithms summarized here can be found in Xu and Wang [13], Tian et al. [10], Jaeger et al. [32], and

Prakash et al. [33]. Only algorithmic principles and computational considerations relevant to affordable implementation are outlined below.

The DAS algorithm remains the cornerstone of affordable reconstruction owing to its simplicity and negligible memory overhead. It computes the propagation delays between detectors and voxels and coherently sums received signals. Although real-time and easily parallelized, DAS provides limited contrast and lateral resolution under sparse sampling [13, 20, 33].

5.3.2.1 Fourier-Domain (FD) Reconstruction

The FD approach reformulates image recovery in the frequency–wavenumber space [32]. Using the Fourier slice theorem, FD reconstruction replaces time-domain delay operations with one or two inverse FFTs, eliminating interpolation errors and drastically reducing computational complexity. In a systematic comparison of ten algorithms, Prakash et al. [33] found that FD, DAS, UBP, and pDAS achieved the shortest runtimes for a 1834 $\times$ 1334-pixel image on a typical CPU, while maintaining robust signal-to-background ratios (SBR) even at -10 dB noise. FD also exhibited the best overall noise tolerance, outperforming adaptive beamformers such as DMAS and SLSC. Because it maps directly onto highly optimized FFT libraries (cuFFT, OpenCL FFT), FD reconstruction enables video-rate display on mid-range GPUs and underpins real-time commercial systems such as AcousticX. These characteristics like speed, robustness, and hardware efficiency make FD reconstruction particularly suitable for low-cost, handheld PAI, where immediate visualization is required and computation per frame must remain minimal.

5.3.2.2 Adaptive Beamforming Extensions

Adaptive extensions such as delay-multiply-and-sum (DMAS) and minimum-variance (MV) algorithms enhance contrast and resolution by dynamically weighting detector signals. Eigen-based (MV-EIG) and coherence-based (SLSC, HRCF)** techniques further suppress sidelobes and incoherent noise [20, 34]. Prakash et al. [33] reported that DMAS and pDAS yielded the highest in-vitro target detectability (SBR $\approx$ max), while FD retained the strongest performance under high noise. These results indicate that DAS, FD, and DMAS represent the most practical operating range for affordable systems balancing quality, noise robustness, and speed.

Representative reconstructions obtained with these algorithms are shown in Fig. 5.2, demonstrating characteristic differences in contrast, noise behavior, and spatial resolution. Analytic methods such as DAS and FD yield uniform but lower-contrast images, while adaptive and coherence-based approaches (DMAS, MV, EIGMV, SLSC) produce improved target detectability at increased computational cost.

Table 5.2 summarizes the comparative performance of ten widely used reconstruction algorithms, emphasizing affordability-related metrics. FD and DAS achieve the

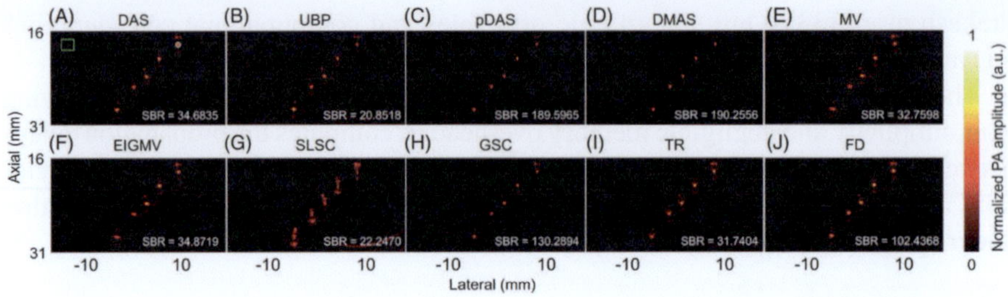

Fig. 5.2 Representative reconstructed images using ten widely used algorithms for linear-array photoacoustic imaging. The compared methods include delay-and-sum (DAS), universal back-projection (UBP), partial DAS (pDAS), delay-multiply-and-sum (DMAS), minimum-variance (MV), eigen-based MV (EIGMV), short-lag spatial coherence (SLSC), generalized spatial coherence (GSC), time-reversal (TR), and Fourier-domain (FD) reconstruction. The images illustrate variation in contrast, resolution, and noise characteristics across algorithmic families. FD, DAS, and UBP provide the fastest reconstructions suitable for real-time and low-cost systems, while adaptive and coherence-based approaches yield higher contrast at greater computational cost. Reprinted with permission from [33]

fastest runtimes, enabling real-time operation on embedded GPUs or low-power processors. DMAS and pDAS provide superior target detectability (SBR) but incur higher computational cost. Coherence-based and adaptive beamformers (SLSC, MV, EIGMV) improve contrast at the expense of processing time, while generalized spatial coherence (GSC) and time-reversal (TR) remain unsuitable for portable, cost-constrained platforms. These results confirm FD reconstruction as the most balanced solution for affordable PAI.

For most research prototypes and low-cost clinical platforms, reconstruction is still executed on general-purpose CPUs or modest embedded processors, where algorithmic efficiency directly limits achievable frame rate. Consequently, the relative execution times reported in Fig. 5.3 provide a realistic measure of affordability: algorithms that complete within a few seconds on a CPU are readily scalable to real-time operation on GPUs or SoCs, whereas those requiring hundreds of seconds remain impractical for point-of-care use. Figure 5.3 quantitatively illustrates this disparity, confirming FD reconstruction as the most computationally efficient option for affordable systems.

5.3.3 GPU and FPGA Acceleration

Hardware acceleration provides the most direct route to affordable real-time imaging.

The GAPAT framework [8] achieved up to 871× speedup over CPU reconstruction using Taichi-based GPU parallelization, enabling full 3-D PACT at video rates on consumer GPUs. Similarly, Gao et al. [11] implemented total-variation-regularized compressed sensing on GPUs, obtaining 20–30× acceleration with negligible accuracy loss [11]. For battery-operated or portable devices, FPGA architectures offer

Table 5.2 Comparison of reconstruction algorithms for affordable photoacoustic imaging

Algorithm	Relative computation time †	Lateral resolution	SBR ‡	Noise tolerance (−10 dB)	Implementation feasibility	Affordability remarks
FD (Fourier-domain)	1.0× (fastest)	High–moderate	High	Excellent	Real-time on mid-range GPU/ Jetson; FFT libraries	Best speed–robustness balance; used in *AcousticX*
DAS (Delay-and-sum)	1.1×	Highest	Moderate	Good	FPGA/GPU-ready; minimal memory	Widely adopted baseline; simplest to implement
UBP (Universal back-projection)	1.2×	Highest	Moderate	Fair	CPU/GPU feasible	High accuracy; slightly higher latency
pDAS (Partial DAS)	1.3×	High	Very high	Good	Embedded GPU feasible	Good trade-off of SBR vs speed
DMAS (Delay-multiply-and-sum)	4–5×	Moderate	Highest	Good	GPU-only	Superior contrast; moderate computational load
SLSC (Short-lag spatial coherence)	6–8×	Moderate	High	Good	Embedded GPU	Effective noise suppression; slower
MV (Minimum variance)	8–10×	High	Moderate	Moderate	Desktop GPU	High quality; heavy matrix ops
EIGMV (Eigen-based MV)	9–11×	High	Moderate	Moderate	Desktop GPU	Improved stability; costly compute

(continued)

Table 5.2 (continued)

Algorithm	Relative computation time †	Lateral resolution	SBR ‡	Noise tolerance (–10 dB)	Implementation feasibility	Affordability remarks
GSC (Generalized spatial coherence)	25–40×	Moderate	High	Fair	High-end GPU/ workstation	Excellent quality; unsuitable for portable units
TR (Time reversal)	> 50× (slowest)	High	Moderate	Poor	HPC/cluster	Quantitative accuracy; impractical for affordable PAI

† Computation times normalized to FD = 1 × (runtime for 1834 × 1334 pixels on standard CPU, ‡ Signal-to-background-ratio (SBR) ranking based on in-vitro phantom study [33]

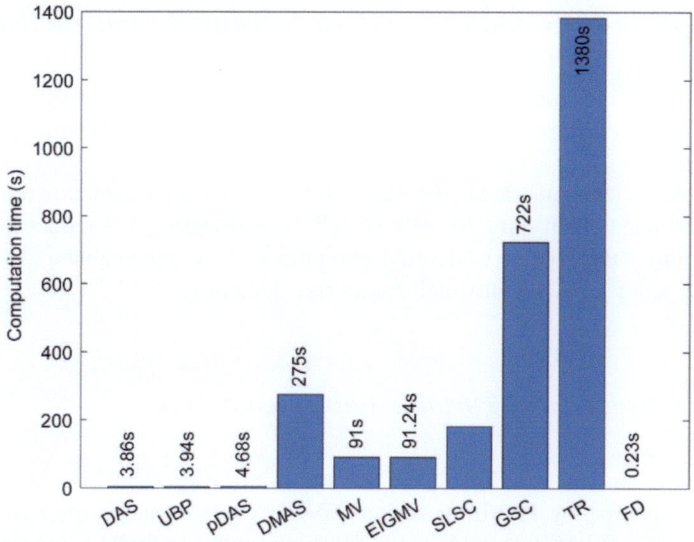

Fig. 5.3 Relative computation times for ten photoacoustic image reconstruction algorithms (CPU benchmark). A 1834 × 1334-pixel photoacoustic image was reconstructed to determine execution time on a standard CPU. Fourier-domain (FD), delay-and-sum (DAS), universal back-projection (UBP), and partial DAS (pDAS) are the fastest, with sub-5 s runtimes. Adaptive and coherence-based algorithms—DMAS, MV, EIGMV, and SLSC—require 90–275 s, whereas generalized spatial coherence (GSC) and time-reversal (TR) are two to three orders of magnitude slower. These results highlight the computational advantage of FD-based methods for affordable, real-time, and embedded PAI implementations. Reprinted with permission from [33]

deterministic latency at power levels below 10 W; Gao et al. [12] demonstrated real-time FPGA-based PACT [12]. While less flexible than GPU code, FPGA pipelines integrate naturally with ultrasound front-ends already present in low-cost probes.

5.3.4 Model-Based and Iterative Reconstruction

Model-based reconstructions incorporate acoustic physics to correct for limited view, speed-of-sound, and attenuation effects. Time-reversal and finite-element solvers yield high accuracy but are computationally intensive. Hsu et al. [9] introduced a learned-physics iterative model achieving faster convergence with retained interpretability [9], while Tian et al. [10] summarized GPU-parallel implementations of conjugate-gradient and ADMM schemes [10]. In affordable systems, such solvers are valuable when cloud or high-end GPU access is available but remain secondary to analytic or FD methods for embedded operation.

5.3.5 Sparse-View and Compressed-Sensing Reconstruction

Under-sampling is common in low-channel affordable setups. Compressed-sensing (CS) reconstruction recovers images from partial data by exploiting sparsity constraints.

Gao et al. [11] demonstrated GPU-accelerated CS-TV reconstruction reproducing acceptable images with only 30–50% of full measurements. CS reduces hardware channel count at the expense of iterative computation—an acceptable trade-off when energy efficiency and cost outweigh frame-rate demands.

5.3.6 Embedded and Portable Implementations

Embedding reconstruction directly on SoCs or mini-GPUs closes the loop between acquisition and display. Platforms such as NVIDIA Jetson and Xilinx Zynq achieve 10–20 fps for 2-D PACT at < 15 W [8, 12]. Fixed-point arithmetic and lightweight CUDA/OpenCL kernels further reduce power and cost. Open-source frameworks like GAPAT or CuPy facilitate deployment by small laboratories without proprietary toolchains.

5.3.7 Cost–Computation Trade-Off

As illustrated in Fig. 5.3, the computation time difference across reconstruction methods spans over three orders of magnitude, underscoring the decisive impact of algorithmic choice on affordability. Reconstruction strategy selection in affordable PAI is defined by the interplay between image quality, computational demand, and hardware cost.

Analytic (DAS, FD) methods occupy the low-cost real-time corner of this triangle, while adaptive and model-based solvers offer higher fidelity at higher computational cost.

CPU benchmarks from Prakash et al. [33] reveal that FD, DAS, UBP, and pDAS are over $100 \times$ faster than GSC and TR for identical image sizes, emphasizing the decisive role of algorithm choice in achieving affordability.

As hardware prices continue to decline and open GPU/FPGA frameworks mature, reconstruction efficiency, not laser power, will define the achievable performance of next-generation affordable PAI. Section 5.4 will address complementary denoising and artifact-suppression strategies for further quality improvement.

5.4 Denoising and Artifact Suppression

5.4.1 Introduction and Motivation

Affordable PAI systems based on LEDs, LDs, or VCSELs inherently operate at lower optical fluence compared with nanosecond-pulsed solid-state lasers [1–5, 21]. This reduced excitation energy results in lower SNR and higher susceptibility to electronic and acoustic noise sources. While Sect. 5.3 described computational reconstruction strategies to compensate for limited sampling and bandwidth, signal- and image-domain enhancement remains equally crucial for restoring diagnostic image quality. In low-fluence systems, denoising and artifact suppression effectively substitute for expensive hardware upgrades, thereby forming a central pillar of affordability. Noise in PAI is multi-origin and broadband. Electronic noise arises from front-end amplifiers and digitizers; stochastic fluctuations stem from source jitter and shot noise; and additional interference originates from surface reflections, acoustic reverberations, and probe motion [35]. These distortions are often non-Gaussian and spatially correlated, making standard linear filtering sub-optimal. Affordable implementations must therefore employ methods that maximize SNR improvement per unit of computational cost, using lightweight algorithms that can run in real time on CPUs, GPUs, or embedded SoCs.

Typical low-fluence PAI acquisitions demonstrate that the dynamic range of photoacoustic signals may fall below the quantization noise floor of low-cost ADCs [35]. In such regimes, software-based denoising and artifact mitigation outperform any practical hardware redesign [36]. Techniques such as temporal averaging, adaptive spectral filtering, wavelet denoising, and reflection suppression (e.g., PAFUSion [22–24], deformation-compensated averaging [26], and LOVIT [29]) have proven effective in recovering meaningful contrast from LED- or LD-excited images. Since these operations are deterministic and parallelizable, they can be implemented efficiently on embedded GPUs (e.g., Jetson series) or FPGAs already present in low-cost systems. From a cost–benefit perspective, computational enhancement provides order-of-magnitude improvements in perceived image quality at negligible additional hardware expense. Thus, the role of denoising in affordable PAI extends beyond aesthetic enhancement, it enables clinical interpretability, particularly for vascular, musculoskeletal, and dermatologic applications where low-contrast features must be visualized in real time.

Subsequent subsections describe noise sources (Sect. 5.4.2), low-complexity filtering strategies (Sect. 5.4.3), and advanced reflection-artifact suppression (Sect. 5.4.4), concluding with lightweight AI-assisted enhancement.

5.4.2 Noise Sources and Statistical Characteristics

Low-cost photoacoustic imaging (PAI) systems experience complex noise behavior that differs substantially from the relatively stationary noise seen in high-fluence laser-based platforms. Because excitation energies are several orders of magnitude lower, shot noise, electronic fluctuations, and reverberant acoustic interference become comparable to or even exceed the amplitude of the desired photoacoustic signal. Understanding these noise components and their statistical properties is essential for designing effective and affordable denoising strategies.

5.4.2.1 Electronic and Instrumental Noise

Electronic noise originates from pre-amplifiers, digitizers, and power supplies within the detection chain. It is typically broadband and exhibits a non-white, frequency-dependent spectrum, often shaped by the amplifier response and impedance matching of the transducer array. Low-cost analog front-ends generally employ higher-gain, lower-cost components that increase thermal and flicker noise. In compact portable systems, power-supply noise and electromagnetic coupling can introduce line-frequency and switching artifacts.

This component of noise is usually additive and can be effectively attenuated by spectral filtering, Wiener filtering, or wavelet-based suppression methods Sect. 5.4.3.

5.4.2.2 Optical and Source-Related Fluctuations

LD and rarely LED sources show intrinsic pulse-to-pulse variation in optical output energy and timing jitter [1–5, 21]. These fluctuations create stochastic amplitude modulation of the acoustic signal, resulting in non-stationary noise that degrades the consistency of reconstructed images. Because the LED/LED emission duration (tens to hundreds of nanoseconds) often overlaps with the acoustic propagation time, the generated pressure wave is partially smeared, leading to temporal decorrelation between consecutive frames.

Frame averaging or adaptive normalization methods are typically employed to mitigate this effect. In severe cases, illumination instability dominates the variance of the recorded signals and cannot be compensated without active source control.

5.4.2.3 Acoustic and Propagation-Related Interference

In handheld and surface-coupled probes, the strong photoacoustic response from the skin–gel interface produces secondary reflections that propagate into deeper tissue, creating false absorbers or ghost structures in the reconstructed image [22–24]. These reflection artifacts appear even in homogeneous phantoms and can extend well into the region of interest.

The problem is accentuated when illumination is delivered just below the ultrasound probe to maximize fluence, a configuration common in portable LED/LD systems.

The high skin signal reflects from acoustic impedance mismatches deeper in tissue and returns along paths indistinguishable from true sources, leading to characteristic arc-shaped or ring-like artifacts [26–30]. Such reflections can dominate the low-frequency spectrum and bias intensity-based quantification.

The noise associated with multipath propagation is coherent and spatially correlated, meaning that simple averaging or spectral filtering is insufficient.

Advanced suppression strategies such as PAFUSion, deformation-compensated averaging, or vibration-tagging approaches (LOVIT) have been developed to model and remove these reflected components Sect. 5.4.4. From a practical standpoint, these methods are particularly important in affordable handheld systems, where probe geometry, limited bandwidth, and variable coupling exacerbate acoustic reverberation.

5.4.2.4 Motion and Environmental Perturbations

Portable PAI platforms often suffer from inter-frame motion due to handheld operation or subject movement. Even submillimeter displacements cause phase and amplitude inconsistencies that degrade coherent averaging and spatial compounding.

Thermal drift and ambient vibration can also introduce low-frequency baseline shifts.

These perturbations manifest as correlated noise rather than random variance, and thus require registration or deformation-compensated averaging before denoising.

5.4.2.5 Summary

Noise in affordable PAI is multifactorial and non-stationary, combining additive electronic noise with multiplicative and correlated acoustic interference. Effective enhancement therefore demands multi-domain filtering (temporal, spatial, and spectral) implemented efficiently on available hardware resources. The next Sect. 5.4.3 presents low-complexity filtering approaches tailored to these characteristics and suitable for real-time embedded operation.

5.4.3 Temporal and Spatial Filtering Approaches

5.4.3.1 Frame Averaging and Compounding

In low-fluence LED- and LD-based photoacoustic imaging, the simplest and most direct means of improving image quality is temporal frame averaging. When N successive frames are averaged, the effective SNR increases approximately as $\sqrt{N}$, assuming uncorrelated noise. This principle is especially effective when the excitation repetition rate is high, such as in systems operating at several kilohertz pulse repetition rates allowing multiple frames to be accumulated without significant motion artifacts. Figure 5.4 illustrates this relationship for an LED-based small-animal tumor imaging system (850 nm LED array, 7 MHz transducer, 4 kHz pulse repetition rate). At lower frame rates, corresponding to higher numbers of averaged frames, the reconstructed image exhibits markedly improved contrast and vessel delineation. Quantitative analysis (Fig. 5.4b and c) shows that SNR increases rapidly with frame averaging up to several thousand frames, beyond which further improvement saturates.

At acquisition rates above 15–30 Hz, the SNR decline reflects the reduced number of averages per displayed frame. Such results confirm that software-level averaging effectively compensates for low pulse energy in affordable LED-based platforms without modifying hardware [37].

Frame compounding can also be extended spatially by averaging adjacent reconstructions or angular views to suppress uncorrelated speckle-like noise. For portable or handheld systems, averaging must be balanced against motion and latency constraints. Real-time averaging pipelines implemented on GPUs or SoCs can maintain frame rates of 5–10 Hz while improving SNR by 10–15 dB, adequate for vascular and musculoskeletal applications. When combined with adaptive filters Sect. 5.4.3.2, averaging provides a baseline denoising mechanism compatible with almost all affordable implementations.

5.4.3.2 Adaptive Spectral and Spatial Filtering

After temporal averaging, more selective filtering can further enhance signal visibility while minimizing loss of resolution. Affordable PAI systems benefit most from low-complexity adaptive filters that can operate in real time and tolerate the broad, non-stationary noise typical of LED and LD excitation [36, 38]. The objective is to preserve vascular or boundary detail while suppressing background noise across the acoustic frequency band.

5.4.3.3 Bandpass and Wiener Filtering

The frequency response of the detected signal is largely defined by the transducer bandwidth and the frequency content of the absorbed optical pulse. Therefore, simple

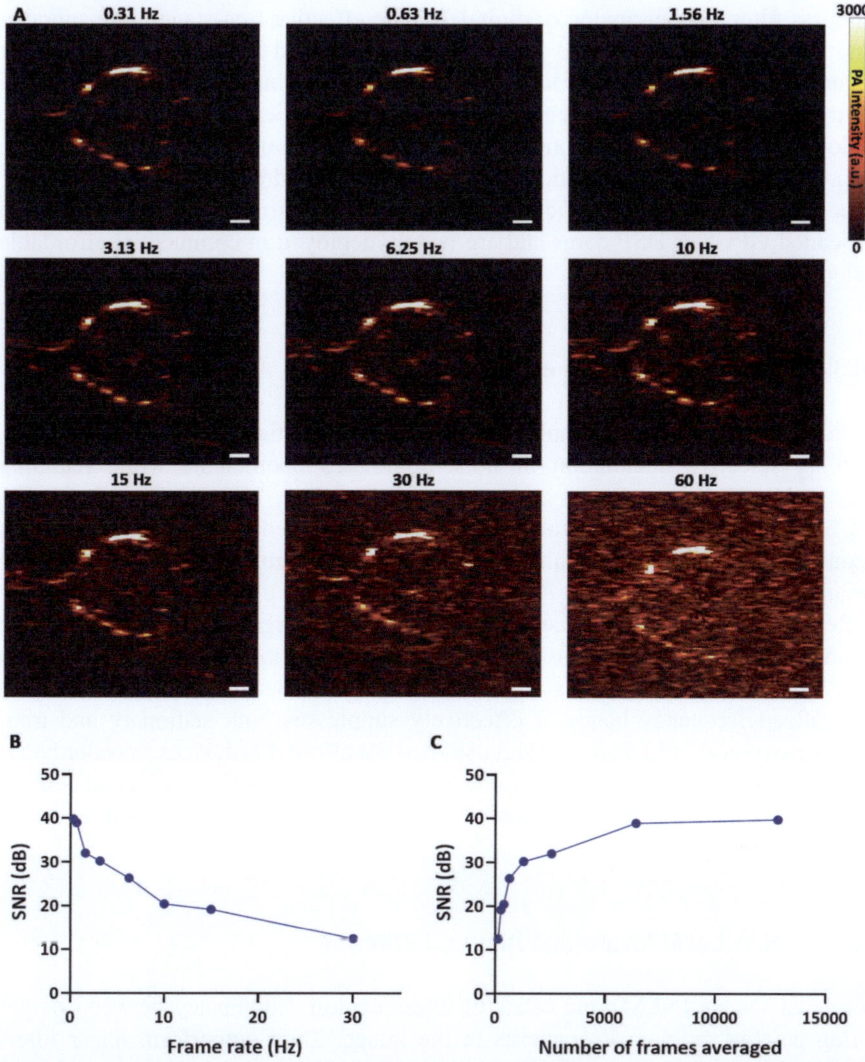

Fig. 5.4 Enhancement of photoacoustic image quality by temporal averaging in an LED-based system. **a** Reconstructed cross-sectional images of a small-animal tumor acquired at different effective frame rates, corresponding to varying numbers of averaged frames (7 MHz transducer, 850 nm LED arrays, 4 kHz pulse repetition rate). **b** Dependence of SNR on frame rate. **c** Dependence of SNR on number of averaged frames. Averaging several thousand frames increases SNR by more than 20 dB, while excessively high frame rates ($\geq$ 30 Hz) reduce SNR due to limited accumulation. Scale bar = 2mm. Reprinted with permission from [37]

bandpass filters, either analog or digital, remain effective first-stage noise suppressors. In practice, narrow-band filtering centered around the dominant transducer frequency (e.g., 5–8 MHz for low-cost linear arrays) removes both low-frequency electronic drift and high-frequency quantization noise. Wiener filters further adapt to the local signal-to-noise environment, estimating the optimal balance between noise suppression and edge preservation in the spectral or time domain [36, 38]. Both techniques can be implemented efficiently using fast Fourier transform (FFT) kernels on embedded GPUs or DSP cores and are widely deployed in commercial affordable systems.

5.4.3.4 Wavelet-Based Denoising

Wavelet thresholding offers multi-resolution decomposition of the acoustic signal, enabling selective attenuation of noise-dominated coefficients while retaining structural detail.

The discrete wavelet transform (DWT) can be applied to individual A-lines or reconstructed images, with hard or soft thresholding guided by estimated noise variance.

Recent adaptive variants use gradient-based or energy-adaptive thresholds to accommodate spatial variations in noise level, yielding improvements of 5–10 dB in SNR without compromising edge fidelity [36, 38]. Because the DWT operates in localized frequency bands, it effectively suppresses both stationary and transient noise typical of LED-driven acquisitions. For affordable devices, wavelet-based filtering is attractive due to its favorable trade-off between performance and computational cost, and can be implemented using open-source libraries optimized for real-time operation.

5.4.3.5 Non-Local Means and Bilateral Filtering

Non-local means (NLM) and bilateral filters exploit redundancy between neighboring patches or intensity patterns in the image. They outperform linear filters in maintaining vascular continuity and are particularly useful when speckle-like noise dominates the reconstructed image. Although originally computationally intensive, simplified fixed-window or lookup-table implementations enable deployment on embedded GPUs or ARM processors, achieving near–real-time performance. In LED-based systems, bilateral filtering has been shown to improve CNR by 30–40% while maintaining fine structural detail [21, 36, 38].

5.4.3.6 Spatiotemporal and Singular-Value Decomposition (SVD) Filtering

For dynamic or high-frame-rate imaging, spatiotemporal filtering such as SVD-based background removal exploits temporal correlation between frames to separate coherent signals from random fluctuations. This method decomposes the data matrix into singular components, retaining only those associated with meaningful photoacoustic signal variance.

When applied over short frame sequences, SVD filtering effectively suppresses low-fluence and electronic noise while preserving motion or hemodynamic signals [39].

Recent studies have demonstrated GPU-accelerated SVD denoising achieving sub-100 ms processing per frame stack, confirming its suitability for portable, low-power PAI platforms.

From an affordability standpoint, these filtering methods offer incremental yet cumulative improvements in image interpretability. Their modular nature allows sequential application (temporal averaging followed by wavelet or Wiener filtering) without increasing hardware cost. A concise comparison of their performance, processing requirements, and suitability for embedded implementation is summarized in Table 5.3.

Among the methods compared, frame averaging and Wiener filtering offer the highest benefit-to-cost ratio and are already integrated into most portable LED-based PAI systems.

Wavelet and non-local means filters provide stronger suppression of broadband and structured noise but require moderate computational resources. Spatiotemporal SVD achieves the greatest contrast recovery when sufficient temporal data are available, yet its matrix decomposition cost limits real-time use on low-power devices.

Table 5.3 Performance of denoising and filtering methods for affordable PAI systems

Method	SNR gain (dB)	Typical latency/frame	Hardware	Complexity	Suitability for affordable systems
Frame averaging	10–25	< 5 ms (GPU); < 50 ms (CPU)	CPU/GPU	Very low	★★★★★
Bandpass/Wiener	5–10	< 5 ms	CPU/FPGA/GPU	Low	★★★★★
Wavelet (adaptive)	8–12	10–50 ms	CPU/GPU	Moderate	★★★★☆
Non-local means/Bilateral	10–15	50–100 ms	Embedded GPU	Moderate	★★★★☆
Spatiotemporal SVD	10–20	50–150 ms (GPU)	GPU/SoC	Mod.–high	★★★☆☆
Multi-stage (combined)	15–25	100–200 ms	GPU/SoC	Moderate	★★★★☆

Hybrid multi-stage filtering like averaging combined with adaptive or wavelet processing represents an emerging compromise that maintains near-real-time performance while approaching the image quality of high-end systems. The relative performance of conventional averaging, discrete wavelet denoising, and spatiotemporal SVD filtering is illustrated in Fig. 5.5. The results confirm that STSVD provides superior noise suppression and contrast recovery while maintaining axial resolution, an attractive balance for LED- or LD-based systems where real-time processing resources are limited.

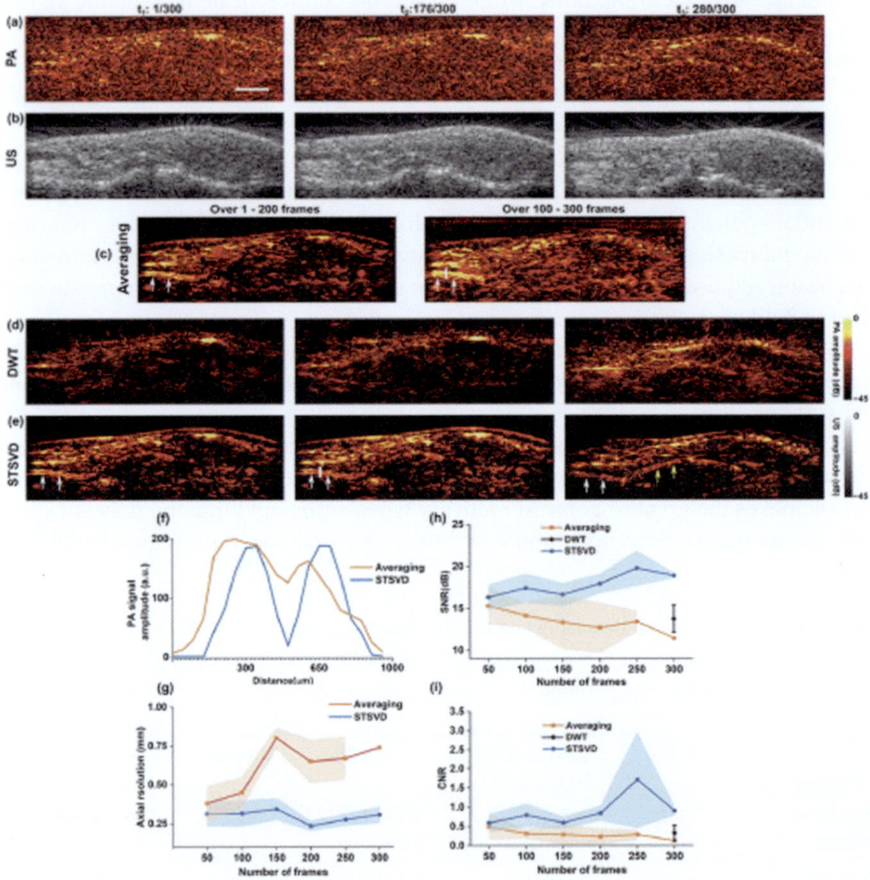

Fig. 5.5 In vivo comparison of denoising approaches for LED-based photoacoustic imaging of a human finger. **a** Representative noisy PA frames; **b** corresponding co-registered ultrasound (US) images; **c** frame averaging; **d** discrete wavelet transform (DWT) denoising; and **e** spatiotemporal singular-value decomposition (STSVD) denoising. Quantitative evaluation of **f** axial signal profiles, **g** axial resolution, **h** signal-to-noise ratio (SNR), and **i** contrast-to-noise ratio (CNR) demonstrate that STSVD yields superior noise suppression and vessel delineation compared with conventional averaging and DWT filtering. Scale bar = 5 mm. Reprinted with permission from [39]

5.4.4 *Reflection Artifact and Out-of-Plane Clutter Reduction*

Reflection artifacts and out-of-plane clutter are coherent distortions that persist in low-fluence, epi-mode PAI. They originate from secondary acoustic propagation and scattering rather than from random noise, producing structured artifacts that can closely resemble true absorbers. In-plane reflections are mainly caused by strong near-surface signals that re-enter the imaging plane after reflecting from deeper impedance boundaries, while out-of-plane clutter arises from absorbers outside the imaging slice whose signals are projected into the reconstruction. Because these mechanisms differ, their correction requires specialized computational strategies. The following sections describe the principal computational and hybrid approaches developed to mitigate such artifacts in affordable handheld photoacoustic systems, including PAFUSion, deformation-compensated averaging (DCA), localized vibration tagging (LOVIT), spectral correction, and AI-based reflection and clutter reduction.

5.4.4.1 PAFUSion: Photoacoustic-Guided Focused Ultrasound

Reflection artifacts caused by acoustic inhomogeneities are a major limitation in epi-mode PAI, particularly in handheld LED- and LD-based systems where high fluence below the probe generates strong surface transients. These surface-generated photoacoustic waves propagate into the tissue and are partially reflected by echogenic interfaces such as fascia, tendon, or bone. The returning echoes overlap with genuine signals, producing mirror-like ghost structures that obscure deeper absorbers and reduce image contrast. The PAFUSion technique was introduced to identify and suppress these reflections by exploiting the reciprocity between photoacoustic emission and ultrasound propagation [22]. In this method, focused ultrasound pulses are transmitted toward bright photoacoustic sources to replicate their inward-travelling acoustic wavefield. By comparing the recorded and simulated reflections, the artifact component can be estimated and subtracted from the original data. This allows selective suppression of reflection artifacts while preserving true absorbers.

To accommodate portable systems that lack programmable ultrasound transmitters, a synthetic-aperture version of PAFUSion was later developed. Here, the reflection paths are modeled numerically by backpropagating the recorded photoacoustic data. Backpropagation is performed in a synthetic way based on the pulse-echo acquisitions after transmission on each single element of the transducer array as shown in Fig. 5.6. Both phantom and in-vivo studies demonstrated artifact reduction exceeding 20 dB, with improved depth visibility and target localization. Because the algorithm is computationally light and parallelizable, it can be executed on embedded GPUs or SoCs, making it practical for real-time handheld operation. An in vivo example of reflection suppression with PAFUSion is shown in Fig. 5.7.

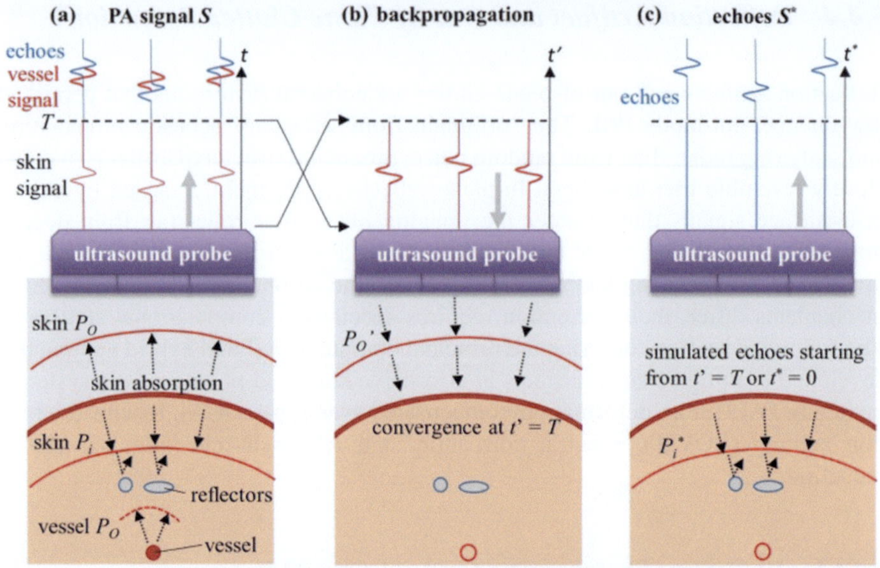

Fig. 5.6 Principle of PAFUSion. **a** Illustration of inward (Pi) and outward-propagating (PO) PA transients, for the special case where they are generated by optical absorption in the skin melanin layer. **b** Time-inversion and backpropagation of the PA signal from interval t = [0, T], towards **c** mimicking the inward-travelling wavefield Pi* and identification of resulting echoes. Reproduced with permission from [23]

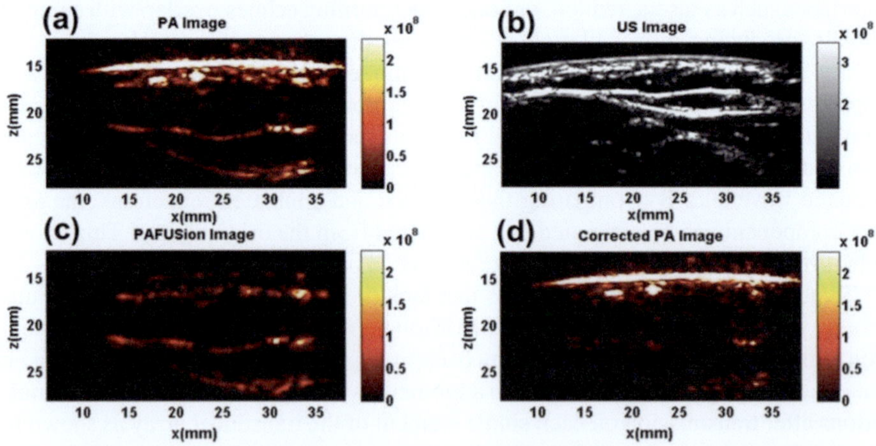

Fig. 5.7 In vivo measurement on human forearm demonstrating the performance of PAFUSion. **a** photoacoustic image, **b** ultrasound image, **c** PAFUSion image with identified reflection artifacts, **d** corrected photoacoustic image after subtraction of reflection artifact. Adapted with permission from [23]

5.4.4.2 Deformation-Compensated Averaging (DCA)

In handheld PAI with LD or LEDs, small probe or tissue motions between consecutive frames can alter the spatial phase of genuine photoacoustic signals while leaving reflection artifacts largely coherent. Simple temporal averaging therefore tends to blur structural detail and only partially reduce reflections. DCA was introduced to overcome this limitation by first aligning consecutive frames to a common reference before averaging [26]. The key concept is that motion-compensated registration preserves real anatomical structures while decorrelating spurious signals arising from reflections or reverberations. The method estimates probe or tissue deformation, typically on the order of tens of micrometres, using cross-correlation between successive PA or ultrasound frames. Once motion fields are determined, each frame is spatially warped to the reference coordinate system and averaged. Because reflection artifacts originate from delayed acoustic paths rather than true anatomical motion, their contribution becomes incoherent and is effectively suppressed. The result is a composite image with reduced clutter and enhanced vascular continuity, particularly valuable in handheld LED or LD-based systems that operate at moderate frame rates. DCA requires only software implementation and modest computation. GPU-accelerated versions achieve near-real-time performance (1–5 Hz) on embedded platforms such as NVIDIA Jetson or Zynq SoCs. Compared with conventional averaging, DCA typically provides an additional 6–10 dB improvement in SNR and visibly stabilizes image appearance under minor hand motion. In phantom and in-vivo LD-PA studies, it has proven effective for suppressing skin-induced reflections and enhancing depth visibility without much hardware modification. An example of reflection reduction using DCA to visualize carotid artery is shown in Fig. 5.8, where motion-corrected averaging eliminates coherent artifacts while preserving fine vascular features.

Display ranges: 50 dB for US (TGC = 5 dB/cm) and 25 dB for PA (TGC = 12.5 dB/cm). Reprinted with permission from [26].

5.4.4.3 Localized Vibration Tagging (LOVIT)

In PAI, clutter arising from absorbers or reflections outside the imaging slice can obscure in-plane structures, particularly in linear-array geometries with limited elevation focusing. The Localized Vibration Tagging (LOVIT) technique was developed to distinguish true in-plane absorbers from out-of-plane clutter by inducing controlled mechanical modulation of the tissue within the imaging plane [29]. During acquisition, a localized vibration applied using a piezoelectric actuator or probe-mounted driver slightly modulates the position of in-plane structures, producing a characteristic temporal fluctuation in their photoacoustic signals. Out-of-plane or reverberant components, which are not physically displaced, remain unmodulated. By comparing sequential frames acquired with and without vibration, correlated components can be identified as genuine absorbers, while unmodulated signals are suppressed. LOVIT provides powerful suppression of out-of-plane and reflection-induced clutter, with reported contrast improvements of 15–25 dB in deep-tissue imaging and enhanced

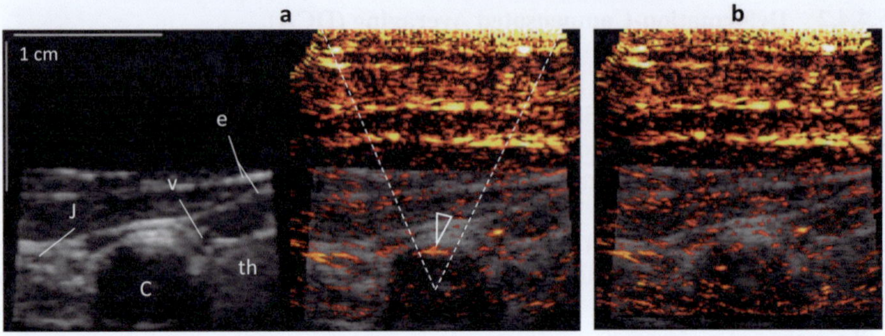

Fig. 5.8 Improved vascular visibility with deformation-compensated averaging (DCA). **a** Side-by-side display of B-mode ultrasound (left) and co-registered photoacoustic/ultrasound overlay (right) of a transverse section through the left carotid artery. US structures are labeled as follows: C—carotid lumen; J—internal jugular vein; v—small vessel; th—thyroid gland; e—muscle epimysia. The photoacoustic overlay (blackbody color scale) shows strong absorption signals at the carotid lumen surface and around vascular structures. Dashed lines indicate the receive angular aperture limiting the lateral extent of the carotid signal. **b** Without motion compensation, simple frame averaging suppresses the carotid signal, whereas DCA preserves vascular contrast and spatial integrity

delineation of vessels embedded in scattering media. However, the approach requires dedicated actuation hardware, precise synchronization between vibration and acquisition, and longer acquisition times for differential processing. These requirements make it more suitable for benchtop or research systems than for compact or low-cost handheld probes. For affordable LED- or LD-based PAI systems, the underlying principle of temporal decorrelation between in-plane and clutter components remains valuable even if full mechanical tagging is impractical. Simplified variants using natural hand-induced micro-motion or probe vibration as a passive tagging source have been explored conceptually, but robust real-time implementation remains an open challenge. As such, LOVIT currently serves as an important conceptual framework for clutter discrimination, rather than a deployable solution for cost-sensitive platforms.

5.4.4.4 Spectral and Multi-Wavelength Correction

Spectral and multi-wavelength correction methods exploit the distinct wavelength-dependent behavior of photoacoustic signals to identify and suppress reflection artifacts and clutter. In epi-illumination geometries, both the skin and its associated reflection artifacts often exhibit similar spectral characteristics because the secondary acoustic wave originates from the same superficial absorber. In contrast, deeper vascular or tissue absorbers display wavelength-dependent absorption spectra characteristic of hemoglobin or other chromophores. By analyzing correlations between signals acquired at multiple wavelengths, components that maintain identical spectral signatures to the skin can be classified as reflections and selectively removed, while true subsurface absorbers are preserved. Dual-wavelength LED/LD arrays, now widely available, make this approach feasible for affordable handheld systems, providing a purely computational means of improving contrast without additional hardware. Such spectral filtering can be implemented using simple correlation or unmixing algorithms and has been shown to substantially reduce superficial reflection artifacts in low-fluence LD-based imaging [27, 28].

5.4.4.5 AI-Based Reflection and Clutter Suppression

Deep learning has recently emerged as a powerful tool for mitigating reflection artifacts and acoustic clutter in PAI, offering data-driven solutions that complement model-based and spectral correction approaches. Rather than explicitly modeling acoustic propagation, convolutional neural networks (CNNs) learn statistical relationships between artifact-contaminated and artifact-free images, enabling automated suppression of structured noise. Allman et al. [31] first demonstrated this concept using simulated and experimental datasets to train a CNN for reflection artifact removal and photoacoustic source localization, achieving substantial improvement in image clarity and depth discrimination [31]. Building on this foundation, Agrawal et al. [40] applied a similar deep-learning strategy to data from an LED-based handheld system (AcousticX, CYBERDYNE Inc.), using realistic synthetic finger models that incorporated optical absorption and acoustic impedance variations [40]. The network, trained on thousands of simulated and in vivo images, effectively suppressed bone- and skin-related reflection artifacts while preserving vascular continuity, marking one of the first demonstrations of deep learning applied to affordable, low-fluence PAI. The method operates entirely in the image domain and can be implemented for real-time inference (<100 ms per frame) on embedded GPUs such as NVIDIA Jetson platforms. AI-based artifact suppression therefore provides a scalable path for improving image quality in portable photoacoustic systems without hardware modification. By combining computational speed, adaptability to different probe geometries, and compatibility with LED/LD illumination, deep-learning approaches such as those of Allman and Agrawal demonstrate that intelligent reconstruction can offset many of the physical limitations inherent to affordable PAI systems.

5.4.4.6 Summary and Comparative Perspective

Reflection artifacts and acoustic clutter remain major challenges in low-fluence, handheld photoacoustic imaging. While their physical origins like secondary acoustic reflections and out-of-plane signal projections are well understood, their correction depends strongly on computational strategy. Model-based and motion-compensated techniques such as PAFUSion and DCA primarily address in-plane reflections; physics-based modulation and spectral methods such as LOVIT and multi-wavelength correction are effective against out-of-plane clutter too; and recent AI-based networks also offer unified suppression of both artifact classes. Each approach carries distinct trade-offs between computational complexity, hardware cost, and achievable image quality. For affordable systems with inherent ultrasound capability, algorithms requiring minimal mechanical or optical modification, particularly PAFUSion, DCA, and lightweight AI inference, represent the most practical near-term solutions. Table 5.4 summarizes their comparative features and implementation suitability.

Table 5.4 Comparison of reflection- and clutter-reduction methods for affordable PAI

Method	Target artifact	Principle/mechanism	Additional hardware	Suitability for affordable systems
PAFUSion	In-plane reflections	Simulates reflected wavefield using focused-US transmission or synthetic back-propagation	Optional US transmit channel	★★★★★
Deformation-Compensated Averaging (DCA)	In-plane reflections, motion-induced clutter	Frame alignment and averaging after motion compensation	None	★★★★★
LOVIT	Out-of-plane clutter, reflection artifacts	Mechanical vibration tags in-plane absorbers for correlation-based discrimination	Piezo actuator or driver	★★☆☆☆
Spectral/Multi-wavelength correction	Mixed reflections and clutter	Identifies reflection components via spectral correlation with skin signals	Multi-LED module	★★★★☆
AI-based suppression	Both reflection and clutter	CNN / U-Net learns mapping from artifact-contaminated to clean images	Embedded GPU (Jetson / SoC)	★★★★☆

5.5 Computational Enhancement for Sparse and Limited-View Systems

5.5.1 *Problem Definition and Relevance to Affordable Systems*

Affordable PAI platforms based on LEDs, LDs, or VCSELs are designed to minimize cost, size, and power consumption. These simplifications inevitably constrain system geometry and acquisition bandwidth, leading to fundamental limitations in data completeness. Two of the most prominent sources of degradation are sparse sampling and limited-view detection, both of which cause characteristic image distortions that cannot be fully resolved by hardware optimization alone. In sparse-sampling systems, the number of detection channels or sequential acquisition positions is reduced relative to the spatial Nyquist requirement.

This condition is common in portable and handheld systems that employ low-element linear arrays or multiplexed receivers to reduce electronics cost and power. Sparse sampling leads to spatial aliasing, streak artifacts, and loss of fine structure, particularly when the object contains multiple small or high-contrast absorbers. Simple interpolation or analytic back-projection cannot restore these missing spatial frequencies, and reconstruction quality becomes highly dependent on prior information or regularization. The limited-view problem arises from geometric constraints of the detector aperture. In many handheld configurations, the transducer array subtends only 60–90° of the full detection circle, causing anisotropic sensitivity: absorbers oriented parallel to the array are reconstructed clearly, whereas those perpendicular to it are blurred or entirely missing as shown in Fig. 5.9. This anisotropy produces incomplete boundary delineation and uneven vascular visibility. The effect is further compounded in LED/LD-based systems, where shallow optical penetration restricts the field of excitation, and in clinical scenarios where the probe cannot fully encircle the tissue.

These two factors including sparse sampling and limited angular coverage represent the core computational challenges for affordable PAI. They define an underdetermined inverse problem in which the available acoustic data are insufficient to reconstruct the true absorption distribution using direct analytic methods. Overcoming these deficiencies requires algorithmic compensation, ranging from model-based iterative solvers and compressed-sensing reconstructions to deep-learning approaches that learn the mapping from incomplete to artifact-free images. The following sections describe computational strategies developed to address these limitations in affordable PAI systems.

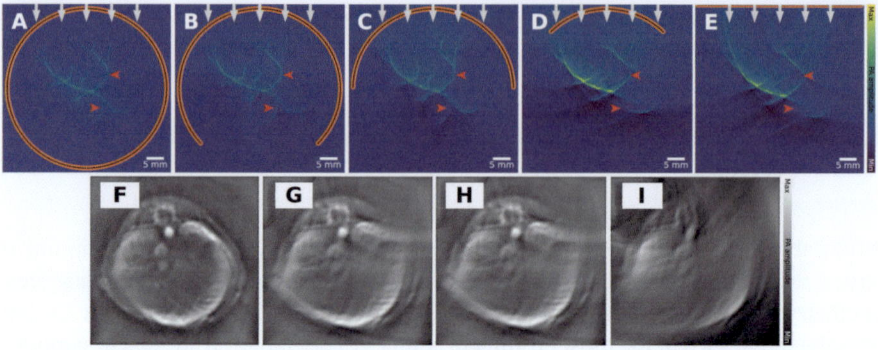

Fig. 5.9 Limited-view artifacts and loss of image fidelity in PAI. Simulated and experimental examples showing degradation as the detector aperture decreases. A vascular network was reconstructed under identical conditions while varying the detection opening angle of a spherical array: **a** 360°, **b** 270°, **c** 180°, **d** 90°, and **e** linear transducer geometry. As the angular coverage narrows, visibility and continuity of vascular structures progressively decline. Panels (**f**–**I**) show corresponding preclinical in vivo data from a mouse, reconstructed with effective apertures of 180°, 120°, 90°, and 70°, respectively, demonstrating similar loss of structural information and increased clutter. Reprinted with permission from [30]

5.5.2 Analytic and Model-Based Reconstruction Strategies

Conventional photoacoustic image reconstruction algorithms such as DAS, universal back-projection (UBP), and filtered back-projection (FBP) assume complete and uniformly sampled acoustic data. In affordable PAI systems, however, limited detector coverage and sparse sampling violate these assumptions, leading to spatial aliasing, streaking, and anisotropic resolution. To recover missing information, analytic reconstructions have been extended and combined with regularization or model-based optimization schemes that incorporate physical priors about acoustic propagation and tissue structure. Early universal back-projection formulations by Xu and Wang [13] established the mathematical basis for accurate reconstruction under ideal full-view conditions [13]. Subsequent optimizations introduced angular weighting and adaptive aperture compensation to reduce limited-view artifacts when using linear or sector transducers. Fourier-domain implementations further accelerated these analytic methods, achieving real-time 2D imaging on GPU hardware [11]. While these analytic approaches remain computationally efficient, their performance rapidly deteriorates as the detection geometry becomes incomplete or sparsely sampled. To address such data incompleteness, model-based iterative reconstruction (MBIR) methods explicitly formulate PAI as an inverse problem governed by the acoustic wave equation.

By minimizing the difference between measured and forward-simulated data, MBIR methods can enforce data consistency while introducing regularization terms such as total variation (TV) or Tikhonov constraints to promote sparsity and suppress streak artifacts [8, 9, 30]. Compressed-sensing (CS) formulations have proven particularly powerful: by assuming that the optical-absorption map is sparse in the gradient or wavelet domain, accurate reconstructions can be recovered from substantially undersampled data [11]. GPU-accelerated and parallelized solvers now enable these iterative methods to operate within seconds per frame, making them viable for portable devices equipped with embedded GPUs [8, 12]. Recent advances have focused on improving both convergence speed and robustness under low-SNR conditions typical of LED or LD-based systems. Hsu et al. [9] introduced a learned physical model that integrates analytical forward modeling with adaptive preconditioning to achieve near-real-time MBIR performance [9]. Rietberg et al. [30] further provided a comprehensive review of algorithmic performance in sparse and limited-view geometries, highlighting the advantages of hybrid methods that combine data-fidelity enforcement with structural priors and spectral weighting [30]. Model-based frameworks yield better depth uniformity and vascular continuity than analytic methods while maintaining interpretability and reproducibility, an essential consideration for clinical translation.

In the context of affordable PAI, model-based iterative approaches represent a practical midpoint between analytic back-projection and fully data-driven (AI) reconstruction.

They require no hardware modification, leverage well-established physical models, and can exploit GPU acceleration for near-real-time operation. Figure 5.9 illustrates the limited-view artifacts these algorithms aim to correct. The next Sect. 5.5.3 describes how learning-based and hybrid approaches extend these concepts further by embedding physical priors within neural-network architectures.

5.5.3 AI-Assisted Sparse and Limited-View Reconstruction

Although MBIR and CS approaches can mitigate streaking and aliasing, their performance depends on the accuracy of the assumed acoustic and optical models. In affordable LED- or LD-based systems, uncertainties in fluence distribution, coupling, and limited bandwidth often cause residual artifacts even after regularized inversion. Recent advances have therefore focused on AI-assisted reconstruction, where neural networks learn mappings between incomplete measurements and artifact-free images, either in the image domain or directly from raw channel data. Early studies employed CNNs to enhance analytically reconstructed images. Guan et al. [41] introduced a fully dense U-Net for two-dimensional sparse-view PAI that recovered missing vascular structures and suppressed streaking compared with CS-TV methods [41]. Reiter et al. [31] used similar concept to channel-domain learning, training a CNN to identify and correct reflection-like artifacts before beamforming, improving generalization and reconstruction speed [31]. More recently, He et al. [42] demonstrated a lightweight deep-learning model for artifact removal in compact PAI systems, achieving substantial enhancement in edge sharpness and contrast under sparse-detection conditions [42]. In parallel, Li et al. [43] proposed a zero-shot, self-incentive framework (Artifact2Artifact) capable of removing artifacts without paired training data, an approach particularly suited to handheld LED/LD-based systems lacking reference datasets [43].

Francis et al. [44] developed one of the earliest GAN-based reconstruction frameworks for photoacoustic imaging, designed specifically for band-limited and limited-view data from handheld LED-based systems [44]. The network was conditioned on measured sinograms and time-reversal reconstructions, allowing compensation for transducer directivity and bandwidth loss. Trained on both simulated and in-vivo datasets, the GAN restored vascular continuity and suppressed directional artifacts compared with DAS reconstruction. This work demonstrated the feasibility of adversarial learning in real, low-fluence PAI and provided a conceptual foundation for later physics-guided and hybrid deep-learning models. Building upon this foundation, Lu et al. [45] introduced a learnable physical deep-learning model that embeds the acoustic forward operator within a neural architecture [45], combining data-fidelity enforcement with a trainable regularizer to recover fine structures from sparse measurements. Such physics-informed networks enhance interpretability and reliability while maintaining real-time performance on GPU hardware. AI-assisted reconstruction therefore represents a continuum from purely data-driven to physically guided learning that complements analytic and iterative methods. By combining physical modeling with neural inference, these hybrid frameworks deliver high-quality reconstructions under the sparse-sampling and limited-view conditions intrinsic to low-cost photoacoustic devices.

5.5.4 Computational Resolution Enhancement and Denoising Under Sparse Sampling

Even after advanced reconstruction, affordable PAI systems remain constrained by the limited bandwidth and sampling density of compact transducers and low-fluence optical sources. These constraints produce spatially blurred, low-contrast images with degraded high-frequency detail, particularly in handheld LED- or LD-based configurations. Hardware upgrades can partly mitigate these effects, but the preferred solution in affordable systems is to compensate computationally—through super-resolution (SR) and denoising strategies that enhance the effective information content of the acquired data without increasing cost or power consumption.

5.5.4.1 Resolution Enhancement Approaches

Model-based resolution enhancement methods apply deconvolution or sparsity-driven priors to recover high-frequency information lost to transducer bandwidth or limited angular sampling. Total-variation and iterative gradient-sparse reconstructions have shown improvements in vessel delineation, but they remain sensitive to noise and require accurate knowledge of the system's point-spread function. In contrast, learning-based super-resolution directly learns mappings between low- and high-resolution image domains. CNN and GAN architectures such as SR-CNN, ESRGAN, and domain-adapted U-Nets have been successfully adapted to PAI, achieving sharper edges and improved vascular continuity. Hariri et al. [46] demonstrated that a deep CNN trained on low-fluence LED-PA data significantly increased contrast and resolution compared with traditional band-limited reconstructions [46]. Subsequent work by Shi et al. [47] used semi-synthetic data generation to train deep networks for enhancing needle visibility in LED-based PAI, overcoming domain gaps between simulations and experimental measurements [47].

5.5.4.2 Learning-Based Denoising and Real-Time Enhancement

Low-fluence systems suffer from stochastic LED/LD intensity variations and electronic noise that cannot be fully averaged out in real-time operation. Deep-learning-based denoising has emerged as an efficient solution that maintains high frame rates while improving signal-to-noise ratio (SNR) and contrast-to-noise ratio (CNR). Paul and Mallidi [48] implemented a real-time U-Net architecture capable of denoising and enhancing LED-based PAI frames during acquisition, resulting in over twofold SNR improvement without perceptible latency [48]. Their follow-up study introduced *DenP2P*, an optimized conditional GAN architecture designed for embedded GPU inference [49]. DenP2P simultaneously enhanced SNR and resolution while preserving vascular geometry, achieving near-video-rate operation on consumer-grade hardware (Fig. 5.10). These frameworks demonstrate that carefully engineered deep networks can act as computational amplifiers, enabling diagnostic-quality visualization from inherently low-energy sources. Collectively, these methods indicate that resolution enhancement and denoising no longer represent offline post-processing steps but integral, real-time components of affordable imaging workflows. Table 5.5 compares different deep learning models used in affordable PAI for enhancing resolution and contrast.

5.5.4.3 Perspective

Resolution enhancement and denoising are now critical computational enablers that transform low-cost PAI from qualitative visualization to quantitative imaging. By combining model-driven priors with learning-based inference, current algorithms recover the resolution and contrast otherwise lost to low fluence, limited bandwidth, and sparse detection. Their implementation on GPUs, SoCs, and embedded AI accelerators makes these methods particularly suited for portable, bedside, and wearable PAI systems where hardware simplicity must be balanced against clinical interpretability.

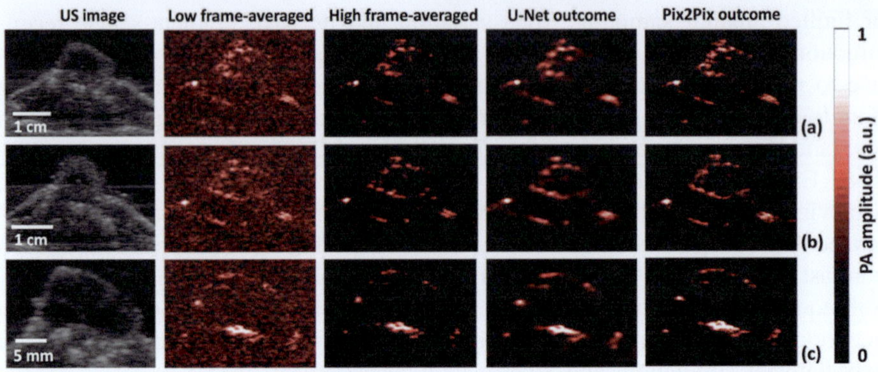

Fig. 5.10 Comparison of U-Net and DenP2P deep-learning networks for noise reduction in LED-based photoacoustic imaging. Representative in-vivo results from three cross-sections of mouse tumors are shown in rows (**a**–**c**). U-Net outputs exhibit mild blurring and partial loss of structural contrast, whereas the proposed DenP2P network effectively recovers fine vascular detail and reproduces image quality comparable to high-frame-average reconstructions, without sacrificing temporal resolution. All scale bars correspond to those shown in the accompanying ultrasound reference images. Reproduced with permission from [49]

Table 5.5 Summary of deep-learning-based computational enhancement approaches for affordable photoacoustic imaging. Each framework targets the restoration of image sharpness, vascular continuity, and SNR in low-fluence LED/LD systems while maintaining real-time feasibility

Study/year	Model type	Application/dataset	System type	Reported improvement	Real-Time Feasibility
Hariri et al. [46]	CNN (contrast and edge enhancement)	Low-fluence vascular imaging	LED-PAI bench setup	↑ SNR × 2; improved edge definition	Offline (GPU)
Shi et al. [47]	U-Net/CNN (semi-synthetic)	Needle visibility improvement	LED linear probe	+ 35% CNR; reduced background clutter	Near-real-time
Paul and Mallidi [48]	U-Net (real-time denoising)	In-vivo LED imaging	Handheld LED system	↑ SNR 2–3 × ; reduced flicker noise	Real-time GPU
Paul and Mallidi [49]	cGAN (*DenP2P*)	End-to-end denoising + resolution recovery	LED portable platform	↑ SNR ≈ 3 × ; ≈ 25% sharper vascular edges	Real-time embedded GPU

5.6 Practical Implementations and Evaluation

The successful translation of computational enhancement methods into affordable PAI systems depends not only on algorithmic performance but also on implementation efficiency, hardware compatibility, and reproducibility. Unlike benchtop systems that rely on high-end GPUs and extensive memory, portable and handheld PAI platforms operate under strict constraints on power, latency, and thermal dissipation. Consequently, the design of the processing pipeline, ranging from reconstruction to denoising and visualization must balance numerical precision with real-time responsiveness [2, 8, 12].

5.6.1 Hardware Acceleration and Embedded Computing

Recent advances in embedded parallel processors and AI accelerators have significantly improved the feasibility of deploying advanced reconstruction and enhancement algorithms on low-cost systems. GPUs remain the primary choice for high-throughput reconstruction and filtering, and libraries such as CUDA, OpenCL, and TensorRT enable optimized execution of delay-and-sum, model-based iterative reconstruction, and CNN inference pipelines [9, 11]. Compact edge modules such as the NVIDIA Jetson Orin, Xilinx Zynq UltraScale + , and Rockchip RK3588 platforms can now execute 2D/3D reconstruction and neural inference within 20–50 ms per frame, sufficient for real-time imaging . Field-programmable gate arrays (FPGAs) further reduce latency and power consumption for deterministic operations such as back-projection, adaptive beamforming, and signal envelope detection [12]. While FPGAs offer high efficiency, their limited flexibility compared with GPUs makes them best suited for fixed-function reconstruction or embedded clinical devices.

5.6.2 Software Integration and Real-Time Workflow

In integrated LED/LD-based systems, reconstruction, denoising, and display must operate concurrently. Pipeline design often follows a producer–consumer architecture, where raw acoustic data are acquired, buffered, reconstructed, and enhanced in parallel threads [5]. Memory-efficient precomputation of delay tables and lookup-based beamforming are critical to sustaining frame rates above 10–20 Hz. Software frameworks such as PyTorch, TensorFlow Lite, and ONNX Runtime allow neural networks (e.g., DenP2P, U-Net) to be deployed in quantized form, achieving speedups of 3–5 $\times$ over full-precision inference [49]. Real-time implementations frequently use mixed-precision (FP16/INT8) arithmetic to reduce computational load while maintaining clinical image quality.

5.6.3 Performance Metrics and Evaluation Criteria

Quantitative evaluation of reconstruction and enhancement algorithms in affordable PAI follows three primary dimensions [46–49]:

Image quality metrics: SNR, contrast-to-noise ratio (CNR), full-width-at-half-maximum (FWHM) of vascular features, and structural similarity index (SSIM).

Computational metrics: frame rate, latency, power consumption, and model size.

Clinical or task-based metrics: target detectability, depth fidelity, and anatomical correspondence with co-registered ultrasound data.

For LED/LD-based systems, performance is best reported as a trade-off between SNR gain and processing time. Typical portable implementations achieve reconstruction and enhancement of 128×128 frames in < 50 ms using edge GPUs, or < 200 ms on low-power FPGAs, meeting the threshold for interactive imaging [12].

5.6.4 Open-Source Deployment and Reproducibility

To ensure transparency and accelerate adoption, several research groups have released open-source PAI reconstruction toolkits and enhancement networks. Frameworks such as k-Wave, SIMPA, PATATO, and PhocoSpace provide standardized simulation and reconstruction environments [50–54]. Public repositories now include pretrained U-Net and GAN models for data, allowing benchmarking across institutions. For regulatory and clinical translation, maintaining traceability of training datasets, hyperparameters, and evaluation scripts is essential. Containerized deployment using Docker or Singularity can further guarantee reproducibility across different computing platforms.

5.6.5 Future Considerations

Continued progress in model compression, neural quantization, and hardware–software co-design will further reduce computational cost and energy demand [45]. In parallel, unified benchmarking protocols reporting frame-rate–SNR trade-offs and hardware power budgets will enable fair comparison among affordable systems. Ultimately, computational efficiency, reproducibility, and clinical robustness will define the next generation of low-cost, high-utility PAI instruments.

5.7 Summary and Outlook

Affordable PAI systems based on LEDs, LDs, and VCSELs are transforming optical–acoustic imaging from laboratory setups into portable and clinically viable tools. However, these compact, low-fluence platforms face inherent challenges of reduced SNR, limited bandwidth, and sparse spatial sampling. This chapter has emphasized that image quality in affordable PAI is now primarily determined by computation, where intelligent reconstruction and enhancement compensate for simplified optics and detection. Three foundational strategies like fast reconstruction, denoising, and artifact suppression together define the computational backbone of modern low-cost systems. From efficient GPU/FPGA-based reconstruction to advanced AI-driven artifact removal and real-time denoising, algorithmic innovation has become the main driver of contrast, depth fidelity, and diagnostic reliability.

Key Takeaways

- **Hardware simplification demands software intelligence**. Affordable LED/LD/VCSEL sources and limited-channel detectors inherently lower fluence and bandwidth; algorithmic correction is essential to recover diagnostic information.
- **Fast reconstruction underpins real-time usability**. GPU- and FPGA-accelerated delay-and-sum, model-based, and Fourier-domain methods enable near-video-rate performance while maintaining low power budgets.
- **Artifact reduction enhances interpretability**. Reflection suppression methods such as PAFUSion, DCA, and LOVIT substantially improve structural fidelity in handheld and clinical imaging.
- **AI has become a practical solution for sparse and noisy data**. CNN, U-Net, and GAN models including physics-informed networks restore vascular continuity, correct for limited-view loss, and enable adaptive enhancement directly on embedded processors.
- **Computational enhancement is cost-neutral**. Improvements in SNR and contrast achieved by deep or hybrid learning frameworks require no additional optical energy, directly addressing safety and cost constraints.
- **Reproducibility and open access are vital**. Public reconstruction toolkits (e.g., *k-Wave*, *SIMPA*) and pretrained models are accelerating cross-lab validation and clinical translation.
- **Integration with ultrasound remains central**. Real-time photoacoustic–ultrasound fusion enhances anatomical localization and provides an interpretable clinical context for low-cost systems.

References

1. Y. Zhu, G. Xu, J. Yuan et al., Light emitting diodes based photoacoustic imaging and potential clinical applications. Sci. Rep. **8**(1), 9885 (2018)
2. P.K. Upputuri, M. Pramanik, Fast photoacoustic imaging systems using pulsed laser diodes: a review. Biomed. Eng. Lett. **8**(2), 167–181 (2018)
3. S. van Heumen, J. Riksen, M.K.A. Singh, G. van Soest, D. Vasilic, LED-based photoacoustic imaging for preoperative visualization of lymphatic vessels in patients with secondary limb lymphedema. Photoacoustics. **29**, 100446 (2022)
4. J. Jo et al., Detecting joint inflammation by an LED-based photoacoustic imaging system: A feasibility study. J. Biomed. Opt. **23**(11), 110501 (2018)
5. M. Erfanzadeh, Q. Zhu, Photoacoustic imaging with low-cost sources: a review. Photoacoustics. **14**, 1–11 (2019)
6. S.K. Kalva, P.K. Upputuri, P. Rajendran, R.A. Dienzo, M. Pramanik, Pulsed laser diode-based desktop photoacoustic tomography for monitoring wash-in and wash-out of dye in rat cortical vasculature. J. Vis. Exp. (147), (2019)
7. M. Seeger, A. Stylogiannis, L. Prade et al., Overdriven laser diode optoacoustic microscopy. Sci. Rep. 13, 19542 (2023)
8. Y. Wang, C. Li, Comprehensive framework of GPU-accelerated image reconstruction for photoacoustic computed tomography. J. Biomed. Opt. 29(6), 066006 (2024)
9. K.-T. Hsu, S. Guan, P.V. Chitnis, Fast iterative reconstruction for photoacoustic tomography using learned physical model: Theoretical validation. Photoacoustics 29, 100452 (2023)
10. C. Tian et al., Image reconstruction from photoacoustic projections. Photonics Insights. **3**(3), R06 (2024)
11. M. Gao, G. Si, Y. Bai, L.V. Wang, C. Liu, J. Meng, Graphics processing unit accelerating compressed sensing photoacoustic computed tomography with total variation. Appl. Opt. **59**, 712–719 (2020)
12. Z. Gao et al., *FPGA acceleration of image reconstruction for real-time photoacoustic tomography* (2022). Preprint arXiv:2204.14084
13. M. Xu, L.V. Wang, Universal back-projection algorithm for photoacoustic computed tomography. Phys. Rev. E **71**, 016706 (2005)
14. C.W. Wei, T.M. Nguyen, J. Xia, B. Arnal, E.Y. Wong, I.M. Pelivanov, M. O'Donnell, Real-time integrated photoacoustic and ultrasound (PAUS) imaging system to guide interventional procedures: ex vivo study. IEEE Trans. Ultrason. Ferroelectr. Freq. Control **62**(2), 319–328 (2015)
15. A. Garcia-Uribe et al., Dual-modality photoacoustic and ultrasound imaging system for noninvasive sentinel lymph node detection in patients with breast cancer. Sci. Rep. **5**, 15748 (2015). https://doi.org/10.1038/srep15748
16. A.A. Oraevsky et al., Clinical optoacoustic imaging combined with ultrasound for coregistered functional and anatomical mapping of breast tumors. Photoacoustics. **12**, 30–45 (2018). ISSN 2213–5979
17. G.-S. Jeng et al., Real-time interleaved spectroscopic photoacoustic and ultrasound (PAUS) scanning with simultaneous fluence compensation and motion correction. Nat. Commun. **12**, 716 (2021)
18. U. Alqasemi et al., Real-time co-registered ultrasound and photoacoustic imaging system based on FPGA and DSP architecture, in *Proceedings of SPIE*, vol. 7899, article 78993S (2011)
19. G. Yang, E. Amidi, W. Chapman, S. Nandy, A. Mostafa, H. Abdelal, Z. Alipour, D. Chatterjee, M. Mutch, Q. Zhu, Co-registered photoacoustic and ultrasound imaging of human colorectal cancer. J. Biomed. Opt. **24**(12), 1–13 (2019)
20. S. Zhang, J. Miao, L.S. Li, Challenges and advances in two-dimensional photoacoustic computed tomography: a review, J. Biomed. Opt. **29**(7), 070901 (2024)
21. M. Kuniyil Ajith Singh, W. Xia, Portable and affordable light source-based photoacoustic tomography. Sensors, **20**(21), 6173 (2020)

22. M. Kuniyil Ajith Singh, W. Steenbergen, Photoacoustic-guided focused ultrasound (PAFUSion) for identifying reflection artifacts in photoacoustic imaging. Photoacoustics. **3**(4), 123–131 (2015)
23. M.K.A. Singh, M. Jaeger, M. Frenz, W. Steenbergen, In vivo demonstration of reflection artifact reduction in photoacoustic imaging using synthetic aperture photoacoustic-guided focused ultrasound (PAFUSion). Biomed. Opt. Express **7**, 2955–2972 (2016)
24. M.K.A. Singh, M. Jaeger, M. Frenz, W. Steenbergen, Photoacoustic reflection artifact reduction using photoacoustic-guided focused ultrasound: comparison between plane-wave and element-by-element synthetic backpropagation approach. Biomed. Opt. Express **8**, 2245–2260 (2017)
25. M.K. Singh, V. Parameshwarappa, E. Hendriksen, W. Steenbergen, S. Manohar, Photoacoustic-guided focused ultrasound for accurate visualization of brachytherapy seeds with the photoacoustic needle. J. Biomed. Opt. **21**(12), 120501 (2016)
26. M. Jaeger et al., Deformation-compensated averaging for deep-tissue LED and laser diode-based photoacoustic imaging integrated with handheld echo ultrasound, in *LED-based photoacoustic imaging. Progress in optical science and photonics*, vol. 7, ed. by M. Kuniyil Ajith Singh (Springer, Singapore, 2020)
27. H.N.Y. Nguyen, W. Steenbergen, Feasibility of identifying reflection artifacts in photoacoustic imaging using two-wavelength excitation. Biomed. Opt. Express **11**(10), 5745–5759 (2020)
28. H.N.Y. Nguyen, A. Hussain, W. Steenbergen, Reflection artifact identification in photoacoustic imaging using multi-wavelength excitation. Biomed. Opt. Express **9**(10), 4613–4630 (2018)
29. M. Jaeger, J.C. Bamber, M. Frenz, Clutter elimination for deep clinical optoacoustic imaging using localised vibration tagging (LOVIT). Photoacoustics. **1**(2), 19–29 (2013)
30. T. Rietberg, J. Gröhl, T.R. Else, S.E. Bohndiek, S. Manohar, B.T. Cox, Artifacts in photoacoustic imaging: Origins and mitigations. Photoacoustics 45, 100745 (2025)
31. D. Allman, A. Reiter, M.A.L. Bell, Photoacoustic source detection and reflection artifact removal enabled by deep learning. IEEE Trans. Med. Imaging **37**(6), 1464–1477 (2018)
32. M. Jaeger, S. Schüpbach, A. Gertsch et al., Fourier reconstruction in optoacoustic imaging using truncated regularized inverse k-space interpolation. Inverse Prob. **23**(6), S51–S63 (2007)
33. R. Prakash, R. Manwar, K. Avanaki, Evaluation of 10 current image reconstruction algorithms for linear array photoacoustic imaging, J. Biophotonics 17(3), e202300117 (2024)
34. S. Paul, S. Mulani, M.K.A. Singh, M.S. Singh, Improvement of LED-based photoacoustic imaging using lag-coherence factor (LCF) beamforming. Med. Phys. **50**, 7525–7538 (2023)
35. T. Agano, N. Sato, K. Awazu, LED-based photoacoustic imaging system: why it achieves the same signal to noise ratio as solid-state-laser-based system: a review, in Proc. SPIE 11240, Photons Plus Ultrasound: Imaging and Sensing 2020 (2020), p. 112401T
36. K. Tang, S. Zhang, Z. Liang, Y. Wang, J. Ge, W. Chen, L. Qi, Advanced image post-processing methods for photoacoustic tomography: A review. Photonics **10**(7), 707 (2023)
37. M. Xavierselvan, M.K.A. Singh, S. Mallidi, In vivo tumor vascular imaging with light emitting diode-based photoacoustic imaging system. Sensors **20**(16), 4503 (2020)
38. R. Manwar, M. Zafar, Q. Xu, Signal and image processing in biomedical photoacoustic imaging: A review. Optics **2**(1), 1–24 (2021)
39. M. Shi, T. Vercauteren, W. Xia, Spatiotemporal singular value decomposition for denoising in photoacoustic imaging with a low-energy excitation light source. Biomed. Opt. Express **13**(12), 6416–6430 (2022)
40. S. Agrawal, K. Johnstonbaugh, T. Suresh, A. Garikipati, M.K.A. Singh, S.P.K. Karri, S.R. Kothapalli, In vivo demonstration of reflection artifact reduction in LED-based photoacoustic imaging using deep learning, in *Proc SPIE 11642, Photons plus ultrasound: Imaging and sensing 2021*, (2021), p. 116421K
41. S. Guan, A.A. Khan, S. Sikdar, P.V. Chitnis, Fully dense UNet for 2-D sparse photoacoustic tomography artifact removal. IEEE J. Biomed. Health Inform. **24**(2), 568–576 (2020)
42. R. He, Z. Liu, Y. Zhao et al., Deep learning realizes photoacoustic imaging artifact removal. Appl. Sci. **14**(12), 5161 (2024)
43. S. Li, Y. Tang, Z. Li, C. Liu, *Zero-shot Artifact2Artifact: Self-incentive artifact removal for photoacoustic imaging without any data*, (2024). arXiv preprint. arXiv:2412.14873

44. K.J. Francis, A. Arora, P. Kancharla, M. Kuniyil Ajith Singh, W. Steenbergen, S.S. Channappayya, Generative adversarial network-based photoacoustic image reconstruction from bandlimited and limited-view data, in *Proc SPIE—photons plus ultrasound: Imaging and sensing*, vol. 11642 (2021), p. 1164235
45. M. Lu, J. Zhang, R. Peng et al., A learnable physical deep-learning model for sparse photoacoustic image reconstruction. Photoacoustics. **30**, 100468 (2025)
46. A. Hariri, K. Alipour, Y. Mantri, J.P. Schulze, J.V. Jokerst, Deep learning improves contrast in low-fluence photoacoustic imaging. Biomed. Opt. Express **11**(6), 3360–3373 (2020)
47. M. Shi, T. Zhao, S.J. West, A.E. Desjardins, T. Vercauteren, W. Xia, Improving needle visibility in LED-based photoacoustic imaging using deep learning with semi-synthetic datasets. Photoacoustics. **26**, 100351 (2022)
48. A. Paul, S. Mallidi, U-Net enhanced real-time LED-based photoacoustic imaging. J. Biophotonics **17**(6), e202300465 (2024)
49. A. Paul, S. Mallidi, Facilitating real-time LED-based photoacoustic imaging with DenP2P: An optimized conditional generative adversarial deep learning solution. AIP Adv. **15**(5), 055227 (2025)
50. B.E. Treeby, B.T. Cox, k-Wave: MATLAB toolbox for the simulation and reconstruction of photoacoustic wave fields. J. Biomed. Opt. **15**(2), 021314 (2010)
51. J. Gröhl, K.K. Dreher, M. Schellenberg, T. Rix, N. Holzwarth, P. Vieten, L. Ayala, S.E. Bohndiek, A. Seitel, L. Maier-Hein, SIMPA: an open-source toolkit for simulation and image processing for photonics and acoustics. J. Biomed. Opt. **27**(8), 083010 (2022)
52. D. O'Kelly, J. Campbell, J.L. Gerberich et al., A scalable open-source MATLAB toolbox for reconstruction and analysis of multispectral optoacoustic tomography data. Sci. Rep. **11**, 19872 (2021)
53. Else et al., PATATO: a Python photoacoustic tomography analysis toolkit. J. Open Source Softw. **9**(93), 5686 (2024)
54. M. T. Graham and M. A. L. Bell, PhocoSpace: An open-source simulation package to implement photoacoustic spatial coherence theory, in *2022 IEEE International Ultrasonics Symposium (IUS)* (Venice, Italy, 2022), pp. 1–5

Chapter 6
Clinical Translation Pathways for Affordable Photoacoustic Imaging

Learning Objectives

After reading this chapter, the reader will be able to:

- **Describe the current state and trajectory of clinical translation in photoacoustic imaging**, from prototype development to regulatory approval and implementation.
- **Explain how affordability and accessibility** influence the pace and scope of clinical adoption across diverse healthcare environments.
- **Recognize the role of standardization and safety frameworks** including IPASC guidelines, CE marking, and FDA pathways in ensuring trust and reproducibility.
- **Identify key clinical applications**, such as breast cancer and vascular imaging, where photoacoustics has demonstrated diagnostic value.
- **Understand how computational intelligence and system integration** enable clinically viable, low-cost imaging solutions.
- **Discuss socioeconomic, workflow, and co-development strategies** that link academia, industry, and clinical practice in advancing affordable photoacoustic medicine.

6.1 The Evolving Landscape of Clinical Translation

Photoacoustic imaging (PAI) has advanced from an experimental optical–acoustic technique to a clinically oriented imaging modality with clear translational momentum. This evolution has been enabled by progress in optical excitation, acoustic detection, and computational reconstruction. Early research focused on small-animal imaging and proof-of-concept studies, whereas recent developments in compact light sources—particularly light-emitting diodes (LEDs) and laser diodes (LDs)—together with hybrid ultrasound integration, have produced portable and affordable systems suitable for human use at the bedside [1]. These technological

M. Kuniyil Ajith Singh, *Photoacoustic Imaging Using Affordable Resources*, Progress in Optical Science and Photonics 40, https://doi.org/10.1007/978-981-95-7175-8_6

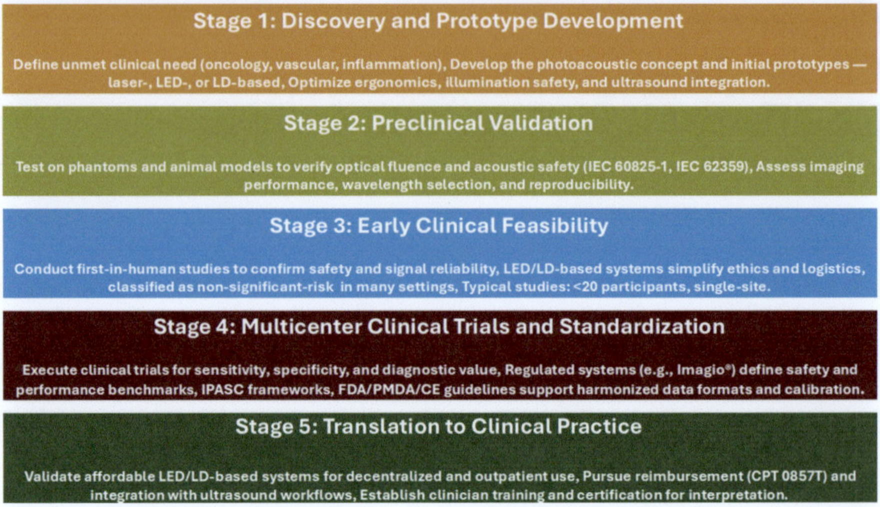

Fig. 6.1 Condensed translational pathway for photoacoustic imaging (PAI). Five key phases summarize the progression from discovery to clinical practice. PAI currently spans Stages 3–5, combining regulated laser-based systems (Imagio®, MSOT) and affordable LED/LD-based platforms for accessible clinical deployment

improvements have been accompanied by a shift in how success in translation is defined: beyond image quality, it now includes affordability, accessibility, workflow integration, and clinical value [2]. As summarized in Fig. 6.1, the translational pathway for PAI spans five critical stages—from discovery and prototype development through preclinical validation and early clinical feasibility to large-scale trials and, ultimately, clinical adoption. At present, PAI technologies are concentrated in Stages 3–5, where both regulated laser-based systems and affordable LED/LD-based devices are demonstrating clinical value.

Every major medical-imaging modality has found its niche by meeting specific unmet needs rather than replacing earlier tools. Computed tomography complemented radiography; magnetic-resonance imaging expanded diagnostic precision; and ultrasound became indispensable for real-time structural visualization. PAI appears poised to follow this same pattern [3]. Its adoption will depend on where it uniquely contributes such as label-free vascular mapping, oxygenation and metabolic imaging, and non-invasive monitoring that complements ultrasound guidance [3, 4]. In dermatology and vascular medicine, multispectral photoacoustic mesoscopy has revealed microvascular and inflammatory signatures inaccessible to purely optical techniques [4]. In oncology, handheld hybrid PAI–ultrasound probes are emerging as adjuncts for tumor characterization, therapy monitoring, and intraoperative guidance [3, 4].

Affordability is a central driver of this transition. LED- and LD-based sources have reduced system cost, footprint, and maintenance burden, enabling clinical deployment in outpatient and resource-limited environments [1, 5]. This democratization mirrors the evolution of portable ultrasound from a specialized diagnostic instrument to a standard component of point-of-care practice. When combined with real-time reconstruction and machine-learning-based enhancement, low-cost PAI systems can deliver functional and molecular information at a fraction of the cost and logistical complexity of MRI or PET, without ionizing radiation [5].

Clinical translation also requires reproducibility and regulatory alignment. The International Photoacoustic Standardization Consortium (IPASC) has established harmonized performance metrics, calibration procedures, and data-format standards to support multi-centre validation [6]. In parallel, photonics organizations such as SPIE and regulatory bodies including the International Electrotechnical Commission (IEC) are developing exposure and metrology standards specific to PAI [7]. These coordinated frameworks are critical for building trust among clinicians, regulators, and industry stakeholders.

Between 2018 and 2025, the field achieved pivotal regulatory milestones that underscore its growing maturity. The Imagio® system (Seno Medical, USA) received FDA Premarket Approval (PMA) in 2021–2022, becoming the first photoacoustic–ultrasound device authorized for breast imaging and establishing a formal regulatory pathway in the United States [3, 8]. Around the same period, Europe granted CE marking (2021) to the MSOT Acuity Echo (iThera Medical, Germany), the first handheld multispectral optoacoustic tomography system certified for clinical use [3]. Shortly thereafter, Japan's Luxonus LME-01 received PMDA approval (2022)—the first such authorization in Asia—demonstrating the modality's expanding global regulatory acceptance [3, 9].

Together with international standardization initiatives and the emergence of affordable LED- and LD-based platforms, these achievements signify the transition of photoacoustic imaging from experimental promise to an established, regulated, and clinically relevant technology. Table 6.1 summarizes these key developments chronologically, highlighting the interplay between technological innovation, regulatory validation, and affordability that defines the current phase of PAI clinical translation.

6.2 Regulatory and Standardization Landscape

The clinical translation of PAI is inseparable from the evolving regulatory and standardization frameworks that define medical-device safety, performance, and reproducibility. As a hybrid modality combining optical and acoustic subsystems, PAI presents regulatory challenges that do not align neatly with existing classifications for either ultrasound or optical devices. Conventional optical standards emphasize laser-safety exposure limits, whereas acoustic frameworks govern mechanical and thermal indices. The integration of both domains requires harmonized evaluation criteria for optical fluence, acoustic output, and patient safety [3, 7].

Table 6.1 Representative milestones in clinical translation of PAI

Year	Milestone/study	Key contribution	Clinical relevance
2018–ongoing	International Photoacoustic Standardization Consortium (IPASC) established [6]	Standardized terminology, performance metrics, and protocol harmonization	Enables reproducibility and multi-centre validation
2021–2022	Imagio® (Seno Medical, USA)—FDA Premarket Approval (PMA) [3, 9]	First FDA-approved photoacoustic-ultrasound system for breast imaging; validated in PIONEER and READER trials	Defined U.S. regulatory pathway and enabled CPT code 0857 T
2021	MSOT Acuity Echo (iThera Medical, Germany)—CE marking [3]	First CE-certified handheld multispectral optoacoustic tomography (MSOT) system	Enabled clinical studies in dermatology, oncology, and inflammation
2022–2023	LME-01 (Luxonus, Japan)—PMDA approval [3, 9]	First approved hemispherical-array PAI system in Asia; validated for vascular and lymphatic imaging	Established Japan's regulatory pathway for PAI devices
2018–ongoing (gaining momentum 2022–2025)	Affordable LED/ LD-based handheld systems [13, 14]	Reduced cost and footprint; real-time PA/US fusion and computational enhancement	Expands access to resource-limited and point-of-care environments
2024–2025	Consensus reviews and standardization frameworks published [3–7]	Established global guidelines and evidence-based translation strategies	Supports clinical trial design and regulatory convergence

6.2.1 Regulatory Pathways

The global regulatory landscape for PAI has advanced substantially since 2020. In the United States, the *Imagio®Breast Imaging System* (Seno Medical) became the first FDA-approved photoacoustic–ultrasound device under the Premarket Approval (PMA P200003) pathway in 2021–2022 [3, 8]. This milestone demonstrated that hybrid optical–acoustic systems could meet Class III device standards for safety and efficacy. The supporting PIONEER and READER trials provided pivotal clinical evidence and ultimately led to the introduction of CPT code 0857T, representing an early step toward reimbursement. In Europe, photoacoustic devices are regulated under the EU Medical Device Regulation (MDR 2017/745). The *MSOT Acuity Echo* system (iThera Medical, Germany) achieved CE marking in 2021, becoming the first handheld multispectral optoacoustic tomography platform cleared for clinical use [3].

In Japan, the *LME-01* system (Luxonus Inc.) received approval from the Pharmaceuticals and Medical Devices Agency (PMDA) in 2022–2023, representing the first Asian authorization for a PAI system [3, 9]. The device met domestic requirements for optical-radiation safety and acoustic-output monitoring. Together, these approvals demonstrate the progressive recognition of PAI as a regulated medical-imaging modality. They also reveal the need for greater international alignment to avoid redundancy in clinical validation and certification procedures.

6.2.2 Standardization Initiatives

Beyond regulation, standardization ensures interoperability and reproducibility across instruments and clinical studies. The International Photoacoustic Standardization Consortium (IPASC), established in 2019, has been instrumental in defining terminology, calibration procedures, and performance metrics for PAI systems [6, 10]. IPASC's collaborative framework provides:

- Reference protocols for optical fluence normalization and transducer calibration;
- Unified data formats enabling cross-platform benchmarking;
- Standardized reporting parameters for multicenter clinical trials.

Expert Perspective: Standardization and the Road to Practical Clinical PAI "Robust phantom-based evaluation and standardized protocols are essential for moving photoacoustic imaging from research to routine clinical use. Consistent benchmarking across systems, centers, and studies creates the confidence needed for broader adoption. As the technology evolves toward compact, affordable illumination solutions such as LEDs and VCSELs, there is real potential for point of care photoacoustic assessment, including breast and vascular imaging in general practice settings where laser-based systems are impractical. The future impact of PAI will depend as much on standardization as on innovation."

—Prof. Srirang Manohar, Chair, Multimodality Imaging Group, University of Twente, Netherlands.

Complementary initiatives from SPIE, the International Electrotechnical Commission (IEC), and the International Organization for Standardization (ISO) are extending these principles to optical and acoustic safety regulations [7]. For example, IEC 60,825–1 specifies laser-radiation safety, while IEC 62,359 defines limits for acoustic output and mechanical index, both directly applicable to PAI devices that combine light and sound. Together, these efforts provide a foundation for consistent device characterization and patient protection.

6.2.3 Toward Regulatory Convergence

A clear trend is the movement toward regulatory convergence, in which lessons from early approvals inform global standards. Recent reviews and editorials [3, 6] highlight the importance of reproducible image acquisition, quantitative biomarkers, and standardized reconstruction pipelines. Regulatory agencies increasingly adopt evaluation criteria derived from hybrid and molecular imaging, prioritizing quantitative accuracy, transparency of algorithms, and longitudinal validation. Future regulatory submissions for affordable PAI systems are expected to integrate hardware certification, software validation, and algorithmic explainability into unified documentation. Such convergence will accelerate cross-border approvals, reduce redundant testing, and support equitable access to low-cost PAI technologies worldwide.

Despite increasing regulatory clarity, PAI manufacturers still face a strategic dilemma in defining their approval pathway. Because PAI performance can be tuned through light wavelength, probe geometry, and detector sensitivity, systems may target diverse anatomical sites and diagnostic purposes. This flexibility complicates the choice between pursuing a narrow, indication-specific approval—such as Seno Medical's breast-imaging route, which demands large and costly clinical trials—or seeking a broader classification, for example as a vascular imaging adjunct. The absence of a clear, standardized framework for such "general-use" PAI systems continues to limit industrial confidence and investment momentum.

6.3 Clinical Trials and Validation Studies

The clinical translation of PAI rests on a growing body of human data obtained from both pre-market feasibility studies and regulated clinical trials. These investigations collectively define the safety, diagnostic performance, and usability of PAI systems, while providing the evidence base necessary for regulatory approval and clinical adoption [1–4]. Over the past decade, clinical activity has broadened from single-site feasibility work to multicenter, statistically powered studies spanning oncology, dermatology, vascular disease, and inflammatory disorders [3, 4].

6.3.1 Evolution of Clinical Study Design

Early human studies (2008–2015) established the foundational safety and feasibility of PAI, typically enrolling fewer than 20 participants. These trials validated light-tissue interaction models, demonstrated non-invasive oxygen-saturation mapping, and correlated photoacoustic biomarkers with histopathology [2, 3]. Since 2016, attention has shifted toward indication-driven, hypothesis-based trials. Prospective designs now incorporate quantitative endpoints like sensitivity, specificity, and

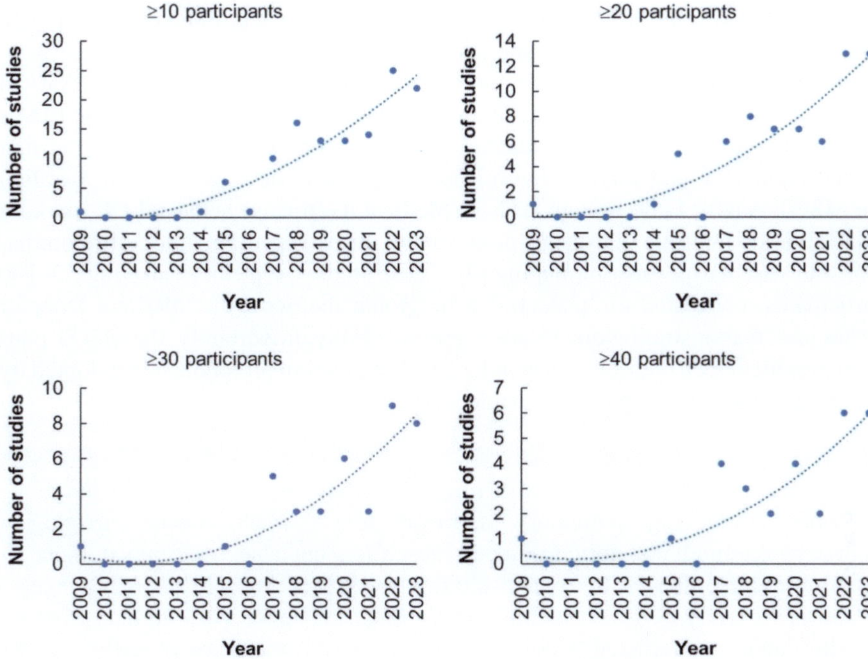

Fig. 6.2 Growth in the number of in vivo clinical photoacoustic studies over time. Graphs show the cumulative number of published optoacoustic clinical studies from 2009 to 2023, categorized by the number of participants per trial (≥10, ≥20, ≥30, and ≥40). The steady increase across all cohorts demonstrates the expanding clinical maturity and acceptance of photoacoustic imaging as a translational modality. Reprinted with permission from [4]

area-under-curve (AUC) and follow Good Clinical Practice (GCP) guidelines. The emergence of standardized reporting frameworks through the International Photoacoustic Standardization Consortium (IPASC) has enabled reproducibility and inter-site comparability [10]. The acceleration of clinical trials reflects the growing confidence in PAI as a safe and reproducible modality. As summarized in Fig. 6.2, the number of in vivo photoacoustic studies has increased exponentially since 2015, with a notable rise in trials enrolling more than 20 participants. This trend corresponds to improved system reliability, standardized imaging protocols, and the emergence of affordable LED- and LD-based platforms that simplify ethical approval and patient recruitment [3, 5, 10].

6.3.2 Large-Scale Clinical Evidence: Commercial Systems

Among regulated devices, two systems have provided the strongest large-scale clinical evidence. The Imagio® Breast Imaging System (Seno Medical, USA) underwent pivotal PIONEER and READER trials enrolling more than 2,500 patients [11, 12]. These studies demonstrated that the addition of photoacoustic information—principally hemoglobin-based vascular and oxygenation maps—improves specificity in breast-lesion assessment while maintaining ultrasound-level sensitivity. The FDA's PMA approval (2021–2022) validated the trial outcomes and established the first formal regulatory pathway for hybrid optical–acoustic devices [3, 8]. In parallel, the MSOT Acuity Echo system (iThera Medical, Germany) achieved CE marking in 2021 after multiple European pilot studies involving dermatology, inflammation, and vascular-perfusion imaging [3]. These studies, typically enrolling 30–100 participants, correlated multispectral hemoglobin metrics with MRI and Doppler ultrasound, demonstrating quantitative reproducibility. Importantly, the MSOT platform provided open datasets and standardized acquisition protocols later adopted by academic groups developing affordable alternatives.

Industry Perspective: Translating Photoacoustics into Clinical Reality "Achieving regulatory approval for a photoacoustic imaging system required not only technical innovation but also a sustained focus on reproducibility, usability, and clinical validation. The greatest challenge and opportunity lies in bridging engineering excellence with clinical trust. Affordable, standardized systems will accelerate this transition globally."

— Bryan Clingman, CTO and VP of Engineering, Seno Medical Instruments, USA

6.3.3 Disease-Specific Validation Pathways

Oncology. PAI is being evaluated for breast, thyroid, prostate, and skin cancer imaging, where oxygenation and angiogenesis serve as biomarkers of malignancy. Multispectral data quantify vascular heterogeneity and hypoxia, providing predictive information for therapy response [3–5]. LED/LD-based prototypes have extended these capabilities to cost-sensitive settings for multiple applications including biopsy guidance and longitudinal monitoring [1, 5, 13, 14].

Dermatology and inflammation. Photoacoustic mesoscopy has enabled high-resolution mapping of cutaneous microvasculature and inflammatory activity in psoriasis, atopic dermatitis, and scleroderma [4, 6]. Standardized wavelength sets allow differentiation between oxygenated and deoxygenated hemoglobin without exogenous contrast, supporting quantitative disease-severity indices.

Vascular and metabolic applications. Feasibility trials in peripheral arterial disease, diabetic microangiopathy, and hemodialysis perfusion have confirmed that PAI can quantify oxygen delivery and perfusion heterogeneity non-invasively [1, 4,

13, 14]. These efforts emphasize safety, repeatability, and integration with handheld ultrasound probes.

6.3.4 Validation Frameworks and Quantitative Metrics

Clinical validation encompasses both device performance and biological correlation. Common protocols include phantom calibration for optical fluence and acoustic sensitivity, intra- and inter-observer repeatability analysis, and correlation with reference modalities such as Doppler US, MRI, or PET [3, 4]. Statistical validation relies on sensitivity/specificity testing, ROC analysis, and Bland–Altman limits of agreement. Machine-learning-based interpretation is being evaluated for automated vascular segmentation and oxygenation estimation, though regulatory guidance now requires algorithmic transparency [3, 6].

6.3.5 Affordable System Pilot Studies

While large commercial trials have defined regulatory acceptance, affordable and LED/LD-based PAI systems have demonstrated the potential for scalable clinical translation. These studies, typically conducted at lower cost and with simplified instrumentation, represent the next frontier of accessibility.

- **Lymphatic and vascular imaging**: Van Heumen et al. used an LED-based PAI system to visualize superficial lymphatic vessels and sentinel pathways in vivo, achieving vascular contrast comparable to laser-based instruments at a fraction of the cost [15].
- **Wound healing and angiogenesis**: Mantri et al. conducted a pilot trial using a dual-LED handheld imager to monitor angiogenic progression in chronic wounds, demonstrating quantifiable increases in hemoglobin-derived contrast over time [16].
- **Inflammatory joint disease**: Janggun et al. applied LED-PAI to rheumatoid-arthritis patients, showing correlation between synovial vascularity and disease activity scores [17].
- **Dermatologic lesions**: Cheng et al. used a multi-wavelength LED system for port-wine-stain mapping, enabling non-invasive visualization of treatment response [18].
- **Hemodialysis and perfusion**: LED-PAI assessed vascular access patency and oxygenation dynamics in dialysis patients, revealing perfusion changes undetectable by pulse oximetry [19].

These pilot studies confirm that LED-based illumination can deliver clinically relevant photoacoustic contrast while dramatically reducing hardware cost, power consumption, and maintenance. Importantly, their imaging biomarkers like oxygenation, vascular density, and tissue perfusion mirror those validated in high-end systems like Imagio® and MSOT, underscoring that affordability need not compromise diagnostic potential. Affordable clinical trials also offer logistical advantages: portable devices simplify ethics approval, patient recruitment, and multicenter replication. When aligned with IPASC calibration standards [10], such studies contribute to a global evidence base spanning both premium and low-cost technologies.

6.3.6 Emerging Trends and Future Directions

Current pilot efforts are converging toward adaptive, multicenter clinical networks that compare LED/LD-based and laser-based platforms under harmonized protocols. Emphasis is shifting toward functional endpoints—for example, monitoring neovascularization during therapy—rather than purely anatomical imaging. The integration of compact light sources, battery operation, and real-time AI reconstruction promises affordable systems capable of operation outside tertiary hospitals. Collectively, the coexistence of high-performance commercial systems (Seno Imagio®, MSOT Acuity Echo) and low-cost LED-PAI platforms defines a continuum of translation. Each contributes complementary evidence: regulated devices validate safety and efficacy, while affordable systems demonstrate scalability and access. This synergy ultimately accelerates the clinical maturity of PAI across economic and geographic boundaries.

The upward trend observed after 2018 coincides with the introduction of compact, affordable LED- and LD-based photoacoustic systems that substantially simplified clinical deployment. These systems operate below laser-safety thresholds and require minimal infrastructure, allowing approval as non-significant-risk (NSR) devices in many ethics frameworks. As a result, investigators could initiate multi-participant trials in outpatient and bedside settings, accelerating data collection and broadening the demographic and geographic scope of PAI research. The figure therefore reflects not only the technological maturity of the field but also the transformative role of affordability and operational simplicity in enabling human studies [1, 3, 5, 13, 14].

6.4 Overcoming Barriers to Adoption in Healthcare Settings

Despite major technological and regulatory progress, the adoption of PAI in routine healthcare remains limited by a combination of technical, clinical, and socioeconomic barriers. Successful translation requires that devices not only perform reliably in controlled research environments but also integrate seamlessly into clinical workflows, meet cost constraints, and comply with safety and training requirements [3–6].

The IPASC has mapped the major translational barriers hindering widespread clinical adoption of PAI. As summarized in Fig. 6.3, these challenges span the entire translational pipeline from discovery to qualification and highlight the need for coordinated efforts in standardization, test-object development, and data harmonization. The following subsections discuss these and other adoption challenges in more detail, focusing on technical, clinical, and socioeconomic dimensions relevant to affordable systems.

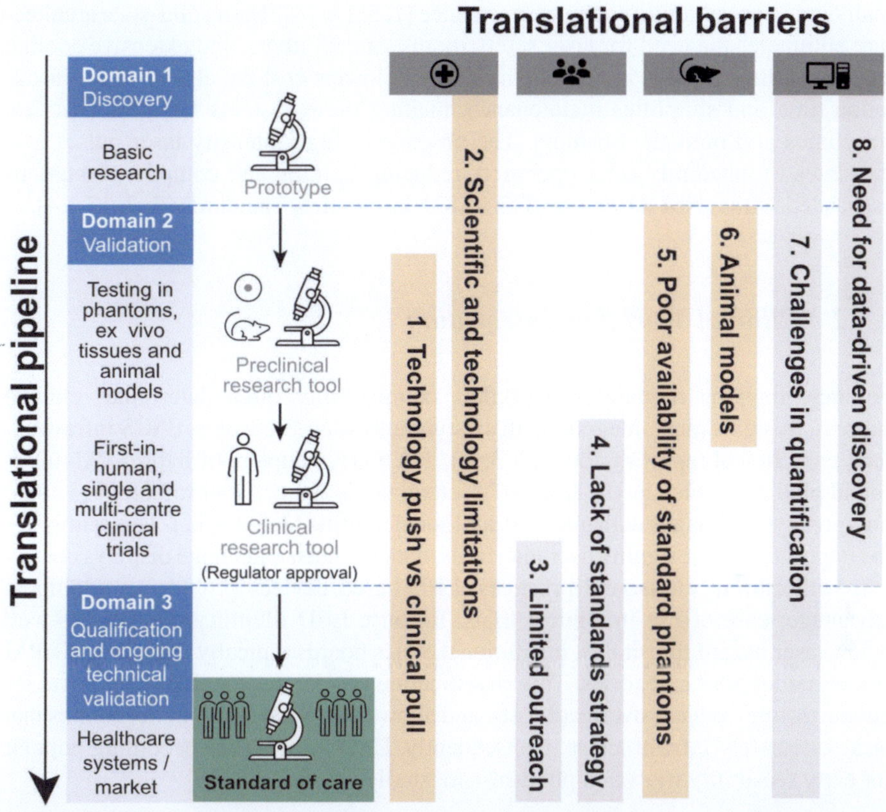

Fig. 6.3 Mapping of translational barriers in photoacoustic imaging (PAI). Adapted from the International Photoacoustic Standardization Consortium (IPASC) translational road-mapping exercise [3, 10]. The framework aligns key barriers with successive phases of discovery, validation, and qualification within the clinical-translation pipeline. Thematic groupings—clinical adoption, standards development, test-object design, and data management—illustrate how challenges extend across both technical and organizational domains, emphasizing the need for coordinated progress toward standardization and reproducible clinical implementation. Reprinted with permission from [21]

6.4.1 Infrastructural Barriers

One of the primary challenges to clinical uptake is economic feasibility. Conventional laser-based PAI systems involve costly components—nanosecond-pulsed lasers, optical delivery assemblies, and high-frequency ultrasound transducers—resulting in capital costs exceeding those of standard ultrasound [3, 4]. Hospitals must justify such investment relative to established modalities like Doppler ultrasound or MRI. In contrast, LED- and LD-based systems dramatically reduce hardware costs while maintaining acceptable imaging performance [1, 5, 13, 14]. Their solid-state architecture eliminates the need for laser safety rooms, optical fibers, and extensive cooling systems. This reduction in complexity not only lowers cost but also shortens installation time and simplifies maintenance, making these systems far more practical for clinics and outpatient centers. The absence of high-intensity laser pulses also translates to inherently safer operation, reducing training and compliance burdens associated with ANSI Z136.1 or IEC 60,825 laser-safety standards [3, 7, 13, 14].

6.4.2 Clinical Workflow Integration

For new imaging modalities, workflow compatibility often determines clinical success. Systems must integrate with existing ultrasound consoles, PACS infrastructure, and clinical reporting tools [2, 3, 13, 14]. Early adopters of hybrid PAI–ultrasound platforms, such as the Imagio® system, demonstrated that radiologists could interpret hybrid images with minimal additional training [3, 8, 11, 12]. This reinforces the importance of designing user interfaces, data formats, and image displays consistent with current ultrasound practices. LED-based devices provide an additional advantage: ease of use in clinical trials**.** Because LED illumination operates well below laser hazard thresholds, institutional ethics boards typically classify LED-PAI studies under NSR categories. This classification accelerates regulatory approval for human testing, reduces overhead costs, and allows deployment in smaller centers that lack laser-safety infrastructure. Consequently, LED systems are becoming popular for early feasibility trials and point-of-care studies [1, 5, 13, 14].

Clinical Perspective: Bringing Photoacoustics to the Bedside "For photoacoustic imaging to become routine, it must integrate seamlessly into clinical workflows, demonstrate clear diagnostic impact, and remain affordable for widespread use. Portable, ultrasound-guided systems that deliver functional information in real time have the potential to transform point-of-care decision making across pediatrics and beyond."

— Prof. *Ferdinand Knieling, Department of Pediatrics and Adolescent Medicine, University Hospital Erlangen, Germany.*

6.4.2.1 Training and User Confidence

A recurring barrier to PAI adoption is the learning curve for clinicians unfamiliar with photoacoustic principles. While ultrasound interpretation is well established, photoacoustic signals introduce novel contrast mechanisms based on optical absorption rather than acoustic impedance [2]. Effective training programs combining simulation, annotated datasets, and standardized reading protocols are critical to building user confidence. Open-access resources developed through IPASC [10] and academic consortia provide reference data and labeling frameworks for education and inter-observer reliability studies.

6.4.3 Economic and Strategic Barriers

Beyond technical and regulatory challenges, economic uncertainty remains one of the strongest deterrents to sustained commercial progress. Several early innovators have faced financial setbacks due to prolonged validation cycles and ambiguous reimbursement prospects. For example, PreXion Corporation transferred its LED-based PAI portfolio to CYBERDYNE about seven years ago, citing investment pressure, and more recently iThera Medical—despite achieving CE marking—declared insolvency in 2025. Such cases illustrate the vulnerability of small and mid-sized enterprises operating in an emerging market with high research costs and delayed clinical returns. The limited number of approved devices reinforces investor caution, contributing to the perception that PAI remains a research-centric rather than commercially mature modality.

> ***Industry Reflection: Sustaining Translation Through Practical Innovation*** "The journey from research to routine clinical use in photoacoustic imaging is shaped not only by scientific and regulatory milestones but also by economic and practical realities. True sustainability in translation depends on aligning innovation with affordability and usability—developing compact, efficient systems, such as those driven by VCSELs or LEDs that inherently support laser class 1 operation. Such systems can deliver clinical value without excessive complexity or cost and remain suitable for use in any environment. As the field matures, the balance between technological ambition and real-world feasibility will ultimately determine how broadly photoacoustics can impact patient care."
>
> — Dr. Patrick Leisching, CTO, iThera Medical GmbH, Germany

6.4.4 Data Management and Interpretation

Another major challenge is data management. Multispectral acquisition generates large datasets requiring complex reconstruction and spectral unmixing. While commercial systems provide embedded reconstruction pipelines, affordable prototypes often rely on open-source algorithms or cloud-based computing [5, 13, 14]. Integration with machine-learning frameworks offers potential automation, but algorithm validation remains a regulatory requirement [3, 6, 13, 14]. The development of lightweight, edge-computing reconstruction strategies compatible with portable devices is an ongoing priority for sustainable deployment.

6.4.5 Reimbursement and Evidence Frameworks

Reimbursement remains a bottleneck for widespread adoption. Despite the introduction of CPT code 0857T for photoacoustic breast imaging in 2024, broader clinical indications lack defined reimbursement pathways [8]. Establishing health-economic evidence—cost per diagnostic gain, reduced biopsy rates, and improved therapy monitoring—will be essential to motivate adoption in hospitals and insurance systems. Affordable systems have an advantage in this context: by lowering equipment cost and operational complexity, they improve the cost–benefit ratio for emerging clinical applications.

6.4.6 Toward Equitable Clinical Integration

Affordable PAI presents a unique opportunity to extend advanced diagnostic capabilities beyond tertiary-care centers. Portable LED-PAI platforms can be deployed in resource-limited regions for vascular, dermatologic, and metabolic disease screening [1, 5, 13, 14]. Their non-ionizing, low-energy illumination and compact form factor make them suitable for bedside, pediatric, and community-health environments. Combined with standardized calibration and safety protocols defined by IPASC [10], these systems can form the foundation of a globally accessible imaging ecosystem.

Ultimately, the clinical translation of affordable PAI depends on continuous and coordinated collaboration between academia, industry, and healthcare systems. Academic groups drive innovation in light sources, imaging algorithms, and phantom-based validation, while industrial partners convert these advances into robust, regulatory-compliant products. Clinicians provide the essential clinical perspective, defining unmet needs and guiding evidence generation through structured trials. A balanced tri-sector ecosystem, supported by regulatory bodies, standardization efforts, and reimbursement frameworks, is therefore fundamental to

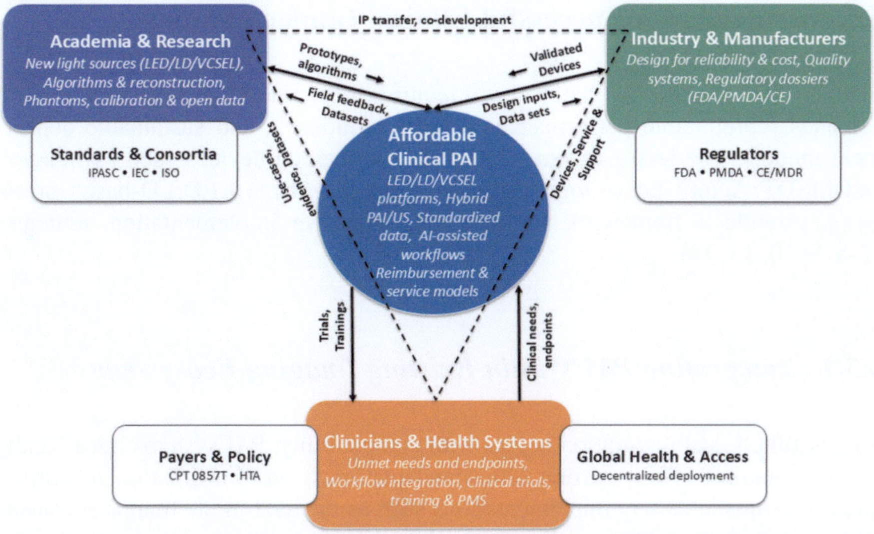

Fig. 6.4 Tri-sector collaboration ecosystem enabling clinical adoption of affordable photoacoustic imaging (PAI). The figure illustrates the multi-stakeholder framework required to translate PAI from research innovation to clinical adoption. Academia, industry, and healthcare sectors interact through a central hub focused on affordable clinical PAI—driven by LED/LD/VCSEL-based platforms, hybrid PAI/US integration, and standardized data frameworks. Outer collaborations with standards bodies, regulators, payers, and global health agencies ensure regulatory alignment, reimbursement, and sustainable deployment. HTA- Health technology Assessment

sustain translation beyond the prototype stage. Figure 6.4 summarizes this collaborative landscape, illustrating how academia, industry, and clinicians interact through a central hub of affordable clinical PAI, with outer partnerships ensuring global access, regulatory alignment, and cost-effective deployment.

An alternative strategic route for affordable LED- or LD-based systems is to emphasize their role as adjunct tools rather than stand-alone diagnostic devices. When integrated with conventional ultrasound scanners, PAI can offer real-time vascular and functional contrast information analogous to an "enhanced Doppler" mode without the need to define a single clinical indication at the outset. Such an approach could accelerate deployment in exploratory clinical settings, allowing clinicians to identify the most impactful use cases organically while manufacturers accumulate safety and performance data for future targeted approvals.

6.5 Strategies for Successful Clinical Implementation

Bringing PAI into routine medical use requires more than regulatory clearance; it demands reproducible performance, clinical confidence, and sustainable deployment models. The lessons from large-scale commercial devices such as Imagio® and MSOT Acuity Echo, together with recent affordable LED/LD-based prototypes, provide a framework for designing effective implementation strategies [1–5, 8–10, 13, 14].

6.5.1 Integrating PAI Within Existing Imaging Ecosystems

Successful clinical integration begins with compatibility. PAI systems must ideally interface naturally with current ultrasound platforms and information infrastructures. The most effective implementations such as the dual-mode Imagio® console leverage familiar workflows, allowing radiologists or dermatologists to acquire and interpret hybrid PAI/ultrasound data without new software ecosystems [3, 8, 11, 12]. Affordable LED/LD-based systems should adopt this model by employing standardized data formats (e.g., DICOM extensions) and interoperable control interfaces to simplify clinical adoption.

6.5.2 Industry–Academia–Clinic Collaboration

Cross-sector collaboration accelerates translation. Academia provides innovation in light-source miniaturization, reconstruction algorithms, and spectral unmixing; industry ensures manufacturability and regulatory compliance; clinicians define relevant diagnostic endpoints [3, 5, 6]. Joint development consortia similar to those seen in ultrasound and MRI community should focus on:

- Translational living labs where prototypes are validated alongside standard imaging tools;
- Shared databases linking optical, acoustic, and clinical metadata for AI training; and
- Open benchmarking supported by initiatives such as IPASC [10].

These frameworks reduce duplication of effort and foster evidence-based optimization across both high-end and affordable device tiers.

6.5.3 Economic Sustainability and Scalable Business Models

Cost remains the strongest determinant of clinical adoption. To ensure sustainability, vendors must balance performance and price through tiered device strategies:

- Premium systems for tertiary hospitals integrating full multispectral capability and research flexibility;
- Mid-range units coupling ultrasound with limited spectral modes for targeted oncology or vascular use; and
- Low-cost portable LED/LD-based devices for community screening and therapy monitoring [1, 5, 13, 14].

Leasing, service-based, or "imaging-as-a-service" models can distribute costs over time, while local manufacturing partnerships in developing regions further enhance affordability.

6.5.4 Education, Training, and Certification

Educational infrastructure is critical. Incorporating PAI modules into radiology and biomedical-engineering curricula will build long-term capacity. Short courses supported by professional societies and online platforms can address the immediate training gap for clinicians. Standardized certification potentially through collaboration with ultrasound credentialing bodies will ensure consistent operator competency and image-interpretation reliability [2, 3].

6.5.5 Clinical Evidence Networks

The next step in implementation is coordinated evidence generation. Multicenter clinical networks can harmonize protocols and share de-identified datasets to build robust validation cohorts. Affordable LED-based systems are ideal for such networks because their safety profile simplifies institutional review and their portability allows outreach to diverse populations. Harmonized data collection under IPASC and GCP guidelines ensures that results from affordable devices are directly comparable to those from established platforms [10, 13, 14].

6.5.6 Digital Integration and AI-Assisted Workflows

The convergence of AI, cloud computing, and edge processing offers transformative potential for clinical PAI. Cloud-based reconstruction pipelines reduce hardware costs and facilitate centralized quality control, while on-device AI supports real-time vascular segmentation and oxygenation mapping [3, 4]. For regulatory compliance, these algorithms must include version tracking, explainability, and bias auditing. Affordable systems benefit most from AI integration, as it compensates for lower signal-to-noise ratios and supports automated interpretation in resource-limited settings.

6.5.7 Toward Equitable Global Deployment

Finally, the strategic goal of clinical implementation is equitable access. Affordable PAI systems which are lightweight, battery-operated, and safe for non-specialist use can extend diagnostic imaging to underserved regions. Partnerships with public-health agencies and NGOs may enable screening programs for vascular, metabolic, and dermatologic diseases where other modalities are unavailable [1, 5, 13, 14]. A globally distributed evidence base will, in turn, reinforce confidence among regulators and clinicians, completing the translational cycle envisioned in this chapter.

In summary, successful implementation of PAI depends on integration, collaboration, affordability, and education. By aligning industry innovation with clinical relevance and public-health priorities, both high-performance and low-cost systems can coexist within a unified, sustainable diagnostic ecosystem.

6.6 The Role of Industry and Academia in Driving Clinical Translation

The successful clinical translation of PAI depends on the sustained partnership between academia, industry, and healthcare institutions. Each sector contributes uniquely to transforming PAI from a research-driven technology into a regulated, accessible, and clinically trusted modality. The recent decade of progress demonstrates that translation thrives not through isolated innovation but through collaborative ecosystems that merge engineering, clinical science, and policy [3–6, 8–10].

6.6.1 Academic Contributions: Discovery and Standardization

Academic laboratories remain the primary source of technological innovation in PAI. Advances in LED- and LD-based illumination, reconstruction algorithms, and multimodal data fusion originated in research environments emphasizing open scientific exchange [1, 2, 5]. Universities and research institutes also play a pivotal role in standardization and benchmarking, with initiatives such as the International Photoacoustic Standardization Consortium (IPASC) providing shared datasets, calibration phantoms, and performance metrics that enable reproducibility across independent centers [10]. By emphasizing affordable design and reproducible science, academic efforts ensure that innovation remains aligned with global accessibility rather than confined to high-resource settings.

6.6.2 Industrial Contributions: Translation, Scaling, and Certification

Industry transforms laboratory prototypes into clinically viable products through design optimization, quality assurance, and regulatory engagement. The successes of Seno Medical's Imagio® and iThera Medical's MSOT Acuity Echo exemplify how sustained industrial investment bridges the gap between feasibility and approval [3, 8]. These companies translated academic discoveries into reliable, user-friendly, and certifiable platforms that could withstand rigorous clinical testing.

Emerging start-ups and mid-tier manufacturers now extend this model to affordable and portable systems, often in partnership with academic groups. Their focus includes scalable manufacturing, modular component design, and software-driven performance enhancement. Collaborative licensing and co development agreements accelerate technology transfer while reducing time-to-market and cost.

6.6.3 Clinician and Healthcare-System Involvement

Clinicians act as the final translators, defining relevant diagnostic endpoints, patient workflows, and usability requirements. Integrating PAI into routine clinical environments requires physicians to contribute actively to study design, ethical review, and post-market surveillance. Early adopters within radiology, dermatology, vascular surgery, and rheumatology have proven that hybrid PAI–ultrasound interfaces can be adopted with minimal workflow disruption [3, 4, 13, 14]. Their feedback informs both academic research priorities and industrial product refinement. Healthcare systems, meanwhile, provide the infrastructure for real-world validation. Hospital consortia can host multicenter studies evaluating both premium and affordable systems under harmonized protocols. These environments generate the high-quality clinical and economic data necessary for reimbursement approval and guideline integration.

6.6.4 Public–Private Partnerships and Funding Mechanisms

A major driver of translational success is the establishment of public–private partnerships (PPPs) that share risk and align incentives across academia, industry, and government. Funding programs in the United States (NIH BRP, SBIR), Europe (Horizon Europe), and Asia (AMED, NEDO) have already supported PAI research consortia linking start-ups with hospitals and universities. Such frameworks ensure that affordable technologies receive the same regulatory and validation attention as high-end systems. To sustain this momentum, future PPPs should prioritize:

- **Low-cost, portable systems** addressing unmet needs in resource-limited regions;
- **Shared manufacturing hubs** to reduce component costs; and
- **Cross-border clinical trial networks** guided by standardized protocols.

6.6.5 Open Data, AI, and the Future of Collaborative Translation

Open-access data and algorithmic transparency are becoming defining features of modern medical-device translation. Shared databases curated by IPASC, the PAI community, and industry partners enable AI-driven innovation without duplicating raw data collection [10].

Affordable LED/LD-based devices can contribute unique datasets from diverse demographics, improving generalizability and fairness of AI algorithms. Collaborative cloud infrastructures further allow continuous learning across sites, creating self-improving clinical systems that evolve post-deployment.

6.6.6 Vision for the Next Decade

The convergence of academic creativity, industrial discipline, and clinical relevance defines the next decade of PAI translation. Academia will continue to refine compact, energy-efficient light sources and AI-enhanced reconstruction. Industry will drive cost reduction, reliability, and global distribution. Clinicians and public-health institutions will validate these systems in real-world populations, ensuring equitable impact.

If sustained, this tripartite collaboration will yield a new class of affordable, standardized, and AI-integrated photoacoustic systems and tools capable of bridging the diagnostic gap between high- and low-resource healthcare environments. The result will not merely be technological success but a paradigm shift in accessibility, fulfilling the translational promise of PAI envisioned at the outset of this chapter.

6.7 Summary and Outlook

The clinical translation of PAI has progressed from proof-of-concept experiments to validated, regulated medical applications. The field now stands at a transition point where technological maturity, affordability, and regulatory acceptance are converging. Affordable LED- and LD-based systems have lowered barriers to entry, enabling safe, decentralized clinical studies and broadening participation beyond specialized research centers. Standardization efforts through IPASC, combined with evidence from large trials such as *Imagio®* and *MSOT Acuity Echo*, have established the technical and clinical credibility needed for global adoption.

The remaining challenges are no longer primarily scientific but organizational, achieving sustainable cost models, ensuring clinician training, and securing reimbursement pathways. Continued collaboration among academia, industry, and healthcare systems will determine how rapidly PAI becomes a routine diagnostic tool. By maintaining affordability, transparency, and interoperability as guiding principles, the community can deliver on the promise of accessible functional imaging across diverse clinical environments. The momentum built in this decade thus defines not the endpoint of translation, but the foundation for a scalable, equitable future for photoacoustic medicine.

Key Takeaways

- **PAI has entered the clinical domain**, supported by successful device trials such as *Imagio®* and *MSOT Acuity Echo*, demonstrating diagnostic performance and regulatory readiness.
- **Affordability and safety innovations**—especially LED- and LD-based systems—have enabled decentralized, low-cost clinical studies, expanding access beyond research hospitals.
- **Standardization initiatives (e.g., IPASC) and regulatory engagement (FDA, CE)** are fostering reproducibility, interoperability, and clinician trust, key prerequisites for market adoption.
- **Clinical translation now depends on systems integration and sustainability**, including user training, cost efficiency, and reimbursement pathways, rather than fundamental physics.
- **Collaborative co-development** among academia, industry, and healthcare providers is proving essential for scaling PAI from pilot studies to routine diagnostics.
- **The future of clinical photoacoustics will be guided by openness and equity**, ensuring that next-generation PAI systems remain affordable, standardized, and accessible across diverse healthcare settings.

References

1. Y. Wen, D. Guo, J. Zhang et al., Clinical photoacoustic/ultrasound dual-modal imaging: current status and future trends. Front. Physiol. **13**, 1036621 (2022). https://doi.org/10.3389/fphys.2022.1036621
2. J.J.M. Riksen, A.V. Nikolaev, G. van Soest, Photoacoustic imaging on its way toward clinical utility: a tutorial review focusing on practical application in medicine. J. Biomed. Opt. **28**(12), 121205 (2023). https://doi.org/10.1117/1.JBO.28.12.121205
3. J. Miao, M. Luo, A. Kotwal, E. Hall, D. Oh, P.A. Valdes, L.S. Li, Clinical translation of photoacoustic imaging using exogenous molecular contrast agents [Invited]. Biomed. Opt. Express 16(11), 4353–4379 (2025).
4. F. Knieling, S. Lee, V. Ntziachristos, A primer on current status and future opportunities of clinical optoacoustic imaging. npj Imaging; 3,4 (2025). https://doi.org/10.1038/s44303-024-00065-9
5. M. Kuniyil Ajith Singh, N. Sato, F. Ichihashi, Y. Sankai, Clinical translation of photoacoustic imaging—opportunities and challenges from an industry perspective. in *LED-based photoacoustic imaging. Progress in optical science and photonics*, vol 7, ed. by M. Kuniyil Ajith Singh (Springer, Singapore, 2020)
6. J. Vonk, F. Knieling, S. Kruijff, Collection on clinical photoacoustic imaging (Editorial). Eur. J. Nucl. Med. Mol. Imaging **51**, 3151–3152 (2024). https://doi.org/10.1007/s00259-024-06780-0
7. SPIE Photonics Focus. Bringing new light to imaging device standards. Nov–Dec 2023. Available at: spie.org (News/Photonics Focus).
8. U.S. Food and Drug Administration, Imagio® Breast Imaging System – Premarket Approval (PMA P200003) (2021). Available at: https://www.accessdata.fda.gov/scripts/cdrh/cfdocs/cfpma/pma.cfm?id=P200003
9. Luxonus Inc. LME-01 Photoacoustic imaging system—Approved by the Pharmaceuticals and Medical Devices Agency (PMDA) (2022). https://www.luxonus.jp/product/medical-product-eng/
10. International Photoacoustic Standardization Consortium. Mission & Themes. https://www.ipasc.science/ (Accessed 21 October 2025)
11. G.L.G. Menezes, R.M. Pijnappel, C. Meeuwis, R. Bisschops, J. Veltman, P.T. Lavin, M.J. van de Vijver, R.M. Mann, Downgrading of breast masses suspicious for cancer by using optoacoustic breast imaging. Radiology **288**(2), 355–365 (2018)
12. S.J. Seiler, E.I. Neuschler, R.S. Butler, P.T. Lavin, B.E. Dogan, Optoacoustic imaging with decision support for differentiation of benign and malignant breast masses: A 15-reader retrospective study. AJR Am. J. Roentgenol. **220**(5), 646–658 (2023)
13. Y. Zhu, T. Feng, Q. Cheng, X. Wang, S. Du, N. Sato, J. Yuan, M. Kuniyil Ajith Singh, Towards clinical translation of LED-based photoacoustic imaging: A review. Sensors, **20**(9), 2484 (2020)
14. M. Kuniyil Ajith Singh, W. Xia, Biomedical photoacoustic imaging and sensing using affordable resources. Sensors, **21**(7), 2572 (2021)
15. S. Van Heumen, J.J.M. Riksen, M.K.A. Singh, G. Van Soest, D. Vasilic, LED-based photoacoustic imaging for preoperative visualization of lymphatic vessels in patients with secondary limb lymphedema. Photoacoustics **29**, 100446 (2022)
16. Y. Mantri, J. Tsujimoto, B. Donovan et al., Photoacoustic monitoring of angiogenesis predicts response to therapy in healing wounds. Wound Repair Regen **30**(2), 258–267 (2022)
17. J. Jo, G. Xu, Y. Zhu, M. Burton, J. Sarazin, E. Schiopu, G. Gandikota, X. Wang, Detecting joint inflammation by an LED-based photoacoustic imaging system: a feasibility study. J. Biomed. Opt. **23**(11), 110501 (2018)
18. Q. Cheng, T. Feng, H. Zhang, Y. Zhu, J. Yuan, X. Wang, *Diagnosis and treatment monitoring of port-wine stain using LED-based photoacoustics: theoretical aspects and first in-human clinical pilot study. In LED-Based Photoacoustic Imaging: From Bench to Bedside* (Springer, Singapore, 2020), pp. 351–377

19. Y. Mantri, T.R. Dorobek, J. Tsujimoto, W.F. Penny, P.S. Garimella, J.V. Jokerst, Monitoring peripheral hemodynamic response to changes in blood pressure via photoacoustic imaging. Photoacoustics **26**, 100345 (2022)
20. B.C. Wilson, M. Jermyn, F. Leblond, Challenges and opportunities in clinical translation of biomedical optical spectroscopy and imaging. J. Biomed. Opt. **23**(3), 030901 (2018)
21. H. Assi, R. Cao, M. Castelino, B. Cox, F.J. Gilbert, J. Gröhl, K. Gurusamy, L. Hacker, A.M. Ivory, J. Joseph, F. Knieling, M.J. Leahy, L. Lilaj, S. Manohar, I. Meglinski, C. Moran, A. Murray, A.A. Oraevsky, M.D. Pagel, M. Pramanik, J. Raymond, M.K.A. Singh, W.C. Vogt, L. Wang, S. Yang, Members of IPASC, Bohndiek S.E**.** A review of a strategic roadmapping exercise to advance clinical translation of photoacoustic imaging: From current barriers to future adoption. Photoacoustics**. 32**, 100539 (2023)

Chapter 7
Future Trends and Innovations in Affordable Photoacoustic Imaging

Learning Objectives

After reading this chapter, reader will be able to:

- **Describe emerging illumination technologies**—LEDs, laser diodes, and VCSELs—and their impact on affordable, compact photoacoustic imaging systems.
- **Explain how component-level innovations** such as transparent ultrasound arrays, flexible substrates, and integrated optical detectors enable simplified, low-cost probe design.
- **Discuss the role of miniaturization and wireless connectivity** in expanding photoacoustic imaging to handheld, wearable, and point-of-care applications.
- **Understand the feasibility of functional and molecular imaging at low optical fluence**, including the use of approved contrast agents such as indocyanine green.
- **Recognize how artificial intelligence and open standards (e.g., IPASC)** support real-time reconstruction, data interoperability, and equitable global access.
- **Assess the transition of photoacoustic imaging from research to routine use**, envisioning its role as an intelligent, noninvasive "optical stethoscope" for vascular and metabolic health monitoring.

7.1 Emerging Technologies for Low-Cost Photoacoustic Imaging

Affordable photoacoustic imaging (PAI) is evolving rapidly through innovations in compact light delivery, acoustic detection, and integrated system electronics. Each of these technological domains contributes to reducing hardware cost, improving portability, and expanding accessibility while maintaining image quality and safety.

M. Kuniyil Ajith Singh, *Photoacoustic Imaging Using Affordable Resources*, Progress in Optical Science and Photonics 40, https://doi.org/10.1007/978-981-95-7175-8_7

The following subsections outline current and emerging directions shaping the next generation of affordable systems.

7.1.1 Next-Generation Light Sources

Light delivery remains the dominant cost and safety determinant in PAI. Conventional Q-switched lasers provide excellent fluence and wavelength tuning but are bulky, expensive, and maintenance-intensive. In contrast, recent developments in high-power light-emitting diodes (LEDs) and laser diodes (LDs) have enabled sub-millimeter packaging and energy efficiencies exceeding 30% [1–5]. Multi-wavelength LED arrays and fiber-coupled LD clusters now achieve pulse energies of tens to hundreds of microjoules sufficient for superficial vascular and musculoskeletal imaging, while remaining eye-safe and inexpensive.

Emerging architectures integrate stacked diode arrays with micro-optical beam shaping, enabling quasi-uniform fluence across large fields of view [2, 5]. Compact driver circuits and field-programmable gate array (FPGA) control allow flexible pulse sequencing and spectral cycling, essential for functional and spectroscopic applications. Early explorations of superluminescent diodes (SLDs) and perovskite-based emitters suggest future opportunities for broader spectral coverage and improved thermal stability [6]. These technologies could eventually bridge the gap between LED affordability and laser coherence, particularly in portable spectroscopic systems designed for quantitative tissue monitoring. A particularly promising light source family is the vertical-cavity surface-emitting laser (VCSEL). VCSEL arrays combine the coherence and narrow linewidth of traditional lasers with the low manufacturing cost and scalability of semiconductor diodes. Their surface emission geometry allows fabrication of compact, 2D addressable arrays capable of delivering uniform illumination across large areas. Recent medical-imaging prototypes have demonstrated pulse energies of several hundred microjoules with minimal thermal drift, making VCSELs attractive for portable and even wearable PAI devices [7–9]. Unlike conventional edge-emitting LDs, VCSELs require no complex fiber coupling and can be directly integrated with optical diffusers or light guides, further reducing system size and cost. The potential for multi-wavelength array fabrication also supports spectroscopic and functional imaging in affordable settings. Figure 7.1 illustrates a representative implementation of this concept, a flexible photoacoustic patch integrating VCSEL emitters with an array of piezoelectric detectors. Such designs demonstrate the feasibility of low-voltage, mechanically compliant light sources co-packaged with acoustic receivers for wearable or body-conformal imaging.

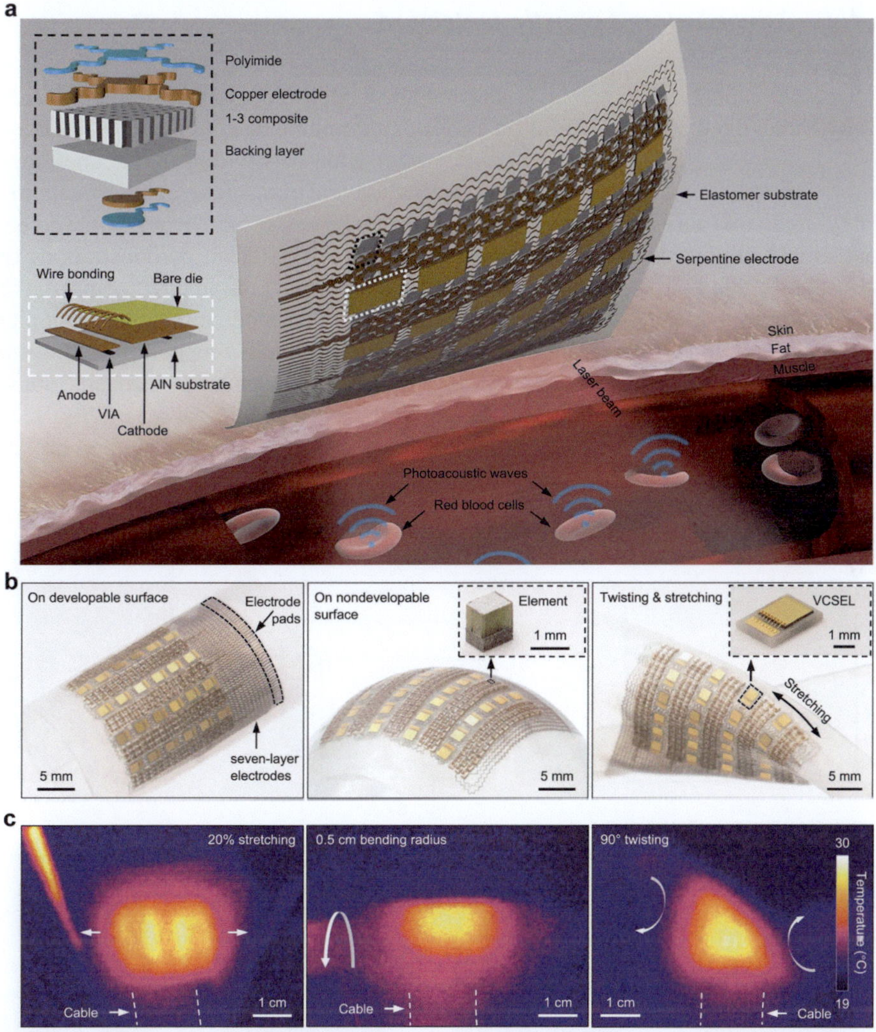

Fig. 7.1 Schematic and operation of a soft photoacoustic patch integrating VCSEL illumination and piezoelectric detection. **a** Structural design showing arrays of vertical-cavity surface-emitting lasers (VCSELs) and piezoelectric transducers interconnected through serpentine copper traces and encapsulated in flexible Ecoflex. Upon pulsed optical excitation, hemoglobin absorption induces thermoelastic expansion, generating acoustic waves that are received by the transducer array. **b** Photographs of the soft patch under various deformation modes—bending, wrapping, and twisting—demonstrate mechanical robustness and conformal adaptability. **c** Infrared images recorded during VCSEL operation (850 nm) confirm uniform illumination across the patch under deformation. Reprinted with permission from [9]

7.1.2 Advances in Acoustic Detection

On the acoustic side, the field is shifting from piezoelectric transducers toward CMOS-integrated and MEMS-based arrays that combine detection and digitization on a single chip [10, 11]. These detectors offer wide bandwidths, small apertures, and compatibility with low-voltage electronics. Optical ultrasound sensors, including Fabry–Pérot and polymer microring resonators, promise transparent or flexible detectors suitable for wearable or multimodal designs [12]. Recent demonstrations of acoustic detection through transparent materials allow integration directly in front of the light source opening up the possibility of even quad-mode ultrasound, Doppler ultrasound, photoacoustic, and fluorescence imaging, simplifying co-alignment and reducing system volume (Fig. 7.2) [13]. This work confirms that acoustic transparency can be achieved without sacrificing impedance matching or sensitivity. Such compact transducer technologies are central to the emergence of handheld probes and head-mounted or skin-conformal devices.

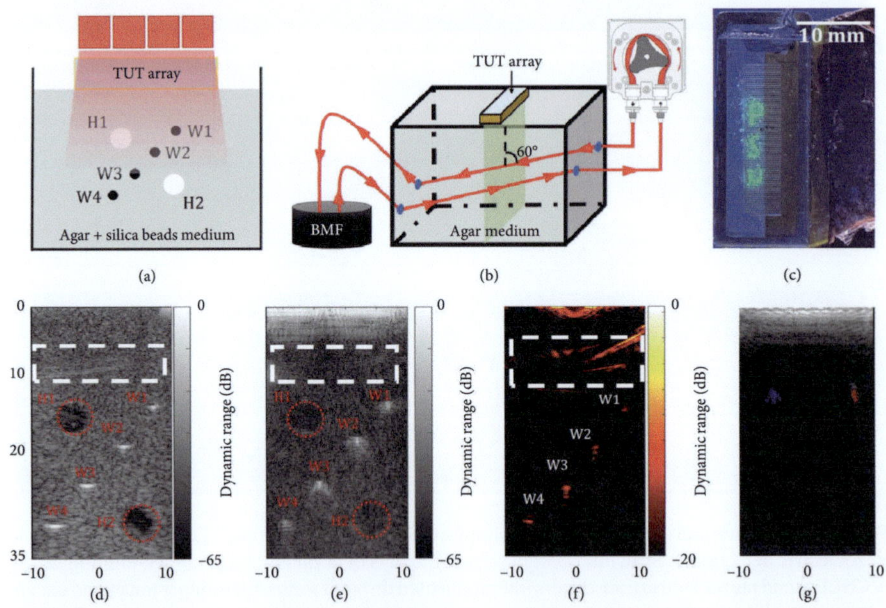

Fig. 7.2 Transparent ultrasound transducer (TUT) array enabling co-registered multimodal imaging. **a, b** Schematics of phantom setups for ultrasound, photoacoustic, and Doppler imaging validation. **c–g** Experimental images acquired through a transparent lithium niobate–based TUT array demonstrate simultaneous transmission of optical and acoustic signals. The same array enables photoacoustic, B-mode, and color Doppler imaging, confirming its capability for co-registered multimodal operation. Reprinted with permission from [13]

7.1.3 Integrated Electronics and Miniaturized Architectures

The transition toward low-cost, field-deployable imaging is accelerated by advances in embedded acquisition and control electronics. System-on-Chip (SoC) platforms combining analog front-ends, data acquisition (DAQ), and edge processors now allow real-time beamforming and reconstruction within palm-sized devices [14]. Wireless data transmission over Wi-Fi or Bluetooth Low Energy (BLE) networks enables connection to tablets or smartphones, reducing cabling and improving mobility [11]. Additive manufacturing and 3D printing further reduce production costs, enabling modular housings and disposable coupling components [15]. These compact architectures, coupled with improved power management, pave the way for wearable and point-of-care PAI units that can operate on rechargeable batteries [16].

7.1.4 Contrast-Enhanced Strategies for Low-Fluence PAI

Low-fluence illumination, while safer and more affordable, can limit imaging depth. The incorporation of clinically approved optical dyes provides a powerful route to enhance contrast and depth performance without sacrificing safety. Indocyanine Green (ICG), Methylene Blue (MB), and emerging agents such as Clofazimine (CFZ) have shown strong absorption peaks compatible with LED or LD wavelengths [5, 17–19]. These agents are particularly stable under low-energy illumination [20], avoiding the photobleaching observed with high-power laser systems. When used judiciously, they can boost photoacoustic signal-to-noise ratios several-fold while maintaining ANSI fluence compliance. Future research may expand this palette toward visible- and NIR-II-range chromophores optimized for LED, LD or VCSEL arrays and repeatable, long-term use. Contrast-enhanced low-fluence PAI thus represents a pragmatic path toward deeper, safer, and more quantitative molecular imaging.

While a wide range of photoacoustic contrast agents has been explored experimentally, only a few have reached true clinical relevance. For low-fluence systems such as LED-, LD-, or emerging VCSEL-based platforms, agents with established safety records and spectral compatibility are especially valuable. Among these, ICG and MB remain the most widely adopted, while CFZ and ICG-loaded nanocarriers offer promising extensions toward functional and longitudinal imaging. Their optical absorption bands align closely with the emission spectra of low-cost light sources, providing practical and regulatory-friendly pathways to boost signal strength and imaging depth without increasing optical fluence. The most representative examples are summarized in Table 7.1.

Table 7.1 Clinically established photoacoustic contrast agents compatible with low-fluence illumination. Among currently used exogenous agents, ICG and MB dominate clinical translation, while CFZ and liposomal ICG formulations offer extended functional imaging capability. Their absorption peaks align with the spectral ranges of affordable light sources, providing a practical route to enhance signal-to-noise ratio and depth without exceeding safety limits

Agent/class	Peak Absorption (nm)	Regulatory/ clinical status	Typical or emerging PAI application	Relevance to affordable/ low-fluence (LED, LD, VCSEL) systems
Indocyanine Green (ICG)	780–805	FDA-approved (diagnostic fluorescence; widely used in clinical PAI)	Vascular and lymphatic mapping; hepatic and tumor perfusion imaging	Strong NIR absorption matching LED/LD/ VCSEL wavelengths; photostable at low fluence; low-dose use (0.5–2 mg kg^{-1}) enables safe bedside and wearable imaging
Methylene Blue (MB)	650–670	FDA-approved (SLN mapping; parathyroid identification)	Lymphatic and endocrine imaging; oxygenation mapping	Absorbs in visible range ideal for compact visible LEDs and emerging red-VCSEL arrays; inexpensive and globally available; minimal photobleaching under low-energy excitation
Clofazimine (CFZ)	495–700	FDA-approved (antimicrobial; not yet approved for imaging)	Reticuloendothelial/ macrophage imaging in inflammatory disease	Passive accumulation allows longitudinal studies; stable signal at low fluence; potential adjunct for functional monitoring with portable systems
ICG-Loaded Liposomal Nanocarriers	780–810	Pre-clinical/ pilot human trials	Liver and sentinel-node imaging; prolonged circulation	Leverages ICG's regulatory status while improving photostability; compatible with LED/ LD/VCSEL sources for deeper penetration at safe fluence levels

7.2 Integration with Artificial Intelligence and Machine Learning

Affordable PAI systems increasingly rely on AI and ML not only for data processing but also for operational intelligence. In contrast to the high-end laser-based research instruments that can rely on abundant signal and expert supervision, low-fluence

LED/LD/VCSEL systems demand algorithms that maximize information extraction under constrained hardware and data conditions. The role of AI is therefore expanding beyond traditional image reconstruction, encompassing adaptive acquisition, workflow automation, and clinical decision support at the point of care.

7.2.1 *From Reconstruction to Intelligent Acquisition*

While deep-learning-based reconstruction and denoising have already demonstrated substantial gains in image quality [21–24], the next step is real-time intelligent acquisition. Drawing inspiration from portable ultrasound systems, modern PAI prototypes are beginning to use AI-guided firing logic that selectively triggers wavelengths or acquisition frames based on live signal feedback. Such closed-loop control can improve frame rate, reduce power consumption, and dynamically optimize SNR without user intervention.

Similar adaptive strategies are already deployed in handheld ultrasound devices, where AI continuously adjusts gain, focal depth, and compression curves [25]. Translating this approach to PAI could allow inexpensive systems to achieve consistent diagnostic performance despite variations in probe placement or optical coupling.

7.2.2 *AI-Enabled Workflow and User Experience*

A persistent barrier to clinical adoption of new imaging modalities is the dependence on expert operators. Borrowing from the AI-assisted ultrasound field, where handheld systems such as *Seno Medical's Imagio*® and iThera Medical's MSOT employ real-time algorithms for lesion characterization and image interpretation [26, 27], affordable PAI systems can similarly benefit from integrated workflow intelligence. AI-driven modules could provide:

- **Automatic region-of-interest identification** (vessels, lymphatic channels, or lesions),
- **Quantitative feature extraction** (sO_2, perfusion indices), and
- **Decision support overlays** for quick interpretation in point-of-care environments.
- This layer of guidance will be critical for resource-limited settings where radiological expertise is scarce but portable imaging tools are accessible.

7.2.3 *Edge-AI and Cloud-Assisted Collaboration*

For affordable systems, computational efficiency is as important as accuracy. The growing availability of edge-AI accelerators (embedded GPUs, FPGAs, or neural processing units) allows real-time inference directly on probe-mounted processors or smartphones. Such configurations minimize data transfer, enhance privacy, and reduce latency, enabling on-device reconstruction and interpretation even in offline conditions. At the same time, federated and cloud-learning architectures can connect distributed devices across clinics or countries. By aggregating anonymized model updates rather than raw data, large-scale population learning becomes feasible without centralized infrastructure. This is particularly powerful for low-resource networks where bandwidth and regulatory hurdles limit conventional data sharing. Edge–cloud synergy could also enable global “AI mentorship” networks where algorithms continuously refine themselves through feedback from geographically diverse usage, similar to tele-ultrasound guidance platforms now emerging in digital health.

A particularly promising direction involves leveraging the vast existing repositories of clinical ultrasound data to accelerate PAI model development. Large hospitals and imaging networks already host millions of annotated ultrasound studies—especially in breast, thyroid, and vascular imaging—which encode well-established anatomical and pathological markers. By linking these datasets with newly acquired photoacoustic information, multimodal AI frameworks can learn to correlate traditional grayscale or Doppler features with novel PAI-derived biomarkers such as oxygenation, hemoglobin distribution, or perfusion dynamics. Such hybrid training paradigms can drastically shorten the learning curve for affordable PAI systems, enabling rapid clinical validation through the synthesis of “known” and “new” markers. In the long term, shared US–PAI databases could form the backbone of integrated diagnostic networks that unify structural and functional imaging under a single intelligent platform. The prospective role of AI in future affordable PAI platforms is summarized schematically in Fig. 7.3, highlighting potential integration pathways across acquisition, reconstruction, learning, and clinical interpretation.

Although not yet realized in commercial devices, the integration of artificial intelligence across the affordable PAI workflow shown in Fig. 7.3 represents a compelling vision for future system development. In this conceptual framework, AI could contribute at multiple levels from adaptive illumination control and intelligent reconstruction to data enhancement, distributed learning, and clinical interpretation. Each domain interacts dynamically with the core LED/LD/VCSEL-based imaging platform, potentially enabling real-time optimization, reduced operator dependence, and improved diagnostic confidence. The clockwise flow depicts a continuous cycle of data acquisition, model learning, and clinical feedback, illustrating how affordable PAI systems may evolve toward autonomous and intelligent diagnostic platforms.

7.2.4 Toward Autonomous and Assistive Imaging

Ultimately, AI integration will redefine what affordability means: not merely low manufacturing cost, but low cognitive cost for users. In the near future, portable PAI devices may self-calibrate, self-diagnose hardware faults, and suggest imaging protocols automatically. Predictive algorithms could even assess patient motion or

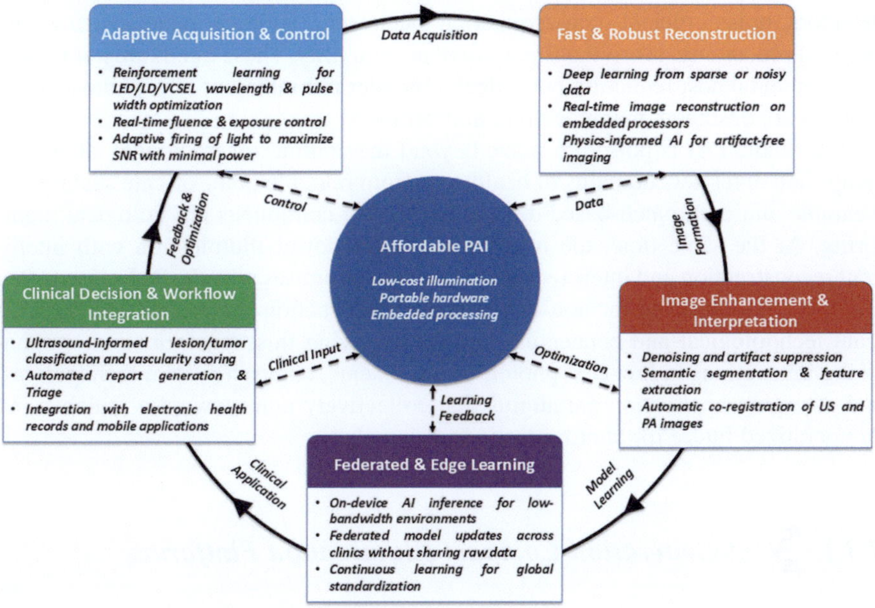

Fig. 7.3 AI empowerment across the affordable photoacoustic imaging pipeline

signal quality and prompt the operator for re-positioning, similar to auto-quality feedback in handheld ultrasound apps [14]. Coupled with advances in voice-guided interfaces and cloud-linked diagnostic support, these systems could extend advanced optical diagnostics beyond hospitals—into community clinics, mobile health vans, and home-care monitoring.

The combination of intelligent acquisition, on-device analytics, and cloud collaboration marks a shift from static imaging devices to adaptive diagnostic ecosystems. Building upon this foundation, the next frontier lies in application-specific innovations, where advances in component design and miniaturization redefine how and where affordable PAI can be used.

7.3 Convergence and Deployment Trajectories for Affordable PAI

The next phase in the evolution of affordable PAI is being shaped by a convergence of photonics, materials science, and artificial intelligence. While Chap. 6 examined the pathways and barriers to clinical translation, the current section looks forward to focusing on technological innovations and conceptual advances that could define the next decade of progress. The recent emergence of compact light sources such as high-efficiency LDs, LED arrays, and VCSELs, together with developments in flexible detectors, noise efficient data acquisition electronics, additive manufacturing, and edge-AI processing, is driving a new era of accessibility. These trends are not limited to incremental cost reduction but reflect a broader rethinking of how photoacoustic systems are designed, manufactured, and deployed.

Affordable PAI is poised to move beyond the confines of traditional laboratory setups and enter new domains of healthcare from portable point-of-care scanners to wearable and even patch-based devices capable of continuous physiological monitoring. At the same time, the integration of low-power illumination with intelligent reconstruction and interpretation algorithms promises to make PAI safer, faster, and more interpretable for non-expert users. This section surveys the most significant technological and conceptual trends underlying this transformation, emphasizing innovations in low-cost photonic components, compact system architectures, and emerging application paradigms that collectively point toward a scalable and democratized future for photoacoustic imaging.

7.3.1 Next-Generation Compact Illumination Platforms

The foundation of affordable PAI lies in continued progress toward compact, energy-efficient, and cost-effective illumination technologies. Early LED- and laser-diode-based implementations demonstrated that meaningful photoacoustic contrast could be achieved even at millijoule-level pulse energies [1–5]. Today, a new generation of semiconductor light sources is closing the performance gap between inexpensive continuous-wave emitters and laboratory-grade OPOs. Among these, VCSEL arrays have emerged as particularly promising devices for portable, real-time PAI.

Two recent reports by Eliason et al. [7] and Konradl et al. [8] presented the first high-power VCSEL subsystems specifically designed for PAI [7, 8]. Their designs integrate the laser array, driver electronics, and thermal management within a compact module small enough to fit inside a standard ultrasound probe housing. Each subsystem produced optical pulses of several kilowatts peak power at 940 nm, with pulse widths of 10–100 ns and repetition rates up to 1 kHz. The wall-plug efficiency exceeded 25%, and the surface temperature during continuous operation remained below 43 °C permitting laser-class 1 operation without active cooling. When coupled with standard ultrasound transducers, these VCSEL-integrated probes achieved image quality comparable to benchtop OPO systems while consuming < 10 W electrical power. A particularly striking feature of the VCSEL-based PA/US probe is its exceptionally broad acoustic response, with a measured fractional bandwidth of approximately 153%. This wideband detection enables simultaneous capture of high- and low-frequency components, enhancing axial resolution and quantitative accuracy. Figure 7.4 summarizes three representative hybrid PA/US probes employing affordable light sources, illustrating the diversity of architectures currently driving miniaturization and cost reduction.

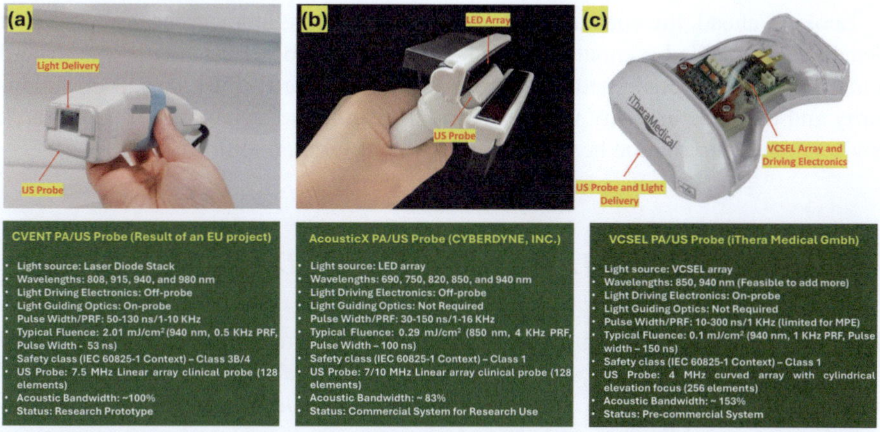

Fig. 7.4 Comparison of representative affordable PA/US handheld probes. Three representative hybrid systems developed using cost-efficient light sources illustrate the ongoing evolution toward portable and clinically viable photoacoustic imaging. **a** The CVENT probe integrates a laser-diode stack (808–980 nm) developed under an EU project. **b** The AcousticX LED-based system (Cyberdyne Inc.) demonstrates multi-wavelength operation with no fiber coupling, achieving imaging depths up to ~2 cm. **c** The VCSEL-PA/US probe (iThera Medical GmbH) represents the next generation of solid-state illumination, achieving high acoustic bandwidth and compact integration. Together, these platforms highlight the transition of affordable photoacoustic imaging from laboratory prototypes to clinically adaptable devices. Image courtesy of Min Wu (Eindhoven University) and Patrick Leisching (iThera Medical Gmbh)

Such integration represents a conceptual inflection point for affordable PAI: the light source becomes a solid-state component rather than an external subsystem, eliminating the need for optical fibers and complex alignment. The combination of LED-level safety with laser-grade coherence and beam uniformity makes VCSELs ideal for regulatory approval and clinical scaling. Moreover, the underlying fabrication leverages mass-production infrastructure from LiDAR and 3-D-sensing industries, suggesting rapid cost reduction and wavelength diversification.

In parallel, advances in high-efficiency LDs and improved LED arrays continue to expand the affordable imaging spectrum. Modern LD modules offer multi-wavelength stacking, <10 ns pulse control, and high repetition rates that support frame-averaged signal enhancement [4–6]. High-power LEDs have benefited from optimized cooling architectures and improved epitaxial efficiency, enabling fluences up to an order of magnitude higher than early prototypes [6]. Recent advances in commercial LED-based PAI systems, exemplified by AcousticX, highlight that tuning the optical pulse width and acoustic reception bandwidth enables imaging depths of up to approximately 2 cm, sufficient for many superficial clinical applications [28]. These developments collectively enable flexible system designs from hand-held probes to wearable patches that deliver clinically meaningful penetration depths while remaining inexpensive and robust.

Looking ahead, the convergence of these light-source innovations with low-cost detection, embedded computation, and AI-based control (Sect. 7.2) points toward a new class of self-contained, intelligent, and affordable PAI devices. Their modularity and manufacturability will be central to scaling the technology from research environments into routine healthcare and resource-limited settings.

7.3.2 Innovations in Low-Cost Component Design

The affordability and scalability of PAI systems depend not only on the illumination source but also on the sophistication, manufacturability, and performance of the acoustic detection chain. Recent advances in transducer materials, optical ultrasound sensors, and flexible integration techniques are redefining how compact and cost-efficient PAI systems can be realized. These innovations aim to preserve broadband sensitivity and resolution while minimizing fabrication cost and form-factor constraints.

A major trend is the emergence of transparent ultrasound transducers (TUTs), which allow optical illumination and acoustic detection to occur through the same aperture. TUT arrays based on indium-tin-oxide (ITO) electrodes and transparent piezoelectric materials, such as lithium niobate or aluminum nitride, have been demonstrated to achieve co-registered ultrasound and PAI within a single field of view [13]. Such designs simplify optical alignment, eliminate bulky side-illumination fibers, and are inherently compatible with miniaturized probes or endoscopic configurations. In comparative studies, prototype TUT arrays produced ultrasound and photoacoustic image quality comparable to that of conventional transducers while maintaining near-infrared transparency above 80%. These developments mark an important step toward monolithic, front-illumination PAI probes suitable for handheld and wearable devices. Parallel progress has been reported in optical ultrasound sensors, which replace piezoelectric elements with interferometric detection mechanisms—such as fiber-optic Fabry–Pérot cavities or micro-ring resonators [12]. These detectors offer broadband frequency response (>30 MHz), small form factors, and electromagnetic immunity, all of which support high-resolution imaging in compact geometries. While traditionally associated with benchtop systems, recent work is exploring their integration with LED or LD excitation modules, where the low noise floor of optical detection can compensate for reduced photoacoustic signal amplitudes.

In addition to detection improvements, the mechanical and electronic integration of these sensors is evolving rapidly. The use of flexible substrates and stretchable interconnects allows ultrasound arrays to conform to curved anatomy, enabling comfortable imaging of peripheral vessels or joints [10, 11]. Advanced materials such as silver nanowire conductors, polymer piezoelectrics, and encapsulating elastomers (e.g., Ecoflex, PDMS) have demonstrated mechanical durability and acoustic transparency suitable for long-term wearable use. Additive manufacturing techniques and printed electronics further promise low-cost, customizable transducer fabrication, aligning well with distributed or point-of-care deployment models [15]. Together, these component-level innovations illustrate a fundamental design shift from rigid, monolithic probes toward integrated, flexible, and potentially wearable architectures. By combining broadband acoustic detection, optical transparency, and scalable manufacturing, these technologies are poised to extend PAI from specialized laboratories to everyday clinical and community health environments.

7.3.3 Miniaturization and Multimodal Integration for Point-Of-Care Imaging

The miniaturization of PAI systems has become a defining trend in their transition from research instruments to practical diagnostic tools. As both light generation and acoustic detection components shrink in size and power consumption, the possibility of integrating photoacoustic capability into portable and even handheld ultrasound scanners is rapidly becoming a reality. For affordable PAI, portability is not merely a convenience—it is central to accessibility, especially in decentralized healthcare environments.

Early progress in this direction was achieved with LD and LED-based probes, which demonstrated the feasibility of combining illumination and detection within compact housings. Commercially available LED-based systems, such as *AcousticX*, have shown that careful tuning of pulse width and acoustic reception parameters can achieve clinically relevant imaging depths of around 2 cm, suitable for vascular and superficial tissue assessment. The CVENT prototype, developed under the EU Horizon 2020 program, further validated that multi-wavelength laser-diode stacks can deliver adequate pulse energy for reflection-mode PAI when coupled to a standard ultrasound array. More recently, VCSEL-integrated probes have pushed this miniaturization further by embedding the light source directly into the ultrasound housing, eliminating external optics and fiber routing. These integrated platforms share a common goal: to enable dual-mode PA/US imaging in a device no larger than a conventional ultrasound probe. Their lightweight form factor facilitates ergonomic use at the bedside or in point-of-care settings, while maintaining full ultrasound functionality. Importantly, because photoacoustic signals are inherently co-registered with ultrasound data, clinicians can interpret functional optical contrast in the familiar anatomical context provided by B-mode ultrasound, an approach that greatly reduces the learning curve and accelerates potential adoption.

Earlier developments in TUTs, such as those pioneered at Penn State University [13], demonstrated not only the feasibility of optical–acoustic integration but also high-quality imaging using optically transparent arrays. These efforts established the foundation for compact, coaxially aligned photoacoustic and ultrasound imaging. Building upon this groundwork, a recent study from Pohang University of Science and Technology illustrates the convergence of compact multimodal imaging within a single handheld platform [29]. By integrating ultrasound, photoacoustic, and fluorescence modalities into TUT array, the system enabled real-time visualization of vascular and lymphatic structures in both volunteers and microsurgical patients. This triple-modal design represents a major step toward multifunctional diagnostic tools that combine structural, molecular, and functional contrasts within a lightweight form factor, aligning with the broader trend toward portable, point-of-care photoacoustic imaging devices. The system architecture and imaging performance are illustrated in Fig. 7.5, which shows the handheld triple-modality opto-ultrasound probe integrating ultrasound, photoacoustic, and fluorescence capabilities through a TUT array. Looking ahead, similar multimodal concepts may be extended to fiber-based or endoscopy-based systems, enabling minimally invasive access to internal organs or confined anatomical regions. Such configurations could support molecular or microvascular imaging during surgical or interventional procedures, further broadening the clinical relevance of integrated photoacoustic technologies. In future implementations, substituting the laser source with compact emitters such as high-power LEDs, LDs, or VCSEL arrays could simplify driver electronics, reduce thermal load, and substantially lower cost, making similar multimodal platforms more accessible for widespread clinical and field use.

Beyond handheld devices, several groups are exploring wearable and patch-based photoacoustic systems. Recent demonstrations include soft, stretchable patches integrating VCSEL arrays and piezoelectric transducers encapsulated in elastomers [9]. These patches can conform to non-planar body surfaces, enabling continuous monitoring of hemoglobin concentration or vascular dynamics under movement. Recent advances from Qualcomm Research have demonstrated the integration of VCSEL excitation with acoustic detection for compact, wearable photoacoustic sensing modules. These studies showcased miniaturized wrist-worn prototypes for blood pressure monitoring, radial artery imaging, and vascular dynamics assessment, highlighting a pathway toward continuous, non-invasive physiological monitoring through photoacoustics [30, 31]. Similarly, prototype photoacoustic wrist-watches

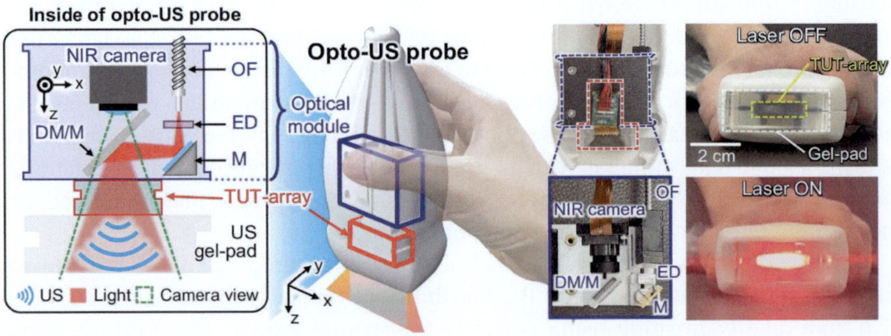

Fig. 7.5 Handheld opto-ultrasound probe integrating ultrasound, photoacoustic, and fluorescence imaging modalities through a transparent ultrasound transducer array. The design enables coaxial optical and acoustic alignment for compact multimodal imaging. Adapted with permission from [29]

have been reported, using compact laser diodes and micro-transducers for real-time assessment of peripheral oxygen saturation and microcirculation [16]. Such systems exemplify the shift toward continuous, non-invasive physiological monitoring using photoacoustic principles.

Looking ahead, the next phase of wearable photoacoustic systems will likely focus on deeper integration, lower power requirements, and algorithmic intelligence. Affordable light sources such as high-power LEDs, pulsed LDs, and VCSEL arrays could enable battery-operated modules capable of safe, continuous use. Similarly, emerging flexible transducer materials ranging from piezocomposites to optical ultrasound detectors may allow curved or textile-based form factors that conform seamlessly to the body. Combined with on-device reconstruction and machine-learning–based signal interpretation, future systems may support longitudinal monitoring of cardiovascular status, microcirculation, wound healing, or tissue perfusion in daily life rather than only clinical settings. These directions suggest that wearable photoacoustics could evolve from early prototypes into practical, autonomous health-monitoring tools, provided that challenges in power efficiency, thermal management, motion robustness, and long-term skin–device interaction are addressed.

In parallel, the proliferation of wireless ultrasound platforms—for example, *Clarius* and *Butterfly* probes—provides a natural foundation for integrating photoacoustic functionality. LED, LD, or VCSEL modules could be coupled to these wireless transducers to create fully untethered PA/US systems, supported by mobile-device processing and AI-based reconstruction pipelines. Recent studies by Periyasamy et al. [32] demonstrated the first wireless, ultra-compact dual-mode ultrasound and photoacoustic imaging system built on an FDA-approved clinical platform (Fig. 7.6) [32, 33]. Using a Clarius L15 wireless ultrasound probe synchronized with a pulsed laser diode (813 nm), the team achieved dual-mode imaging with ~300 μm resolution, fully controlled through iOS or Android devices. Although this was a feasibility study, this prototype provides a clear technological roadmap toward truly handheld, app-based PAI systems. Future integration of LED or VCSEL excitation modules could further reduce cost and weight, enabling wireless PA/US imaging suitable for point-of-care and global-health applications. Such progress underscores how affordable photoacoustic imaging can evolve naturally from the wireless ultrasound ecosystem already familiar to clinicians. The combination of wireless connectivity, compact photonic modules, and on-device computation represents the next major leap toward portable, intelligent, and affordable photoacoustic imaging.

7.3.4 Contrast Enhancement and Functional Imaging in Affordable PAI

As discussed earlier, clinically approved dyes such as ICG have emerged as practical agents to enhance photoacoustic contrast, particularly under the low optical fluence typical of VCSEL-, LED- and LD-based systems. Building on these foundations,

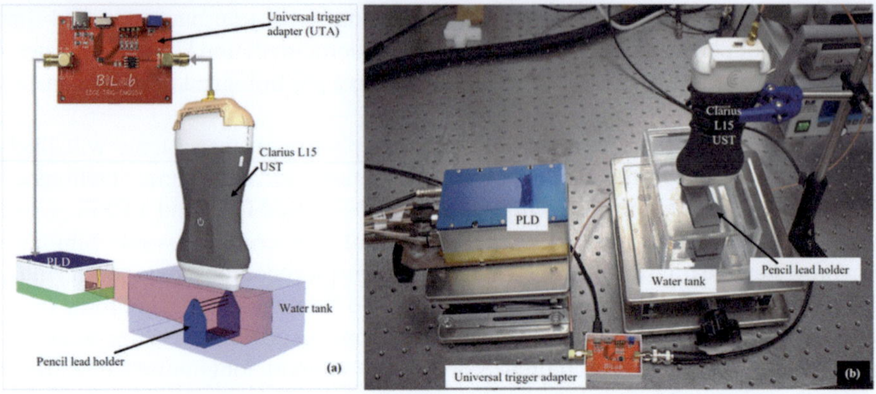

Fig. 7.6 Wireless handheld dual-mode ultrasound and photoacoustic imaging prototype. **a** Schematic diagram and **b** photograph of the experimental setup integrating a Clarius L15 wireless ultrasound transducer (UST) with a pulsed laser diode (PLD, 813 nm) for photoacoustic excitation. The universal trigger adapter (UTA) synchronizes optical and acoustic acquisitions via a mobile-controlled interface. This proof-of-concept system demonstrates the feasibility of fully wireless, compact dual-mode PA/US imaging. Reprinted with permission from [32]

recent studies are exploring how molecular contrast can expand the functional reach of affordable PAI beyond purely structural visualization. As illustrated in Fig. 7.7, exogenous contrast agents open functional imaging pathways which, when combined with affordable hardware (see Sects. 7.3.1–7.3.3), hold great promise for accessible PAI. In low-fluence systems, exogenous agents can offset reduced penetration depth and signal strength by amplifying optical absorption within targeted regions. Because dyes like ICG absorb strongly near 800 nm—well within the emission range of compact LEDs and VCSELs—they are ideal for integration with low-cost light sources. Moreover, their established safety record in clinical fluorescence and angiography applications significantly lowers the regulatory barrier for translation to photoacoustics. Several recent pilot studies using LED-based illumination have demonstrated successful lymphatic and vascular mapping with ICG enhancement, reaffirming its compatibility with compact excitation modules [20].

An additional advantage of such dyes in the affordable-imaging context is their photostability at low fluence levels. While conventional high-energy laser systems may bleach chromophores or induce nonlinear effects, LED and VCSEL excitations operate well below these thresholds, allowing repeatable imaging over longer sessions. This opens opportunities for dynamic monitoring such as perfusion tracking, pharmacokinetic analysis, or sentinel node assessment using entirely non-invasive, portable hardware. Looking ahead, a promising direction lies in the design of next-generation molecular probes tailored for the visible and near-infrared windows, where low-cost emitters are most efficient. Engineered nanoparticles and organic dyes with enhanced quantum yields and tunable absorption spectra could serve as dual fluorescence–photoacoustic agents. These developments align with

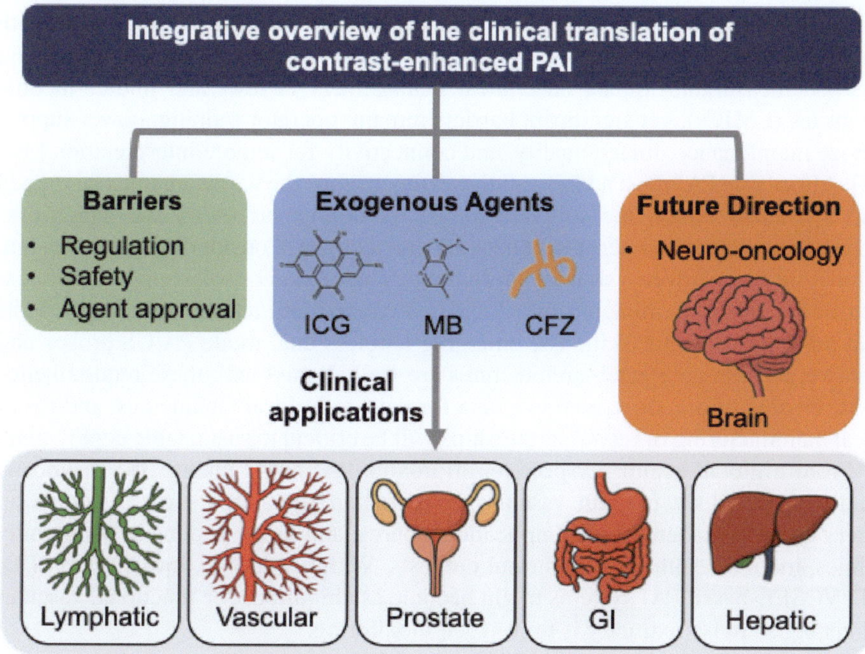

Fig. 7.7 Overview of the clinical translation of contrast-enhanced PAI (GI: gastrointestinal). Reprinted with permission from [17]

the broader movement toward functional photoacoustic imaging, in which multi-wavelength illumination and computational spectral unmixing enable quantitative mapping of tissue oxygenation, perfusion, and metabolic state.

From an accessibility standpoint, the ability to achieve functional contrast without complex laser systems represents a paradigm shift. In the long term, affordable PAI devices integrated with simple dye-based or endogenous contrast mechanisms could perform many of the roles now reserved for high-end tomographic systems, bridging the gap between functional ultrasound and molecular optical imaging.

7.3.5 The Future of PAI in Resource-Limited Settings

The innovations in light source design, acoustic detection, miniaturized architecture, and artificial-intelligence integration, all discussed in prior sections, converge toward a transformative vision: enabling PAI in settings far beyond tertiary hospitals. In resource-limited environments, where MRI or CT may be unavailable or unaffordable, an affordable PAI device offers an opportunity to deliver functional imaging (vascularity, oxygenation, perfusion) in a portable, ruggedized, and low-cost form.

Lessons from the global deployment of portable ultrasound devices highlight both the potential and the challenges. Systematic reviews show that handheld ultrasound has had measurable impact on clinical management in low- and middle-income countries (LMICs), yet significant barriers remain: operator training, power supply, device maintenance, image quality, and connectivity for remote interpretation [34–36]. Affordable PAI must address all these dimensions: the hardware must be rugged and low-maintenance, the software robust for non-expert users, and data workflows compatible with intermittent network connectivity. Considering system design, affordable PAI platforms should aim for battery-operation or solar support, wireless connectivity for tele-imaging or AI-assisted acquisition, and simplified protocols that minimize operator skill. The emerging wireless dual-mode PA/US probes and lightweight wearable patch-architectures are well aligned with these requirements. Additionally, efforts to standardize data formats, train local technicians, and foster local manufacturing or repair partnerships will be critical to sustainable deployment.

From a global-health perspective, affordable PAI could change the diagnostic landscape for diseases with vascular or perfusion signatures such as peripheral artery disease, diabetic foot complications, burns, and breast lesions in community clinics, mobile health vans, and rural outposts. Within a decade, low-fluence LED/LD/VCSEL-based PAI systems might become as ubiquitous for functional vascular screening as pulse oximetry is for oxygen monitoring.

Affordable LED-, LD-, and VCSEL-based PAI systems are also well positioned for deployment in primary-care and mobile screening environments, where their compact, safe, and infrastructure-light design enables functional assessment without the need for specialized imaging facilities. In the context of breast health, these systems could support early triage and referral in settings where mammography is unavailable or inaccessible. Beyond this, their low cost and operational simplicity open the door to additional high-impact applications such as monitoring wound healing, assessing peripheral vascular function, or evaluating inflammatory joint conditions, where accessible functional imaging can meaningfully guide community-level care. Together, these scenarios illustrate where affordable PAI could offer particularly high-impact clinical benefits and align with the kinds of transformative applications envisioned for LED/LD/VCSEL-based systems.

The path ahead is not without risk, regulatory hurdles, device reliability in harsh environments, and evidence of clinical utility must all be addressed. Nonetheless, the foundation of scalable, intelligent, and affordable PAI has been laid. The next generation of systems must prioritize accessibility, ease of use, and integration into real-world clinical workflows if they are to fulfil the promise of bringing advanced optical diagnostics to the point of need.

7.3.6 Predictions for the Next Decade

The coming decade will likely witness the maturation of PAI from a specialized research tool to a widely available clinical and point-of-care modality. This transition will be driven not by single breakthrough technologies but by the cumulative refinement and convergence of affordable components, intelligent algorithms, and standardized clinical workflows. Based on the trajectory of recent advances, several developments can be reasonably anticipated.

- **Integration of solid-state illumination within standard ultrasound probes**

Compact light sources like LED, VCSEL and LD are expected to become natively embedded into ultrasound transducers, enabling true dual-mode imaging in a single device. By the early 2030s, such systems may become as commonplace in vascular or dermatologic diagnostics as Doppler ultrasound is today.

- **Expansion of wearable and continuous-monitoring platforms**

Flexible photoacoustic patches and wrist-mounted sensors will evolve from research prototypes into regulated medical devices and smart watches. Their ability to monitor hemodynamic parameters, perfusion, and oxygenation in real time could transform chronic disease management, critical-care monitoring, and rehabilitation medicine.

- **AI-driven automation and decision support**

Advances in embedded computing and federated learning will allow real-time image reconstruction, artifact suppression, and diagnostic classification directly on handheld or mobile devices. Automated region-of-interest detection and quality control will make the technology accessible to non-specialists and community health workers.

- **Democratization through open standards and modularity**

Initiatives such as IPASC and open-source reconstruction frameworks will accelerate the adoption of standardized data formats and calibration protocols. This will enable interoperability among devices and facilitate cloud-based data sharing, multi-center studies, and AI training pipelines even in resource-limited settings.

- **Hybrid and multimodal convergence**

The boundaries between ultrasound, photoacoustic, and optical imaging will continue to blur. Affordable systems will increasingly combine PAI with elastography, Doppler, fluorescence, and thermal sensing, allowing comprehensive assessment of tissue physiology with minimal hardware complexity.

- **Regulatory and economic evolution**

As more affordable systems demonstrate clinical equivalence, regulatory frameworks will expand to include modular, low-fluence devices. The market will likely shift from capital-equipment sales toward service and subscription models, lowering entry barriers for smaller clinics and mobile-health operators.

- **Broader societal impact**

The combination of affordability, portability, and safety will enable PAI to contribute meaningfully to global-health equity. Just as pulse oximetry became ubiquitous in a generation, PAI could become a universal adjunct to ultrasound, providing optical contrast, vascular mapping, and functional assessment in virtually any clinical environment.

Collectively, these trends point toward a decade of consolidation and expansion, in which the principles of affordability, accessibility, and intelligence guide the evolution of PAI. The innovations described throughout this chapter ranging from LED and VCSEL illumination to flexible detectors and AI-enabled reconstruction are not isolated advances but interconnected steps toward a single goal: making functional optical imaging available to every patient, everywhere.

7.4 Summary

The trajectory of affordable PAI reflects a profound shift in how medical imaging is conceived, designed, and delivered. From solid-state light sources and transparent transducers to AI-enabled reconstruction and wireless integration, the field is converging toward systems that are compact, intelligent, and accessible. These technologies are not merely scaled-down versions of laboratory devices; they represent a fundamentally new philosophy of imaging: one that prioritizes safety, cost-efficiency, and usability at the point of care.

As illumination modules shrink to chip scale and acoustic detectors become flexible and transparent, PAI may soon evolve from a specialized diagnostic tool into an everyday clinical companion. In this sense, affordable PAI has the potential to become an "optical stethoscope", a universal, non-invasive instrument capable of revealing microvascular and metabolic health in real time. Such a transformation would democratize advanced optical diagnostics, bringing the power of functional imaging from research hospitals to community clinics and even the home environment.

Key Takeaways

- **Affordable illumination is maturing rapidly**. LED, LD, and VCSEL technologies now deliver sufficient fluence for clinical-scale PAI, enabling compact, fiber-free designs compatible with handheld and wearable devices
- **Component-level innovation drives system integration**. Advances such as transparent ultrasound arrays, optical detectors, and flexible substrates are simplifying probe architecture and lowering manufacturing costs
- **Miniaturization and wireless connectivity redefine accessibility**. Handheld and patch-based PA/US systems, coupled with mobile-device control, promise real-time functional imaging in decentralized and global-health settings

- **Functional and molecular imaging at low fluence is feasible**. Clinically approved dyes like ICG, and future visible-range agents, can enhance contrast and depth performance while maintaining safety under LED or VCSEL excitation
- **AI and open standards will democratize PAI**. On-device reconstruction, federated learning, and standardized data protocols (IPASC) will accelerate global deployment and clinical adoption
- **A new paradigm is emerging**. Affordable PAI is evolving from a laboratory instrument into an "optical stethoscope", a universal, intelligent tool for assessing vascular and metabolic health at the point of care

References

1. H. Zhong, T. Duan, H. Lan, M. Zhou, F. Gao, Review of low-cost photoacoustic sensing and imaging based on laser diode and light-emitting diode. Sensors **18**(7), 2264 (2018)
2. B. Huang, T.T.W. Wong, Review of low-cost light sources and miniaturized designs in photoacoustic microscopy. J. Biomed. Opt. **29**(Suppl 1), S11503 (2024)
3. M.K.A. Singh, N. Sato, F. Ichihashi, W. Xia, Y. Sankai, The evolution of LED-based photoacoustic imaging: from labs to clinics, in W. Xia (Ed.), *Biomedical Photoacoustics: Technology and Applications* (1 ed., pp. 573–608). Springer Nature (2024)
4. J.J. García-Garrigós, A. Cebrecos, J.A. Navarro-Calvo, F. Camarena, A fiber-coupled laser diode design for reflection mode optical resolution photoacoustic microscopy. Ultrasonics **132**, 107008 (2023)
5. M. Kuniyil Ajith Singh, W. Xia, Biomedical photoacoustic imaging and sensing using affordable resources. Sensors **21**(7), 2572 (2021)
6. C. Wang, Q. Yu, M. Li et al., Challenges and opportunities in next-generation LED therapeutic devices. Light Sci. Appl. **14**, 319 (2025)
7. B. Eliason, A. Longo, C. Dehner, L. Lilaj, S. Gronenborn, C. Klett, A. Weigl, D. Dörich, U. Malzahn, G. Zahnd, J. Konradl, P. Leisching, Performance characterization of a high-power, compact, low-cost, and high-efficiency VCSEL subsystem for photoacoustic imaging. *Proceedings SPIE 13319, Photons Plus Ultrasound: Imaging and Sensing 2025*, 133190S (2025)
8. J. Konradl, S. Gronenborn, A. Weigl, C. Klett, D. Dörich, U. Malzahn, A. Longo, C. Dehner, B. Eliason, C. Reichl, G. Zahnd, P. Leisching, High-power compact VCSEL subsystem as game-changer for photoacoustic medical imaging and sensing. *Proceedings SPIE 13384, Vertical-Cavity Surface-Emitting Lasers XXIX*, 133840G (2025)
9. X. Gao, X. Chen, H. Hu et al., A photoacoustic patch for three-dimensional imaging of hemoglobin and core temperature. Nat. Commun. **13**, 7757 (2022)
10. X. Xue, H. Wu, Q. Cai, M. Chen, S. Moon, Z. Huang, T. Kim, C. Peng, W. Feng, N. Sharma, X. Jiang, Flexible ultrasonic transducers for wearable biomedical applications: a review on advanced materials, structural designs, and future prospects. IEEE Trans. Ultrason. Ferroelectr. Freq. Control **71**(7), 786–810 (2024)
11. S. Zhou, G. Park, M. Lin et al., Wearable ultrasound technology. Nat. Rev. Bioeng. **3**, 835–854 (2025)
12. L. Zhu, H. Cao, J. Ma, L. Wang, Optical ultrasound sensors for photoacoustic imaging: a review. J. Biomed. Opt. **29**(Suppl 1), S11523 (2024)
13. H. Chen, S. Agrawal, M. Osman, J. Minotto, S. Mirg, J. Liu, A. Dangi, Q. Tran, T. Jackson, S.R. Kothapalli, A transparent ultrasound array for real-time optical, ultrasound, and photoacoustic imaging. BME Front. **8**(2022), 9871098 (2022)

14. Y. Xu, T.M. Khan, Y. Song et al., Edge deep learning in computer vision and medical diagnostics: a comprehensive survey. Artif. Intell. Rev. **58**, 93 (2025)
15. Z. Yin, K.M. Clark, T.R. Ray, Emerging additive manufacturing methods for wearable sensors: opportunities to expand access to personalized health monitoring. Adv. Sensor Res. **3**, 2300137 (2024)
16. T. Zhang, H. Guo, W. Qi, L. Xi, Wearable photoacoustic watch for humans. Opt. Lett. **49**(6), 1524–1527 (2024)
17. J. Miao, M. Luo, A. Kotwal, E. Hall, D. Oh, P.A. Valdes, L.S. Li, Clinical translation of photoacoustic imaging using exogenous molecular contrast agents [Invited]. Biomed. Opt. Express 16(11), 4353–4379 (2025)
18. Y. Wen, D. Guo, J. Zhang et al., Clinical photoacoustic/ultrasound dual-modal imaging: current status and future trends. Front. Physiol. **13**, 1036621 (2022)
19. J. Park, S. Choi, F. Knieling et al., Clinical translation of photoacoustic imaging. Nat. Rev. Bioeng. 3, 193–212 (2025)
20. A. Hariri, E. Zhao, A.S. Jeevarathinam et al., Molecular imaging of oxidative stress using an LED-based photoacoustic imaging system. Sci. Rep. **9**, 11378 (2019)
21. A. Hariri, K. Alipour, Y. Mantri, J.P. Schulze, J.V. Jokerst, Deep learning improves contrast in low-fluence photoacoustic imaging. Biomed. Opt. Express **11**(6), 3360–3373 (2020)
22. M. Shi, T. Zhao, S.J. West, A.E. Desjardins, T. Vercauteren, W. Xia, Improving needle visibility in LED-based photoacoustic imaging using deep learning with semi-synthetic datasets. Photoacoustics. **26**, 100351 (2022)
23. A. Paul, S. Mallidi, U-Net enhanced real-time LED-based photoacoustic imaging. J. Biophotonics **17**(6), e202300465 (2024)
24. A. Paul, S. Mallidi, Facilitating real-time LED-based photoacoustic imaging with DenP2P: an optimized conditional generative adversarial deep learning solution. AIP Adv. **15**(5), 055227 (2025)
25. K. Song, J. Feng, D. Chen, A survey on deep learning in medical ultrasound imaging. Front. Phys. **12**, 1398393 (2024)
26. S.J. Seiler, E.I. Neuschler, R.S. Butler, P.T. Lavin, B.E. Dogan, Optoacoustic imaging with decision support for differentiation of benign and malignant breast masses: a 15-reader retrospective study. AJR Am. J. Roentgenol. **220**(5), 646–658 (2023)
27. C. Dehner, G. Zahnd, V. Ntziachristos et al., A deep neural network for real-time optoacoustic image reconstruction with adjustable speed of sound. Nat. Mach. Intell. **5**, 1130–1141 (2023)
28. A. Thomas, M. Kuniyil Ajith Singh, N. Sato, F. Kalloor Joseph, Extending the depth of LED-based photoacoustic imaging for carotid and breast applications (medRxiv Preprint 2025.11.13.25339983) (2025). medRxiv
29. J. Park, D. Oh, J. Yoo et al., Clinical ultrasound, photoacoustic, and fluorescence image-guided lymphovenous anastomosis microsurgery via a transparent ultrasound transducer array. Nat. Commun. **16**, 9853 (2025)
30. H.V. Panchawagh, S. Agrawal, B.H. Soukup, K. Djordjev, Compact photoacoustic imaging module for wearable health monitoring devices. *2025 IEEE International Ultrasonics Symposium (IUS)*, Utrecht, Netherlands, pp. 1–4 (2025)
31. S. Agrawal, H.V. Panchawagh, K. Djordjev, Miniaturized photoacoustic wrist sensor with beamforming for radial artery imaging and vascular dynamics study. *2025 IEEE International Ultrasonics Symposium (IUS)*, Utrecht, Netherlands, pp. 1-4 (2025)
32. V. Periyasamy, A. Das, M. Pramanik, Wireless ultracompact handheld dual-mode ultrasound and photoacoustic imaging. *Proceedings SPIE 13319, Photons Plus Ultrasound: Imaging and Sensing 2025*, 133191L (2025)
33. V. Periyasamy, A. Das, M. Pramanik, Feasibility of an ultra-compact, wireless ultrasound and photoacoustic imaging system (wUS-PA). Opt. Lett. **50**, 6915–6918 (2025)
34. D.M. Becker, C.A. Tafoya, S.L. Becker, G.H. Kruger, M.J. Tafoya, T.K. Becker, The use of portable ultrasound devices in low- and middle-income countries: a systematic review of the literature. Trop. Med. Int. Health **21**, 294–311 (2016)

35. X. Gao, Q. Lv, S. Hou, Progress in the application of portable ultrasound combined with artificial intelligence in pre-hospital emergency and disaster sites. Diagnostics (Basel). **13**(21), 3388 (2023)
36. M. Haji-Hassan, R.D. Capraş, S.D. Bolboacă, Efficacy of handheld ultrasound in medical education: a comprehensive systematic review and narrative analysis. Diagnostics (Basel). **13**(24), 3665 (2023)

Chapter 8
Practical Resources and Tools for Developing Affordable Photoacoustic Imaging Systems

Learning Objectives

After reading this chapter, the reader will be able to:

- **Identify accessible optical, acoustic, and electronic components** suitable for designing and building affordable photoacoustic imaging systems.
- **Explore open-source software frameworks and data standards** that support reproducible image reconstruction, simulation, and analysis.
- **Recognize key collaborative initiatives and community platforms** (e.g., IPASC, SPIE, ESMI, GitHub) that enable shared progress in photoacoustic imaging research and education.
- **Understand the importance of data integrity and FAIR principles**, including standardized metadata, harmonized reporting, and transparent data exchange.
- **Discuss strategies for sustainable global engagement**, including mentorship, regional hubs, and training programs that strengthen capacity in emerging research environments.
- **Formulate a roadmap for open and inclusive innovation**, combining technology, standards, and community collaboration to advance the affordability and reach of photoacoustic imaging.

8.1 Accessing Affordable Components and Materials

Affordable photoacoustic imaging (PAI) relies not only on theoretical or algorithmic advances but on practical access to dependable, cost-effective hardware. This section provides guidance for identifying and sourcing optical, acoustic, mechanical, and electronic components that form the backbone of affordable PAI systems. The emphasis is on verified suppliers, reproducible configurations, and open documentation—helping researchers and system developers balance cost, performance, and scalability.

M. Kuniyil Ajith Singh, *Photoacoustic Imaging Using Affordable Resources*, Progress in Optical Science and Photonics 40, https://doi.org/10.1007/978-981-95-7175-8_8

The discussion begins with illumination technologies, covering laser diodes (LDs), light-emitting diodes (LEDs), and vertical-cavity surface-emitting lasers (VCSELs)—the three most accessible and rapidly evolving light-source families for low-cost PAI. Subsequent subsections address transducers, coupling materials, and synchronization electronics, with attention to procurement strategies that support reproducibility and sustainable system development.

8.1.1 Light Sources

Light sources form the foundation of any PAI system, providing the optical energy that is converted into acoustic waves within the target medium. Their characteristics—wavelength, pulse energy, temporal width, and repetition rate—directly determine image contrast, spatial resolution, and penetration depth [1, 2]. Selecting an appropriate source therefore represents one of the most critical and cost-defining steps in the development of affordable and portable PAI platforms. Traditional research-grade systems have relied on high-energy nanosecond lasers such as Q-switched Nd:YAG or optical parametric oscillator (OPO) sources. These systems offer excellent spectral versatility and pulse energy but are expensive, bulky, and maintenance-intensive. Their cost and operational complexity have historically limited the accessibility of PAI to well-funded laboratories.

Recent advances in semiconductor optoelectronics have opened new pathways toward compact and economical illumination. Modern LDs, LEDs, and VCSELs can now deliver optical pulses in the nanosecond-to-microsecond range with sufficient power for shallow to moderate imaging depths [2–5]. These sources are compact, electronically triggered, and often available as fiber-coupled or modular units, simplifying alignment and integration with ultrasound transducers.

Each of these emitter types presents a distinct balance of performance and affordability:

- LDs provide high optical power density and nanosecond pulsing capability at relatively low cost compared with solid-state lasers. They are well suited for portable and handheld systems requiring moderate penetration depths [6].
- LEDs are the most cost-efficient and robust sources. Although their pulse energies are lower, modern high-current drivers and array configurations allow practical imaging of superficial tissues with excellent stability and eye safety [7].
- VCSELs represent an emerging class that combines the array scalability of LEDs with the coherence and beam quality of laser diodes, offering new potential for multi-element and wide-field excitation [3–5].

For researchers and system developers, understanding the trade-offs among these technologies is essential. The following subsections present detailed discussions of each source type, together with verified component suppliers, selection guidelines, and procurement considerations relevant to affordable PAI research.

8.1.1.1 Laser Diodes (LDs)

LDs remain one of the most widely used excitation sources for affordable PAI systems [2, 6]. They offer compactness, high wall-plug efficiency, and flexible wavelength coverage from the visible to near-infrared spectrum. Compared with bulky solid-state lasers, LDs provide lower per-pulse energy but excellent electrical-to-optical conversion efficiency, allowing compact and portable implementations when combined with appropriate driving electronics.

Most LD-based PAI systems employ nanosecond-pulsed LDs, where short electrical pulses generate optical transients satisfying both thermal and stress confinement. Commercially available pulsed LD modules can produce microjoule-level optical energies at repetition rates up to tens of kilohertz, sufficient for superficial and mesoscopic imaging. Their small form factor and fiber-coupled configurations make them attractive for integration with ultrasound transducers in handheld or probe-based geometries.

Although many LD sources are sold as continuous-wave (CW) devices, numerous PAI and optical-resolution photoacoustic microscopy (OR-PAM) studies have shown that CW-rated LDs can be overdriven using nanosecond electrical pulses. In this mode, the diode is operated far above its CW current limit but for extremely short durations (typically <100 ns, <1% duty cycle), producing transient peak powers, an order of magnitude higher than CW ratings [2]. This approach—often called pulsed-CW operation—has become common in low-cost PAI prototypes and visible-wavelength PAM systems. When using CW diodes in pulsed mode, careful control of peak current, duty cycle, and junction temperature is essential to ensure device longevity and consistent optical fluence.

Pulsed LD systems typically employ driver circuits that deliver tens of amperes with sub-100 ns pulse widths, synchronized precisely to the ultrasound acquisition clock [2, 6]. Fiber coupling, microlens collimation, or simple free-space focusing optics are used to illuminate the imaging region uniformly. The resulting acoustic amplitudes scale with pulse energy and beam profile, so multiple LDs are often arranged in bar or array configurations to increase fluence while maintaining compactness. Table 8.1 lists representative manufacturers producing pulsed LD modules or stacks directly relevant to PAI, while Table 8.2 summarizes commonly used visible CW-rated LDs that can be safely overdriven for nanosecond pulsed operation in laboratory systems.

Following the evolution of compact PAI systems, it becomes apparent that only a small number of manufacturers provide pulsed LD modules suitable for biomedical PAI use [2]. This concentration is largely a result of technical and regulatory constraints. Generating nanosecond pulses with microjoule-level energy requires specialized packaging, thermal management, and driver circuitry that exceed the scope of most general laser-diode suppliers. Moreover, eye-safety certification, beam-quality control, and reproducibility standards applicable to biomedical research further limit the number of companies willing to qualify such products.

Consequently, suppliers like Lumibird (Quantel) have maintained a leading position because they either developed LD illuminators specifically for photoacoustics or

Table 8.1 Representative manufacturers of photoacoustic LD sources (2024–2025). Verified manufacturers offering pulsed or fiber-coupled LD modules relevant to PAI. Technical notes summarize wavelength coverage and application suitability. Cost tiers are qualitative indicators only. The listed manufacturers are verified producers of high-power LD sources relevant to PAI

Supplier/ manufacturer	Example product or family	Wavelength range (nm)	Typical use or PAI relevance	Indicative cost tier[a]
Lumibird/Quantel (France/USA) https://www.lumibird.com	Nanosecond PLD and illuminator modules	800–980	mJ-class pulsed LD systems documented in compact and handheld PAI; widely cited in literature [8–10]	High (OEM)
Laser Components (Germany/USA) https://www.lasercomponents.com	Pulsed LDs and arrays (850, 905, 1550 nm)	850–1550	Manufacturer of pulsed LDs for industrial and biomedical use; 905 nm devices used in the first LD-based in vivo PAI demonstration and cited in reviews as cost-efficient sources [11]	Low-Moderate
ams OSRAM (Germany) https://www.osram.com	Pulsed LD SPL UL90AT08 (905 nm)	905	Manufacturer of pulsed LD for industrial and biomedical use. Potentially suitable for PAI when multiple elements are used as an array	Low
Jenoptik (Germany) https://www.jenoptik.com	JOLD diode stacks/bar modules	800–1060	Manufacturer of high-power LD modules suitable for PAI integration (*no brand-specific PAI paper identified by the author*)	High

[a] Qualitative indicator based on module complexity and market segment; actual pricing varies with configuration, wavelength, and integration level

already supplied to the LIDAR and defense markets where similar pulse characteristics are required. Other major photonics firms focus on continuous-wave (few can be overdriven as mentioned before) or telecom-grade diodes, which are unsuitable for the short-pulse, high-peak-power demands of PAI. As a result, the global supply chain for PAI-capable LD sources remains small but stable, with most research groups relying on a few proven OEMs or in-house assemblies.

Despite the limited number of specialized LD manufacturers, diode-based excitation remains one of the most efficient routes toward compact and high-performance PAI. Yet the dependence on custom-packaged, nanosecond-pulsed emitters has kept system costs relatively high compared with solid-state alternatives. To further reduce entry barriers, many research groups have turned to high-power LEDs, which offer

Table 8.2 Visible laser-diode components used or suitable for PAI/OR-PAM (Build-Level).

Supplier/ manufacturer	Example part (as cited/sourced)	Wavelength (nm)	Notes for PAI use	Evidence/ source
Lasertack (Germany) https://www.lasertack.com	PD-01236 (6 W, 445 nm); PD-01289 (1 W, 520 nm); PD-01229 (0.7 W, 638 nm)	445/520/638	CW modules overdriven with ns pulses (≤10 ns, 50–60 A) for PA microscopy; compact heads with external drivers	Explicitly listed in PAM study [12]
ams-OSRAM (Germany/Austria) https://www.ams-osram.com	PLPT5-447KA blue LD (TO-56)	447	Supports CW and pulsed operation with high modulation bandwidth; common in visible PAM builds	Datasheet and explicit listing in PAM study [13]
Thorlabs (USA) https://www.thorlabs.com	L450G2 (450 nm, 100 mW)	450	Easily available TO-can LD; used with external ns pulsers; monitor junction temperature under overdrive	Product page and explicit listing in PAM study [14]
Monocrom S.L. (Spain) https://www.monocrom.com	Custom diode-laser modules/mounts	400–1000+	Manufacturer of bar and stack LD modules; suitable for custom ns-pulsed drivers	Manufacturer data and explicit listing in PAM study [15]

Caution: Overdriving CW-rated LDs shortens lifetime if peak current, junction temperature, or duty cycle exceed safe limits. Use proper heatsinking, pulse-width control, and electrical interlocks

lower energy per pulse but exceptional affordability, robustness, and safety. The next section reviews the progress, design considerations, and current sources of LED-based excitation modules for PAI.

8.1.1.2 Light Emitting Diodes (LEDs)

LEDs represent a fundamentally different illumination approach from LDs. While LDs provide coherent, high-intensity optical pulses through stimulated emission in a confined cavity, LEDs rely on spontaneous emission within a semiconductor junction, producing broad-spectrum, low-divergence light. The absence of an optical cavity makes LEDs intrinsically simpler, more robust, and less costly to manufacture, but also limits their peak power and energy density. These contrasting characteristics define the primary trade-off between LD- and LED-based excitation in PAI—coherence and fluence versus cost and scalability. Over the past decade, improvements in high-current drivers and semiconductor packaging have significantly extended the capabilities of LEDs for PAI applications [2, 7]. Modern high-power LED arrays can deliver nanosecond-to-microsecond optical pulses with fluences sufficient to image superficial vasculature, skin lesions, and small-animal tissues. While the achievable imaging depth is typically limited to less than 2 cm, the high repetition rate (up to tens of kHz) and excellent stability of LEDs enable rapid averaging and frame accumulation, producing clear and reproducible signals under safe illumination conditions.

The main advantages of LEDs in PAI stem from their affordability, compactness, and safety. LED systems are eye-safe even at close working distances, do not require laser-class safety enclosures, and can be battery-powered or integrated with portable ultrasound scanners. Their broad emission spectra are well suited for multispectral and functional imaging, especially when using arrayed emitters of different center wavelengths (e.g., 450, 530, 630, 690, 750, 850, and 940 nm). Moreover, their long operational lifetime (>10,000 h) and simple electronic triggering make them ideal for low-cost, educational, and point-of-care PAI systems [2, 7].

Despite these strengths, LED-based excitation imposes some design considerations. To maximize acoustic signal strength, LEDs must be driven by short, high-current pulses (typically 20–200 A) while maintaining thermal stability. Efficient heat sinking and synchronization with ultrasound acquisition are crucial to prevent timing jitter and intensity drift. In practical systems, multiple LED boards or bars are often combined to form composite illuminators, each contributing a few microjoules of energy per pulse, achieving total fluences on the order of 0.1–0.3 mJ/cm^2 at the imaging plane. Commercially available LED arrays are typically packaged with integrated drivers and cooling modules, making them straightforward to align and couple optically to the target.

Compared to LDs, LEDs trade coherence and peak fluence for simplicity, robustness, and low cost. Modern high-power LED arrays, when driven with short, high-current pulses and averaged at kilohertz repetition rates, enable superficial vascular and dermal imaging with excellent portability and safety, often integrated into handheld ultrasound/PAI probes [2, 7]. Multiple peer-reviewed demonstrations (preclinical and early clinical) now use commercial LED-PA systems rather than bespoke benches, with most citations referencing the AcousticX platform (CYBERDYNE, INC.), including feasibility in arthritis, needle guidance, tumor vasculature, and oximetry studies [16]. Table 8.3 summarizes representative LED sources and systems explicitly cited in photoacoustic imaging literature, including both commercial platforms and component-level emitters suitable for laboratory prototyping.

The LED-based approach has now matured from laboratory proof-of-concept to reliable clinical and educational instrumentation. While commercial LED-PAI platforms such as AcousticX have validated feasibility in humans, many research laboratories continue to construct customized systems using readily available high-power LEDs like those listed in Table 8.3. For superficial vasculature, wound assessment, and dermatologic imaging, visible or near-infrared LEDs (620–940 nm) offer optimal tissue contrast and penetration. Adequate synchronization and overdrive control of LED currents (typically 20–200 A pulses) are critical for maintaining consistent fluence and safe operation. As diode and driver technology continues to evolve, LED excitation is expected to remain the most accessible route to affordable and reproducible PAI across both academic and clinical settings.

Table 8.3 Representative LED sources and systems cited in photoacoustic imaging research (2016–2025). Verified LED components and systems whose brand or model are explicitly named in published PAI studies

Supplier/ manufacturer	Example product/system	Wavelength range (nm)	Typical use or PAI relevance	Indicative cost tier[a]
Luminus (USA) https://www.luminous.com	**SST-90** high-power visible LED	620–630	Used by UCL for pulsed-LED excitation in biomedical PAI (2016); over-driven 200 ns pulses (~9 μJ per pulse) [17]	Low—Moderate
ams-OSRAM (Germany/ Austria) https://www.ams-osram.com	**LZC-03MA07** four-channel RGBA LED package	460–630	Demonstrated by UCL for multi-wavelength functional PA measurements (2016); compact multispectral excitation source [17]	Moderate
Vishay (Germany/ USA) https://www.vishay.com	**VSMA1085250X02** high-power 850 nm IR LED	850	Reported by Iowa State University (2025) in a 37-LED array (0.19 mJ per pulse total); nanosecond pulsing up to 20 A drive current [18]	Low—Moderate
ams-OSRAM (Germany/ Austria) https://www.ams-osram.com	**SFH 4171S (OSLON P1616)** IR emitter	850–870	Used by ETH Zurich (2025) for full-view LED-based tomographic PA in mice; 160-LED array (~0.5 mJ per pulse total) [19]	Moderate
CYBERDYNE Inc. (Japan) https://www.cyberdyne.jp	**High power LED arrays** sold along with**AcousticX system**	690–940	Widely cited commercial system (2017–2024) for multiple preclinical and clinical applications (1 mJ/ pulse for 850 nm–100 ns pulse width) [2, 7]	High (OEM system)

[a] Qualitative indicator based on emitter power and integration complexity; actual prices depend on packaging, driver design, and quantity

8.1.1.3 Vertical-Cavity Surface-Emitting Lasers (VCSELs)

VCSELs represent a rapidly emerging class of light sources bridging the gap between traditional LDs and LEDs. Unlike edge-emitting LDs, VCSELs emit light perpendicular to the wafer surface, allowing wafer-scale fabrication, excellent beam uniformity, and dense array integration. Each emitter produces a low-divergence, near-Gaussian beam, which can be combined across many elements to achieve high optical fluence while maintaining compact form factors and low cost per channel [4]. In PAI, VCSELs are particularly appealing because they can be driven in pulsed or quasi-continuous modes at high repetition rates with low threshold currents. Arrays comprising tens to hundreds of emitters can deliver micro- to millijoule-class pulses while maintaining stable emission and eye-safe operation in the 850–940 nm range, ideal for hemoglobin-sensitive imaging. Their high electrical-to-optical efficiency and surface-mount architecture make them excellent candidates for portable, wearable, or probe-integrated PAI systems [4, 5].

The first clinical-relevant demonstration of VCSEL-based PAI was reported in 2022 by Gao et al. [5], where a flexible wearable patch incorporated an array of 24 $\times$ 850 nm VCSEL chips (Ace Photonics, Japan). Each die delivered approximately 40 W peak optical power when driven with 200 ns pulses at 3 kHz, enabling 3D vascular and thermal mapping of deep tissue (>2 cm). More recently, TRUMPF Photonic Components and iThera Medical jointly introduced a 940 nm VCSEL subsystem optimized for PAI [3, 4]. These efforts demonstrate that VCSELs can combine compactness, low energy consumption, and scalability, key ingredients for the next generation of affordable PAI instrumentation (Table 8.4).

Practical Guidance

When selecting VCSELs for PAI, devices near 800–940 nm provide an optimal balance between hemoglobin contrast and optical penetration depth. Arrays should deliver tens of watts peak power per chip, driven with 100–500 ns pulses at kHz PRFs. Effective thermal management, typically using planar heat spreaders or TEC plates is crucial for maintaining stability under repeated pulsing. VCSELs dissipate heat laterally across the substrate, so uniform heatsinking improves reliability. Their low divergence and wafer-scale fabrication make them ideal for collimated or fiber-coupled excitation directly integrated with ultrasound probes. As fabrication costs fall and driver technology matures, VCSELs are expected to bridge the performance gap between traditional LDs and LEDs, offering an energy-efficient, digitally addressable light source for the next generation of compact and wearable PAI devices.

Table 8.4 Representative VCSEL sources and subsystems for PAI (2022–2025). Verified VCSEL arrays, modules, or subsystems whose manufacturer or system name appears in PAI literature and/or public technical material

Supplier/ manufacturer	Example product/System	Wavelength range (nm)	Typical use or PAI relevance	Indicative cost tier[a]
TRUMPF Photonic Components (Germany/ Global) https://www.trumpf.com	940 nm VCSEL subsystem (co-developed with iThera Medical)	760–940	Demonstrated 2025 for deep-tissue PAI; compact, eye-safe subsystem targeting handheld/wearable devices [3, 4]	Moderate-High
Ace Photonics (Japan) https://www.acephotonics.com	850 nm VCSEL dies/arrays (used in PAI patch)	850	Used in Gao et al., *Nat. Commun.* 2022; 24-element array, ~40 W per chip, 200 ns pulses @ 3 kHz [5]	Moderate—High
DoGain (China) https://www.dogain.com	760/800/850 nm VCSEL arrays	760–850 nm	Manufactures and distributes VCSELs for industrial and biomedical use	Low-Moderate
Mesa Quantum (USA) https://www.mesaquantum.com	760/800/850 nm VCSEL arrays	760–850 nm	Manufactures and distributes VCSELs for industrial and biomedical use	Moderate
Sony Semiconductor Solutions (Japan/Global) https://www.sony-semicon.com	905/940 nm VCSEL arrays	905–940 nm	Manufactures and distributes VCSELs for industrial and biomedical use	Low—Moderate
Coherent, Inc. (USA/Global) https://www.coherent.com	905/940 nm VCSEL arrays	905–940 nm	Manufactures and distributes VCSELs for industrial and biomedical use	Low—Moderate
Lumentum Operations LLC (USA/Global) https://www.lumentum.com	905/940 nm VCSEL arrays	905–940 nm	Manufactures and distributes VCSELs for industrial and biomedical use	Low—Moderate

[a] Cost tier indicates relative pricing based on array complexity and market segment; actual values depend on configuration and volume

8.1.2 Driver and Synchronization Electronics for LD/LED/VCSEL PAI

Reliable photoacoustic excitation depends as much on the driving electronics as on the optical source itself. The driver determines pulse energy, temporal precision, and long-term stability, parameters that directly affect the signal-to-noise ratio and reproducibility of PAI measurements. Drivers for LDs, LEDs, and VCSELs must deliver short, high-current pulses with low jitter while protecting the emitter against thermal and electrical stress.

8.1.2.1 Design Principles and Key Parameters

Most time-domain PAI systems require nanosecond optical pulses to satisfy stress and thermal confinement. The corresponding electrical pulse widths typically range from 50 to 300 ns, with peak currents of 10–200 A depending on source type. Key parameters include:

- **Pulse amplitude and duration**: Must match the emitter's transient capability and maintain sub-microsecond confinement.
- **Rise/fall times**: <20 ns preferred to minimize timing uncertainty.
- **Trigger jitter**: <5 ns to ensure synchronization with the ultrasound data-acquisition clock.
- **Thermal management and duty cycle**: Maintain junction temperature below rated limits, usually <1% duty cycle.
- **Safety features**: Over-current limiting, temperature feedback, and interlock control to protect expensive diodes or arrays.

8.1.2.2 Commercial Nanosecond Drivers

A number of compact, reliable nanosecond current sources are commercially available and have been explicitly used in published PAI studies. The LDP-V series (PicoLAS GmbH, Germany) and PCO-7120/7121 (Directed Energy Inc., USA) are among the most cited. For example, the LDP-V 50–100 V3 driver was used in early high-speed LED pulsing research, while LDP-V 240–100 V3.3 modules have been employed for high-current LD excitation in recent PAI systems [2, 6, 7]. Both provide programmable pulse widths from 1 ns to several μs and up to hundreds of amperes peak current. Similarly, the PCO-7120/7121 drivers offer ns-scale current modulation for both LD and LED excitation and remain a common benchmark in published hardware builds.

8.1.2.3 Custom and Academic Drivers

Several research groups have developed fully documented, open driver circuits specifically for affordable PAI. At Helmholtz Munich/TUM, overdriven CW-rated LDs were pulsed at 10 ns width and 50–60 A peak current using a discrete MOSFET design, achieving reproducible μJ-class pulses for mesoscopic imaging [12, 20]. At Iowa State University, a complete nanosecond pulsed current source (NSPCS) for LED-PAI was described, capable of 20 A peak current and 50–100 ns pulses with open schematics and bill of materials [21]. These detailed publications form valuable blueprints for laboratories developing their own low-cost systems.

8.1.2.4 VCSEL-Specific Driver Integration

For VCSEL arrays, commercial integrated driver ICs and control boards have recently appeared. In the TRUMPF–iThera handheld VCSEL probe (2025), two rows of 940 nm triple-junction VCSELs were mounted on a copper heat sink and connected to iC-Haus driver boards (Germany) [3, 4]. The boards incorporated dual-channel iC-HT control ICs and were managed by a custom FPGA timing controller, externally triggered by a Rigol DG1022 signal generator. The system demonstrated reliable nanosecond pulsing, FPGA-synchronized triggering, and DAQ-level averaging, underscoring how digital control can simplify complex multi-source timing in modern probes.

8.1.2.5 Synchronization and Timing

Accurate synchronization between the optical and acoustic subsystems is typically achieved using laboratory delay generators such as the Stanford Research Systems DG535 and Berkeley Nucleonics 577-8C, providing picosecond-level resolution and <200 ps jitter. These instruments allow deterministic sequencing between driver output, transducer firing, and data capture, ensuring phase-coherent averaging and high image fidelity.

The diversity of available driver solutions—from commercial ns modules to open academic circuits illustrates the maturation of affordable excitation control in PAI.

Table 8.5 summarizes all verified drivers and synchronization hardware explicitly cited in published PAI literature or patented system designs, spanning LD, LED, and VCSEL implementations between 2010 and 2025.

Cost and Accessibility

Affordable driver solutions now span more than two orders of magnitude in cost.

- Open or academic designs (TUM, Iowa State, GAC) can be assembled for €200–€500, allowing reproducible PAI excitation in teaching and small research labs.

Table 8.5 Verified drivers and synchronization hardware used in affordable PAI (2010–2025). Entries are drawn exclusively from peer-reviewed PAI papers, patents, or published prototypes. "Cost Tier" indicates approximate affordability at research scale: Low (< €1 000) | Moderate (€1 000–€5 000) | High (> €5 000) | High (OEM) = integrated commercial or clinical subsystems

Supplier/Platform (website)	Model/Type	Emitter Use	Source/Application	Cost tier[a]
Custom (German Aerospace Center (GAC)—FLOAT System)	Custom MOSFET driver (ns, high-current)	LED	Developed for high-speed optical pulsing; later adopted in FLOAT LED-PA system [19, 22]	Low–DIY
CYBERDYNE Inc. Japan) https://www.cyberdyne.jp	Custom "Giant and Ultra-Short-Pulse" LED Driver (MOSFET switch, 300–400 V, 14 A, 30–150 ns)	LED	Proprietary driver powering AcousticX LED-PAI system [23]	High (OEM)
Directed Energy Inc. (USA) https://www.directedenergy.com	PCO-7120/7121	LD/ LED	Widely used commercial ns drivers; cited in PAI driver-development studies [17]	Moderate
PicoLAS GmbH(Germany) https://www.picolas.de	LDP-V 50–100 V3, LDP-V 240–100 V3/V3.3	LD/ LED	High-current ns LD driver benchmark for diodes and diode stacks [5]	Moderate
Brightloop SAS (France) https://www.brightloop.fr	Proprietary low-inductance LD driver(US9819 146B2 patent)	LD	Patented resonant-loop driver (Liffran & Rabot 2017) co-developed with Quantel; used in FULLPHASE/CVENT handheld LD-PA/US systems [9, 24]	High (OEM)
Custom (TUM/ Helmholtz Munich)	Overdriven CW-LD ns driver (MOSFET, 10 ns, 50–60 A)	LD	Published schematic enabling μJ-level ns pulses from CW LDs [12]	Low–DIY
Custom (Iowa State University)	Nanosecond Pulsed Current Source (NSPCS) (20 A, 50–100 ns)	LED	Fully documented open design for LED-PAI [18]	Low–DIY
TRUMPF × iThera Medical https://www.trumpf.com · https://www.ichaus.de	VCSEL module (940 nm) with iC-Haus driver boards	VCSEL	Handheld PAI probe (2025): dual-row 940 nm VCSELs, iC-Haus boards, FPGA control, Rigol DG1022 trigger [3, 4]	High (prototype)

(continued)

Table 8.5 (continued)

Supplier/Platform (website)	Model/Type	Emitter Use	Source/Application	Cost tier[a]
Stanford Research Systems (USA) https://www.thinksrs.com	DG535 Delay Generator	Sync/ Timing	5 ps resolution, < 50 ps jitter; standard optical–acoustic trigger [25]	Moderate
Berkeley Nucleonics (USA) https://www.berkeleynucleonics.com	Model BNC 577 trigger generator	Sync/ Timing	Multi-channel delay generator with 250 ps resolution for multi-illumination control [26]	High

[a] Cost tier indicates relative pricing based on design complexity and market segment; actual values depend on configuration and volume

- Commercial ns modules such as PicoLAS and Directed Energy fall in the €800–€5 000 range per channel—well suited to mid-budget research systems.
- Integrated OEM or clinical-grade subsystems (CYBERDYNE AcousticX, Brightloop–Quantel, TRUMPF–iThera) exceed €10 000, but combine safety, diagnostics, and regulatory compliance. This cost diversity demonstrates that robust, repeatable excitation control for LD-, LED-, and VCSEL-based PAI can now be achieved at virtually any budget level.

8.1.2.6 Practical Considerations

Researchers should match drivers to the electrical and thermal characteristics of their emitters:

- **LEDs**: require high current (20–100 A), moderate voltage, and relatively wider pulses (70 130 ns).
- **LDs**: demand lower capacitance paths and strict current limiting; nanosecond control (<100 ns) is critical.
- **VCSELs**: benefit from low-inductance layouts and can often be driven by compact integrated ICs with FPGA synchronization.

Where possible, record the driver trigger and current-monitor outputs alongside the ultrasound signal to maintain reproducibility. Reliable synchronization ensures that optical, acoustic, and electronic subsystems operate in deterministic phase alignment, a prerequisite for quantitative PAI.

8.1.3 Transducers and Acoustic Modules

The acoustic detection chain in a PAI system determines spatial resolution, imaging depth, and overall sensitivity. Regardless of excitation source (LDs, LEDs, or VCSELs), the fidelity of the photoacoustic signal is ultimately limited by the characteristics of the ultrasonic transducer and its associated front-end electronics. Practical selection therefore requires balancing frequency, bandwidth, aperture size, and geometry against the intended imaging depth and application.

8.1.3.1 Classification and Design Considerations

Transducers for PAI fall into four major categories:

1. **Single-element transducers**
   Used for laboratory prototypes and optical-resolution PAI microscopes (OR-PAM). They offer high sensitivity and broad bandwidths (often > 70%), enabling axial resolutions of a few tens of micrometres. However, point-by-point scanning is required for image formation, limiting frame rate [6].
2. **Linear and curvilinear arrays**
   The most common choice for handheld and clinical PAI systems, typically sharing the same housings and interfaces as conventional ultrasound (US) probes. Typical element counts range from 64–256 with central frequencies between 3–10 MHz and bandwidths of 60–100%. Arrays enable real-time hybrid PA/US imaging and straightforward integration with existing ultrasound front ends [7].
3. **Custom PAI-optimized arrays**
   Research-grade designs using materials and geometries tuned for photoacoustic sensitivity. Examples include cylindrically focused arcs (e.g., Imasonic arrays in the TRUMPF–iThera VCSEL probe, 4 MHz, 256 elements, 153% fractional bandwidth) and custom PZT or PVDF elements designed for optical transparency or wide angular coverage [3, 4].
4. **Capacitive micromachined ultrasonic transducers (CMUTs) and polymer transducers (PMUTs)**
   Emerging microfabricated detectors that enable integration with optical components. CMUTs provide wide bandwidth and are compatible with flexible substrates, making them promising for wearable or miniaturized PAI sensors. Ongoing research explores their co-fabrication with VCSELs or LED arrays for fully integrated photoacoustic-on-chip devices.

Key design parameters are central frequency (f_0), fractional bandwidth ($\Delta f/f_0$), element pitch, and sensitivity (V/Pa). For most biomedical PAI applications, a frequency range of 3–10 MHz offers an effective compromise between resolution ($\propto f_0$) and penetration depth ($\propto 1/f_0$). Broadband transducers with high sensitivity are crucial for capturing the first acoustic cycle generated by nanosecond optical pulses.

In parallel with optical source development, significant progress has occurred in custom transducer manufacturing for affordable PAI. Table 8.6 summarizes verified

Table 8.6 Representative transducer and array developers for photoacoustic imaging (2020–2025). Entities verifiably involved in custom or co-developed transducer design for PAI or hybrid PA/US systems. The "Focus" column highlights core expertise; "Example Use" cites a representative published or documented system. Cost tiers: Low—DIY/Academic, Moderate—OEM/Research, High—Commercial/Clinical (OEM)

Developer/ supplier (website)	Headquarters/ region	Focus/expertise	Example use in PAI/ collaboration	Cost tier[a]
Imasonic SAS https://www.imasonic.com	Besançon, France	Custom PZT and CMUT arrays; broadband, high-sensitivity transducers	Cylindrically focused 4 MHz, 256-element array in TRUMPF–iThera VCSEL probe; numerous hybrid PA/US research systems [3, 4]	High (OEM R&D)
Vermon SA https://www.vermon.com	Tours, France	Custom linear/ curvilinear arrays for photoacoustic integration	Collaborations with multiple EU projects; breast and vascular PAI probe development [27]	High (OEM Clinical)
ULSO Tech Co., Ltd. https://www.ulso-tech.com	Xingtai City, China	Custom linear/ curvilinear arrays for photoacoustic integration	Supplier for linear and curved ultra-high bandwidth probes for PAI research systems [28]	Moderate (OEM)
Suzhou Norayso Technology Co., Ltd. https://www.norayso.com	Suzhou, China	Custom acoustic stacks/ array fabrication	Supplier for prototype probes in Asia–Pacific collaborations	Moderate (OEM)
CYBERDYNE Inc. (AcousticX System) https://www.cyberdyne.jp	Tsukuba, Japan	Commercial LED-PAI platform with 7 MHz and 10 MHz linear probes	Integrated dual-mode LED-PA/US probes used in clinical feasibility studies (arthritis, vascular imaging, etc.) [7]	High (OEM)
Blatek Inc. https://www.blatek.com	Pennsylvania, USA	PZT transducer elements/ acoustic stacks for research arrays	Fabrication partner for North-American PAI prototype probes [29]	Moderate
Olympus Scientific Solutions https://www.olympus-ims.com	Tokyo, Japan/ Global	Broadband immersion and single-element transducers (customizable)	Commonly used in single-element and acoustic-resolution PAI setups [18]	Moderate
Academic Groups (e.g., Twente, POSTECH, Penn State, ETH Zurich)	Europe	Design and open dissemination of probe design and geometries	Open information in publications about PA probe design [30–33]	Low—DIY/ Open

[a] Cost tier indicates relative pricing based on design complexity and market segment; actual values depend on configuration and volume

industry and academic partners actively designing or co-developing acoustic arrays and probe modules, ranging from OEM clinical suppliers to open academic platforms.

8.1.3.2 Integration and Coupling

Acoustic coupling between the transducer and tissue is commonly achieved using aqueous gels, thin membranes, or transparent coupling media such as perfluorocarbon-based oils. In probe-based systems, optical and acoustic paths are co-aligned either through optically transparent apertures, fiber bundles arranged around the aperture, or beam combiners that merge optical and acoustic paths. For flexible or wearable arrays, thin PDMS membranes provide both acoustic matching and mechanical stability.

8.1.4 Data Acquisition, Front-End Electronics, and Signal Conditioning

The DAQ subsystem defines the performance bridge between the acoustic domain and computational reconstruction. Its architecture determines achievable bandwidth, dynamic range, frame rate, and synchronization precision between optical excitation and acoustic detection. For affordable or custom-built PAI systems, DAQ integration remains one of the most critical and cost-sensitive design steps.

8.1.4.1 Functional Overview

A complete PAI DAQ chain typically includes:

- **Low-noise preamplifiers (LNA)**: boost weak microvolt-level acoustic signals while maintaining wide bandwidth and phase linearity.
- **Analog filters**: band-limit the signal to suppress low-frequency drift and aliasing before digitization.
- **High-speed digitizers**: multi-channel ADCs or ultrasound DAQ cards with 8–14 bit resolution, sampling rates from 20–100 MS/s per channel.
- **Timing and synchronization circuits**: ensure precise correlation between light firing, driver triggers, and ADC sampling clocks.
- **Data-transfer interface**: PCIe or USB 3.0 for real-time data streaming to the reconstruction computer.

Typical PAI system requirements lie within 2–20 MHz bandwidths and 60–80 dB dynamic range. In hybrid PA/US research platforms, synchronization can often be achieved through firmware updates or shared trigger lines with existing ultrasound controllers.

8.1.4.2 DAQ and Front-End Solutions

Affordable multi-channel acquisition platforms have evolved rapidly, moving from proprietary ultrasound hardware toward open or semi-open FPGA–ADC architectures. In practice, one of the most direct and economical strategies is to repurpose existing ultrasound research hardware, since most modern systems already provide synchronized multi-channel digitization, programmable gain, and trigger control suitable for PAI. However, it should be noted that ultrasound front ends are optimized for higher acoustic pressures and larger dynamic ranges, while photoacoustic signals—especially those generated by low-cost LEDs or laser diodes—are several orders of magnitude weaker. Consequently, the built-in amplification and noise performance of clinical or preclinical ultrasound systems may not always be adequate for faint PA signals, and external low-noise preamplifiers or dedicated analog front ends are often required to preserve SNR. Table 8.7 summarizes DAQ hardware verified in published PAI literature, highlighting the trend from high-cost research systems to integrated low-cost frameworks.

Key trends: Recent work demonstrates that a complete 30-channel hemispherical PACT can be built with a TI AFE5832EVM front-end coupled to a TSW14J56EVM FPGA capture board, achieving 12-bit, 40 MS/s sampling and 46 dB mean SNR—all for under €2 000 in hardware [41]. This marks the first published framework delivering research-grade performance at a genuinely affordable price point. High-end DAQs (e.g., Verasonics or AcousticX) remain indispensable for advanced beamforming and hybrid PA/US imaging but are cost-prohibitive for open, educational, or portable systems.

8.1.4.3 Front-End and Signal Conditioning Considerations

- **Impedance Matching**: Most piezoelectric transducers present ~ 50 Ω impedance; mismatches introduce reflections and noise. Compact preamps or buffer stages near the probe minimize these effects
- **Noise and Gain**: Published low-cost designs use two or three amplification stages totaling 40–127 dB gain, often employing discrete low-noise op-amps (e.g., LT1226, OP37) or integrated RF LNAs (e.g., Mini-Circuits ZX-series).
- **Integrated AFEs**: The AFE5832 architecture combines programmable gain amplifiers, filters, and ADCs, greatly simplifying analog front-end design for PAI
- **Filtering**: Simple first-order high-pass ($\approx$ 100 kHz) and low-pass ($\approx 0.4 \times$ Fs) filters suffice for 2–10 MHz transducers while preventing DC drift and aliasing.
- **Isolation and Grounding**: Optical or transformer isolation between driver and DAQ grounds prevents coupling of high-current transients, especially in LED/LD-based systems.
- **Proximity and Layout**: Locating preamplifiers close to the transducer connector minimizes cable capacitance and improves effective bandwidth and SNR.

Table 8.7 Commercial and custom-made DAQs suitable for affordable photoacoustic imaging

Platform/supplier	Channel count	Sampling rate/ resolution	Reported use/reference	Approx. cost tier[a]
Verasonics Vantage 64/ 256 https://www.verasonics.com	64–256	62.5 MS/s @ 14 bits	Standard research ultrasound/ PA system; widely cited in hybrid PA/US imaging [34]	High (>€30 000)
CYBERDYNE AcousticX https://www.cyberdyne.jp	128	40 MS/s @ 14 bits, 105 dB gain	Standard LED-PA/US research system; widely cited in hybrid PA/US imaging papers [7]	High (>€30 000)
TsingPAI MarsonicsDAQ https://www.tsingpai.cn	64–1024	40–125 MS/s @ 16 bits, 74 dB gain	PA/US research system recently visible in literature [35]	High (>€30 000)
Opensonics OpenPioneer https://www.opensonics.tech	64–1024	50 MS/s @14 bits, 110 dB gain	US research system launched in 2025	High (>€30 000)
PhotoSound Legion ADC https://www.photosound.com	128–256	40 MS/s @ 12 bits	Customizable PA/US research system [36]	High (>€30 000)
S-Sharp Prodigy https://www.s-sharp.com	128–256	125 MS/s @16 bits, 48 dB gain	Customizable PA/US research system [37]	High (>€30 000)
us4us us4R-lite https://www.us4us.eu	32–64	80 MS/s @ 12 bits	Customizable PA/US research system [38]	High (≈€15 000)
National Instruments PXIe-5170R https://www.ni.com	8	250 MS/s @ 14 bit	Benchmark reference DAQ; used in cost-comparison frameworks [39]	High (≈€15 000)
GaGe CSE1642 https://www.vitrek.com	4	200 MS/s @ 12–14 bits	Single-element PACT and benchtop prototypes [40]	Moderate (≈€12 000)
LED-based PACT (Iowa State University)	1	25 MS/s @ 16 bits, 127 dB	Single-element LED-PACT [18]	Low (≈€10 000 for whole system)
TI AFE5832EVM + TSW14J56EVM (University of Illinois)	16–32	40 MS/s @ 12 bits, 51 dB analog gain	Low-cost open-source hPACT system (30-element array, SNR ≈ 46 dB, 100 ms capture) [41]	Low (<€2 000)
NI PCIe-6320	8	250 kS/s @ 16 bits	PACT and CW/LED-PAI cost studies	Low (≈€800)
Sound-card/Arduino DAQs	1–4	48 kHz @ 16 bits (SC)/10 kS/s (μC)	Continuous-wave and envelope-detection PA sensors	Very Low (<€100)

[a] Cost tier indicates relative pricing based on design complexity and market segment; actual values depend on configuration and volume

8.1.4.4 Synchronization and Timing

Accurate synchronization between optical excitation and data capture is critical for quantitative PAI. Recent low-cost frameworks employ dual-channel function generators to issue synchronized triggers to both the light drivers and FPGA capture. The FPGA generates a stable system clock (typically 4 $\times$ sampling frequency) and records the optical trigger signal on an auxiliary channel for precise alignment during reconstruction. Sub-microsecond jitter ($<$10 ns) has been reported using this configuration.

8.1.4.5 Summary

Affordable, modular DAQ solutions—particularly those integrating FPGA-based digitizers and shared-clock triggering—now make real-time, multi-channel PAI feasible without proprietary ultrasound consoles. Published demonstrations confirm that high-fidelity, 30-element PACT imaging can be achieved with open evaluation boards under €2 000, while retaining bandwidths above 10 MHz and SNRs $>$45 dB. By combining open-hardware digitizers, low-noise amplification, and deterministic timing control, researchers can assemble a complete, high-performance DAQ chain for under €5 000, laying the hardware foundation for a fully open and affordable PAI instrument.

8.1.5 Open-Source Wearable Ultrasound Platforms and Future Integration with Affordable PAI

Open-hardware ultrasound systems have recently begun to bridge the gap between conventional instrumentation and fully wearable diagnostic devices. A notable example is the WULPUS platform (Wearable Ultra Low-Power Ultrasound), an open-source system developed for continuous acoustic monitoring using ultra–low-power electronics, Bluetooth connectivity, and a modular analog front-end [42]. All hardware schematics, firmware, and host-side software are publicly available through its repository, with a total material cost of approximately 150–200 USD [43].

Although WULPUS is designed as an ultrasound-only platform, its architecture including miniaturized electronics, efficient signal chains, and wireless data transmission offers a compelling template for future wearable photoacoustic devices. Integrating a compact excitation module, such as a small LED or VCSEL driver, could enable patch-based or wrist-mounted PAI systems for monitoring vascular dynamics, tissue perfusion, or physiological trends in daily-life settings. The feasibility of combining low-power ultrasound sensing with eye-safe optical excitation illustrates how open hardware ecosystems can accelerate innovation in affordable, body-worn PAI.

8.2 Open-Source Software and Reconstruction Frameworks

This section presents the open and reproducible software ecosystem that complements affordable PAI hardware, covering simulation, reconstruction, AI/ML integration, embedded computing, and data-standardization tools. Readers should come away understanding how to build, analyze, and share PAI data using freely available frameworks.

8.2.1 Reconstruction Frameworks

The growing ecosystem of open-source toolboxes has made PAI reconstruction and simulation accessible to researchers without commercial platforms. These frameworks cover time-domain acoustic propagation, optical fluence modeling, system response simulation, and deep-learning-based inversion. Many of them are interoperable through standard data formats such as IPASC HDF5, enabling reproducible workflows from raw data to final images [44]. A selection of actively maintained open-source simulation and reconstruction toolboxes for PAI is summarized in Table 8.8. These frameworks collectively span acoustic and optical modeling, forward and inverse reconstruction, and system-level simulation. Many support standardized IPASC HDF5 data formats, enabling interoperability between simulation, reconstruction, and deep-learning workflows.

The frameworks listed in Table 8.8 form the foundation for data generation and reconstruction in modern PAI pipelines. When combined with open datasets and deep-learning toolkits (Sect. 8.2.2), they enable fully reproducible, end-to-end development of affordable and quantitative photoacoustic systems.

8.2.2 AI and Machine-Learning Resources

The rise of data-driven methods has opened new directions for PAI, enabling image reconstruction, denoising, artifact suppression, segmentation, and quantitative analysis beyond traditional model-based algorithms [64]. The combination of open datasets, simulation frameworks, and freely available deep-learning toolboxes has made it possible to train and evaluate advanced algorithms even on affordable or custom-built PAI systems.

Table 8.8 Open-source simulation and reconstruction toolboxes for photoacoustic imaging

Framework	Language/ domain	Core features	Website/repository
k-Wave	MATLAB/ Python	Pseudospectral time-domain acoustic simulation; time-reversal reconstruction; GPU acceleration	https://github.com/ucl-bug/k-wave [45]
NIRFAST, NIRFASTer-FF	MATLAB/ Python	FEM based package for modeling Near-Infrared light transport in tissue	https://www.github.com/NIRFAST-admin [46, 47] https://github.com/milabuob/nirfaster-FF [48]
MSOT Toolbox	MATLAB	Fluence correction, multispectral unmixing, and reconstruction for iThera MSOT datasets	https://github.com/juestellab/mb-rec-msot [49, 50]
PATLAB	MATLAB	Delay-and-sum and analytical back-projection algorithms; educational use	https://github.com/Lawson-Optics-Lab/PATLAB [51]
PATATO (PhotoAcoustic Tomography Analysis Toolbox)	Python	Full simulation-to-reconstruction workflow; IPASC-compliant; data pre- and post-processing	https://github.com/BohndiekLab/patato [52]
SIMPA (Simulation and Image Processing for PAI)	Python	Modular hybrid forward modeling combining MCX and k-Wave; supports LED/LD systems	https://github.com/IMSY-DKFZ/simpa [53]
PhocoSpace	Python	Physically consistent optical–acoustic simulation and inversion framework	https://gitlab.com/pulselab/phocospace [54, 55]
LED System Modelling	Python	Simulation of LED-array illumination, transducer response, and noise for low-cost PAI systems	https://www.github.com/jgroehl/led_system_modelling [56]
SKYQUANT 3D	Python	Quantifying Vascular Anatomy With an Open-Source Workflow for Comprehensive Analysis of Volumetric photoacoustic Angiography Data	https://github.com/KAY-imaging/OA-pipeline [57]
IPASC linear array standardized reconstruction	Python	Standardized image reconstruction framework for linear array probes	https://github.com/IPASC/standardised-image-reconstruction [58–63]

8.2.2.1 Public and Synthetic Datasets

Open datasets provide standardized material for developing and benchmarking learning-based reconstruction and classification methods. They include both experimental measurements and synthetic data generated by Monte-Carlo and k-Wave simulations, facilitating transfer learning between modalities and setups. These datasets together with synthetic data produced using SIMPA, PATATO, or k-Wave enable the generation of large, labeled training corpora required for neural-network-based PAI methods. Recently, PhotIQA introduced the first expert-rated photoacoustic image dataset for quantitative image quality assessment (IQA) [65]. It includes over a thousand reconstructed images from tissue-mimicking phantoms, each evaluated by two experts across attributes such as edge visibility, homogeneity, and contrast. This resource enables benchmarking of reconstruction and enhancement algorithms Table 8.9 summarizes publicly available synthetic and experimental datasets suitable for machine learning development in PAI.

8.2.2.2 Deep-Learning Toolboxes and Frameworks

The open release of deep-learning architectures tailored for PAI has accelerated research reproducibility and cross-laboratory benchmarking. Most of these frameworks use PyTorch or TensorFlow back ends and are designed for modular training and inference on commodity GPUs. Table 8.10 shows the latest list (2025) of open source deep-learning toolboxes and frameworks for PAI reconstruction and analysis.

Complementary general-purpose medical AI frameworks such as MONAI (https://www.monai.io) and PyTorch Lightning (https://www.pytorchlightning.ai) can be used to structure training workflows, perform augmentation, and standardize evaluation.

8.2.2.3 Integration and Use

Researchers can combine simulation, reconstruction, and learning resources in an open pipeline:

1. Simulate realistic acoustic and optical data using SIMPA or US-PA Simulation Codes.
2. Format outputs using the IPASC HDF5 standard for interoperability.
3. Train/fine-tune models such as PA OmniNet or Y-Net on real + synthetic data.
4. Evaluate reconstruction accuracy or segmentation metrics using public benchmarks (e.g., OADAT, PATATO).

These resources allow labs with even modest hardware to perform advanced reconstruction and quantitative analysis using low-cost PAI setups.

Table 8.9 Public in vivo, phantom, and synthetic datasets available for machine-learning development in photoacoustic imaging

Dataset	Type/content	Highlights	Access/repository
PhotIQA: Photoacoustic Image Dataset with Image Quality Ratings	Reconstructed PA images with expert-assigned quality ratings	1,134 images from tissue-mimicking phantoms annotated for five attributes; designed for benchmarking image quality assessment and reconstruction algorithms	https://doi.org/10.5281/zenodo.13325196 [65]
OADAT (Photoacoustic Data Set)	Experimental + synthetic raw and reconstructed data	Multiple geometries; vessel segmentation labels; sparse and full-view acquisitions	https://www.github.com/berkanlafci/oadat [66]
OA-Breast Database (UIUC, Anastasio Lab)	Numerical breast phantoms (optical + acoustic)	3D MRI-based breast tissue maps (skin, fat, vessels); binary/HDF5 format	https://www.anastasio.bioengineering.illinois.edu/downloadable-content/oa-breast-database [67]
PATATO Dataset	Phantom and in-vivo PACT data	Quantitative imaging and oxygen-enhancement sequences; HDF5 format	https://www.repository.cam.ac.uk/items/bf455f97-1ff0-4694-87ca-b72c85366dc6 [68]
Dataset: Quantitative PAI using tissue-mimicking phantoms	Experimental phantom datasets (multispectral)	Experimentally well characterized phantom data and their digital twins	https://www.repository.cam.ac.uk/items/c7cc6ab7-9ce6-44b5-9f6f-af241e65ffc9 [69]
Photoacoustic Vascular Dataset (Mendeley)	Experimental microvessel images	Mouse ear vasculature, labeled for segmentation and validation	https://www.data.mendeley.com/datasets/dp5jgıkd6k [70]
Photoacoustic Microscopy Dataset (Figshare)	Experimental PAM images	For deep-learning upsampling and denoising tasks	https://www.figshare.com/articles/dataset/Photoacoustic_Microscopy_Dataset/12871805 [71]

8.2.2.4 Summary

The open availability of PAI datasets and deep-learning toolboxes now allows affordable systems to achieve performance once limited to high-end platforms. By leveraging synthetic data generation, shared benchmarks, and portable networks, researchers can develop, validate, and deploy AI-enhanced PAI systems entirely within the open-science ecosystem.

Table 8.10 Open-source deep-learning toolboxes and frameworks for photoacoustic image reconstruction and analysis

Toolbox/model	Domain/ objective	Core features	Access/repository
PA OmniNet	Generalizable deep-learning based reconstruction	Retraining-free U-Net-derived network; cross-system generalization (OADAT benchmark)	https://www.github.com/olivierstam4/PA_OmniNet [72]
Y-Net	Learned reconstruction + denoising	Hybrid deep network combining CNN and residual paths for in-vivo PAI	https://www.github.com/chenyilan/Y-Net [73]
DeepSino	Limited-view PAT reconstruction	Dual-domain network reconstructing full images from sparse sinograms	https://www.github.com/Gaurav1302/DeepSino [74]
Photoacoustic-FasterRCNN	Source detection and artifact removal	Object-detection-style CNN for PA source localization and reflection suppression	https://www.github.com/derekallman/Photoacoustic-FasterRCNN [75]
US-PA Simulation Codes (Kothapalli Lab)	Co-simulation + dataset generation	Combined ultrasound/ photoacoustic forward modeling using k-Wave + NIRFAST; suitable for AI training	https://www.github.com/KothapalliLabPSU/US-PA_simulation_codes [76]

The growing use of open datasets and deep-learning frameworks has transformed PAI from an offline, reconstruction-intensive modality into a data-driven, near–real-time imaging technique. However, practical translation of these algorithms into portable and affordable PAI systems requires computational efficiency and embedded implementation. Recent advances in edge AI hardware, FPGA-based accelerators, and GPU-enabled single-board computers now allow direct deployment of reconstruction and inference pipelines at the acquisition site. These developments form the basis of the next section, which focuses on embedded and real-time computing platforms for compact and cost-effective PAI instruments.

8.2.3 Embedded and Real-Time Computing

Modern PAI increasingly relies on embedded computing platforms that combine data acquisition, signal processing, and reconstruction in a compact form factor. For affordable and portable systems, these platforms provide the necessary computational power for real-time visualization, deep-learning inference, and hardware-level synchronization without requiring dedicated workstations. Table 8.11 summarizes embedded computing platforms suitable for affordable PAI.

Embedded architectures fall broadly into three categories:

1. GPU-enabled single-board computers—e.g., NVIDIA Jetson series; ideal for deploying trained neural networks and performing k-Wave-based reconstruction in near-real time.
2. FPGA-based DAQ and accelerators—used for high-throughput data streaming and on-board delay-and-sum or beamforming; can share clocks with acquisition electronics for deterministic timing.
3. Lightweight ARM/SBC controllers—used for instrument control, triggering, and simplified envelope-detection or PAI visualization.

Table 8.11 Embedded computing platforms for real-time or portable photoacoustic imaging. Information compiled from official manufacturer documentation and publicly available specifications (accessed Oct 2025)

Platform/device	Processor/ accelerator	Typical use in PAI	Verified reference/link	Approx. cost tier
NVIDIA Jetson Nano/Xavier/ Orin NX	ARM CPU + CUDA GPU (128–1024 cores)	Real-time reconstruction, CNN inference (e.g., Y-Net, PA OmniNet)	https://www.developer.nvidia.com/embedded	€150–800
Xilinx Zynq UltraScale + MPSoC/PYNQ Z2	FPGA + ARM Cortex-A9	FPGA-based beamforming, DAQ synchronization, and LED/LD trigger control	https://www.pynq.io	€150–600
Intel FPGA (Arria 10 GX Dev Kit)	1.5 M logic elements, DDR4 support	On-board reconstruction and PCIe DAQ acceleration (e.g., TI AFE5832 integration)	https://www.intel.com/products/fpga	€3 000–5 000
Raspberry Pi 4/ 5	Quad-core ARM Cortex-A72	Instrument control, LED/laser triggering, envelope-mode display	https://www.raspberrypi.com	<€100
NVIDIA Jetson Orin AGX + TensorRT	275 TOPS AI performance	Embedded deep-learning reconstruction and segmentation	https://www.developer.nvidia.com/embedded/jetson-orin	€1 000–2 000

8.2.3.1 Summary

Compact embedded platforms now deliver performance once reserved for desktop workstations, enabling real-time image formation and AI-based enhancement directly at the point of acquisition. By coupling FPGA-based DAQs with GPU-edge modules such as the Jetson Orin, affordable PAI systems can achieve frame rates above 10 Hz with end-to-end latency below 200 ms, sufficient for dynamic small-animal or handheld imaging. These developments bridge the gap between open research prototypes and deployable, field-ready photoacoustic devices.

As embedded and real-time computing solutions mature, the ability to share, reproduce, and compare results across diverse hardware and software environments becomes increasingly important. Open frameworks and low-cost devices can only achieve their full impact when supported by common data formats, metadata standards, and harmonized reporting practices. The following section therefore focuses on emerging initiatives such as the International Photoacoustic Standardization Consortium (IPASC) data format and the adoption of FAIR principles that enable transparent data exchange, interoperability, and reproducible research in PAI.

8.2.4 Standards, Data Sharing, and Interoperability

The adoption of open hardware and software in PAI must be matched by shared data formats, metadata standards and interoperable workflows to enable collaboration, reproducibility and translation. The IPASC plays a central role in this process [77].

8.2.4.1 Why Standards Matter

- PAI systems span diverse hardware: lasers/LEDs, transducers (single-element, arrays), and varied digitization chains. Without a unified format for raw time-series data and metadata, comparison across systems becomes difficult. The IPASC data format—designed to store raw time-series data (not just reconstructed images), and built on HDF5, addresses this [44].
- Interoperability of data enables:

  1. Shared reference datasets and benchmarks,
  2. Cross-system algorithm validation (e.g., reconstruction, ML),
  3. Transparent reporting of acquisition and system properties (illumination geometry, transducer configuration, digitizer settings).

- Standardization supports open-science goals (data reuse, FAIR principles) and helps pave the path to clinical translation by aligning with regulatory and industrial workflows.

8.2.4.2 IPASC: Core Aims and Themes

According to the IPASC website [77]:

- The goal is to reach consensus on PAI standardization to improve preclinical study quality and accelerate clinical translation.
- Key aims include:
  1. Defining widely accepted phantoms for use with preclinical and clinical PAI systems.
  2. Providing open, publicly available reference datasets for testing data reconstruction and spectral processing algorithms in a standard format.
  3. Agreeing on standardized test methods for new PAI instruments to help comparison of published results.
- Thematic areas: Clinical Adoption, Test Objects & Methods, Standards Development, and Data Management.

8.2.4.3 The IPASC Data Format

- The format is described in detail in the paper "The IPASC data format: A consensus data format for photoacoustic imaging" [44]. Table 8.12 summarizes key metadata modules in the IPASC data format.

Table 8.12 Key metadata modules in the IPASC data format

Module	Purpose/content	Examples of parameters captured
Device	Describes physical system configuration	Transducer geometry, element spacing, bandwidth, illumination wavelength(s), laser repetition rate
Acquisition	Records experimental acquisition conditions	Sampling frequency, digitizer bit depth, trigger timing, number of averages, channel mapping
Processing/reconstruction	Documents processing steps applied to the data	Beamforming or delay-and-sum parameters, reconstruction algorithm name, voxel size, interpolation method
Metadata/general information	Global identifiers and contextual metadata	System manufacturer, software version, experiment date/time, user notes, coordinate system definition

- Highlights:
  1. Uses HDF5 as underlying container.
  2. Captures raw time-series (not only images) to preserve the full signal chain.
  3. Defines device, acquisition, and reconstruction metadata modules (illumination, detection, digitization, coordinate systems).
  4. Accompanied by an open-source API ("PACFISH") for conversion and quality control [78].
- The format is now the recommended baseline for open datasets and toolboxes to support cross-platform sharing and algorithm benchmarking.

8.2.4.4 Data Sharing and Interoperability Practices

- Open toolboxes and datasets (see Sects. 8.2.1 and 8.2.2) increasingly support the IPASC format or export routines to it, a positive sign of ecosystem convergence.
- Researchers are encouraged to publish data with structured metadata (device geometry, excitation parameters, detector specs) to enable reuse, comparison and archival.
- Cross-discipline alignment: IPASC's efforts complement other imaging standards (e.g., the addition of PAI modality to the DICOM standard via Supplement 229) which helps frame clinical translation pathways [79].

Community Perspective: Building Standards for Open and Reproducible Photoacoustics

"Shared data standards and harmonized benchmarking are essential for reproducible photoacoustic imaging. Through collaborative initiatives like IPASC, the community is developing open formats and protocols that enable interoperability, transparency, and equitable access as the field grows globally."

—Prof. *Sarah Bohndiek, Chair—International Photoacoustic Standardization Consortium (IPASC), University of Cambridge, United Kingdom.*

8.2.4.5 Summary

Standards are the glue that hold hardware, software, data, and community together. By adopting the IPASC framework and associated data-sharing best practices, affordable and open PAI systems gain major advantages: reproducibility, scalability, and real-world impact. In the broader context of this book where we emphasize affordable hardware, open software, AI-enabled reconstruction, the standards layer ensures that innovations are not isolated but interoperable, shareable, and future-proof.

The establishment of shared data standards and open repositories has laid the technical foundation for a transparent and collaborative PAI ecosystem. Yet, sustainable progress depends not only on open tools and formats but also on active networks of researchers, educators, and industry partners who maintain and evolve them. The next section therefore explores how coordinated community efforts through consortia, training programs, and regional hubs are fostering a truly global and inclusive PAI community, ensuring that affordable and open innovation continues to thrive beyond individual laboratories.

8.2.5 Building and Sustaining a Global PAI Community

The success of affordable PAI depends not only on open technology but also on an active and connected community that sustains it. Collaborative networks spanning academic institutions, industry partners, and international consortia now play a central role in shaping how PAI evolves, ensuring that innovation remains accessible, reproducible, and globally inclusive.

8.2.5.1 Global Collaboration and Open Exchange

The IPASC serves as the primary forum for harmonizing data standards, performance metrics, and phantom materials across laboratories worldwide. Its community-driven structure and open membership model promote transparent collaboration between academia, clinics, manufacturers, and government regulators. Complementary initiatives by SPIE, Optica (OSA), and the European Society for Molecular Imaging (ESMI) reinforce this culture of openness through conferences, workshops, and dedicated PAI working groups at events such as Photonics West, BiOS, and EMIM.

8.2.5.2 European Research and Innovation Projects

Several European Union–funded programs under Horizon 2020 and the European Research Council (ERC) have advanced affordable and clinically relevant PAI technologies:

- **FULLPHASE**—Pioneered fully integrated real time multi-wavelength LD-based photoacoustics combined with ultrasound for early disease detection. (https://www.fullphase-fp7.eu/index.php?id=1121)
- **CVENT**—Focused on the development of a portable multimodality, multiwavelength photoacoustic imaging system with a hand-held PAI and ultrasound probe for screening, in-depth diagnosis and treatment monitoring of carotid plaque vulnerability (https://www.cvent-2020.eu/)

- **RSENSE**—Revolutionizing disease and environmental detection with portable optoacoustic sensing https://rsense.munichimaging.eu/

Together, these projects demonstrate the European commitment to open, cost-efficient, and interoperable photoacoustic technologies that bridge research and clinical translation.

8.2.5.3 Academic–Industry Partnerships

Joint academic–industrial initiatives are increasingly important for turning open research prototypes into deployable systems. Collaborations between Helmholtz Munich, ETH Zurich, University College London, University of Twente, Erasmus Medical Center, and companies such as CYBERDYNE INC, iThera Medical, Lumibird, and Trumpf have accelerated progress in hybrid PA/US devices, compact diode-based sources, and user-friendly reconstruction software [16]. These partnerships balance commercial viability with open-science principles, ensuring that affordability and reproducibility remain central to translational design.

8.2.5.4 Education, Training, and Skill Development

Capacity building forms the foundation of a sustainable PAI ecosystem. Open documentation and tutorials from toolboxes such as SIMPA, PATATO, and k-Wave, along with hands-on training through IPASC workshops, SPIE short courses, and ESMI summer schools, have made it easier for students and engineers to enter the field. Such initiatives promote cross-disciplinary understanding of optics, acoustics, electronics, and computation skills essential for advancing affordable imaging systems.

8.2.5.5 Standardization and Phantom Development Within IPASC

A recent milestone from IPASC is the standard phantom activity, in which member institutes agreed on a common phantom recipe to benchmark PAI performance. Each laboratory fabricated the phantom locally, performed imaging using its own system, and reported on reproducibility, imaging fidelity, and fabrication ease. The resulting dataset demonstrated that harmonized yet low-cost phantoms can substantially improve cross-laboratory comparability and validation. IPASC membership is open and free of charge, allowing individuals and institutions from academia, healthcare, and industry to participate directly in working groups (https://www.ipasc.science/about/get-involved). In parallel, IPASC has formed a dedicated Industrial Board that collaborates on different translational aspects including defining emerging IEC standards for safety, calibration, and performance assessment, providing a bridge between open research and regulatory pathways.

Table 8.13 Major international collaborations and projects supporting open and affordable photoacoustic imaging (accessed November 2025)

Initiative/organization	Scope/focus	Website/reference
IPASC—International Photoacoustic Standardization Consortium	Data standards, phantoms, reproducibility, industrial board	https://www.ipasc.science
RSENSE (H2020 #862,811)	Disease detection with portable photoacoustic sensing	https://www.cordis.eu.rsense
FULLPHASE (FP7-ICT-2011-8)	Development of compact handheld LD-photoacoustic and ultrasound probe	https://www.cordis.eu.ful lphase
CVENT (H2020 #731,771)	Risk assessment of plaque rupture and future cardiovascular events by handheld multi-spectral photoacoustic imaging	https://www.cordis.eu.cvent
SPIEPAI Working Group	Conferences, education, and industry alignment	https://www.spie.org/photoa cousticimaging
ESMI/EMIM Workshops	Training and molecular imaging community engagement	https://www.e-smi.eu

8.2.5.6 Long-Term Sustainability

To preserve openness beyond individual projects, researchers increasingly release datasets and code through persistent repositories such as Zenodo and GitHub, using DOI-linked versioning and open licenses. Community-led maintenance, issue tracking, and documentation ensure continuity even after formal funding periods end. This distributed governance model rooted in transparency and collaboration has become essential to keeping PAI innovation both affordable and reproducible. Table 8.13 summarizes major collaborations which are relevant for affordable PAI.

8.2.5.7 Summary

A truly sustainable photoacoustic community relies on shared effort across disciplines and continents. By combining open standards, collaborative EU projects, industrial partnerships, and freely accessible education, PAI is evolving into a globally inclusive and interoperable field. This collective approach ensures that affordable, high-quality PAI will continue to expand its reach from research laboratories to clinics and classrooms guided by openness, reproducibility, and shared innovation.

8.3 Networking and Collaboration Opportunities

The sustained growth of PAI has been shaped not only by advances in technology but by the strength of its global research network. International consortia, recurring conferences, and open educational activities have created a collaborative environment in which affordable and reproducible PAI continues to evolve across continents and disciplines.

8.3.1 International Consortia and Working Groups

The IPASC remains the key international body aligning PAI research with shared standards, interoperability goals, and regulatory readiness. Beyond its scientific and industrial members, IPASC includes participation from regulatory stakeholders such as the U.S. FDA, ensuring that future reporting practices and performance metrics align with preclinical and clinical translation needs. Open membership and accessible working groups allow researchers from academic, industrial, and clinical institutions to directly contribute to data-format development, phantom standardization, and quality-control methods. Regular phantom fabrication workshops organized by IPASC provide hands-on training in preparing, imaging, and analyzing standardized test materials which is crucial for reproducibility in both high-end and low-cost PAI systems.

Complementary activities are led by professional societies such as SPIE, Optica (OSA), and the ESMI. The annual SPIE BiOS "Photons Plus Ultrasound: Imaging and Sensing" conference remains the largest international forum dedicated to PAI bringing together hundreds of researchers each year to present advances ranging from instrumentation to clinical applications. These meetings serve as a neutral ground for collaboration between academia, industry, and standards organizations.

8.3.2 Community Platforms and Educational Outreach

Open communication now extends well beyond conferences. Repositories such as GitHub, Zenodo, and ResearchGate have become the backbone of technical exchange, enabling version-controlled sharing of software, datasets, and tutorials that can be cited in publications and reused by others. A representative example of open educational outreach is the AcousticX LED-PAI Webinar Series, freely available on YouTube (playlist link). These publicly accessible sessions, hosted by the developers of the AcousticX system, provide structured demonstrations of LED-based PAI, covering acquisition workflows, reconstruction methods, and biomedical case examples. Presented in an instructional rather than commercial format, the webinars exemplify how manufacturers can contribute to open technical education while promoting reproducible and affordable system design.

LED-driven PAI has also begun entering university teaching laboratories, thanks to its low optical fluence and intrinsic safety. Institutions such as Eindhoven University of Technology have integrated compact PAI systems into undergraduate and graduate physics or biomedical engineering courses, giving students firsthand experience with imaging concepts once limited to advanced research labs. This trend marks a crucial step toward democratizing photoacoustics and cultivating early-career expertise.

8.3.3 Collaborative Research and Data-Sharing Networks

International research programs and consortia continue to demonstrate the value of collective progress in PAI, even as formal data-sharing standards are still evolving. Large-scale European projects such as RSENSE, FULLPHASE, and CVENT established collaborative frameworks in which multiple academic and industrial partners shared simulation methods, system designs, and experimental findings through project websites and joint publications. Although these initiatives predated the formalization of the IPASC data format, they set important precedents for transparency and cross-institutional coordination in PAI research. Each project maintains an accessible web portal—typically hosted on the CORDIS platform—where summaries of results, device concepts, and open-access papers are made publicly available. Future collaborative efforts are expected to build upon these models by adopting standardized data structures such as the IPASC HDF5 format, enabling reproducible analysis and cross-platform validation.

Together, these distributed networks highlight how shared documentation, openly published methodologies, and coordinated reporting already serve as stepping-stones toward full interoperability and open-science compliance in upcoming PAI projects. In parallel, shared benchmark datasets from phantom workshops and simulation challenges are beginning to support cross-validation of reconstruction algorithms and machine-learning models across institutions. These practices embody the FAIR (Findable, Accessible, Interoperable, and Reusable) principles and ensure that affordable imaging systems can achieve scientific rigor comparable to commercial platforms.

8.3.4 Summary

Networking and community engagement remain at the core of sustainable progress in PAI. Through coordinated standards efforts, inclusive conferences, open-access webinars, and educational adoption in teaching laboratories, the field now offers multiple entry points for students, engineers, and clinicians alike. This expanding network ensures that the tools, knowledge, and culture of open and affordable PAI continue to grow collectively, bridging continents, disciplines, and generations of researchers.

8.4 Future Outlook and Skill-Development Pathways

Affordable PAI has progressed from early experimental prototypes to a globally networked field grounded in open hardware, software, and community standards. The next phase will emphasize integration, accessibility, and education, ensuring that the technology's benefits extend to research, clinical, and educational environments worldwide.

8.4.1 Technology Convergence and Affordability

Ongoing advances in affordable light source-based illumination, FPGA–GPU co-processing, and miniaturized transducer arrays are expected to further reduce system cost while improving frame rate and quantitative accuracy. Future devices will potentially integrate AI-assisted reconstruction and embedded processing directly within the imaging probe, enabling real-time analysis without external computing. Interoperability through IPASC-compliant data structures and modular hardware interfaces will allow new systems to interconnect and evolve collaboratively rather than in isolation.

8.4.2 Skill Development and Education

Expanding access to training will be essential for sustaining progress. As systems and open toolboxes become safer and more affordable, PAI is increasingly entering university and teaching laboratories, allowing students to gain practical imaging experience early in their careers. Short courses at SPIE BiOS, IPASC phantom workshops, and online educational series such as the AcousticX LED-PAI webinars demonstrate how technical training can now be delivered globally at minimal cost. Integrating open PAI content into biomedical engineering curricula will help standardize knowledge transfer across regions.

8.4.3 Open-Science Sustainability

The long-term vitality of the PAI community will depend on sustaining open repositories, curated datasets, and transparent code maintenance. Institutional recognition of open-source contributions through citations, DOI-linked datasets, and funding mechanisms for maintainers will be vital to preserving reproducibility. Community governance models, already emerging within IPASC and open toolbox projects, are likely to become the backbone of future collaborative research.

The trajectory of PAI is shifting from innovation through instrumentation to innovation through openness. Affordable light sources, shared datasets, and globally connected training ecosystems are redefining who can contribute to and benefit from the field. By coupling technological accessibility with continued investment in education and open collaboration, PAI is poised to become a standard, inclusive imaging modality across both research and clinical domains.

Key Takeaways

- **Affordable PAI development is now feasible end-to-end**, with cost-effective light sources, transducers, DAQ modules, and embedded computing platforms readily available to researchers.
- **Open-source toolboxes and data formats** including k-Wave, SIMPA, and PATATO have made simulation, reconstruction, and interoperability widely accessible.
- **Community-led initiatives and consortia** such as IPASC, SPIE, and ESMI are shaping standards, datasets, and training frameworks that sustain reproducible research.
- **Adoption of FAIR principles and standardized reporting** ensures that shared data, code, and protocols remain verifiable and reusable across laboratories.
- **Global collaboration and education** from webinars and phantom workshops to mentorship networks are building the skills and infrastructure needed for equitable participation in PAI research.
- **The collective shift toward openness and affordability** marks a defining transition in photoacoustics from specialized instrumentation to a shared, sustainable, and globally connected imaging ecosystem.

Epilogue: Enabling Global Clinical Impact Through Safe and Affordable PAI

The evolution of PAI, as traced throughout this book, reveals a clear truth: the technologies most likely to transform clinical practice are those that remain safe, affordable, and accessible to all. LED-, LD-, and VCSEL-based systems, low-cost detectors, compact DAQs, efficient algorithms, and open-source tools together redefine what biomedical imaging can be—not a luxury confined to advanced hospitals, but a practical modality deployable in any clinic, anywhere in the world.

Affordable PAI is not merely a cost-reduction exercise; it reshapes who can use the technology, where it can be used, and how quickly a diagnostic impact can be made. Safe, laser-class-1 illumination opens the door to imaging in pediatric wards, general practices, community health centers, and resource-constrained environments. Open standards and reproducible methods ensure that devices built continents apart can speak the same scientific language. And accessible educational resources empower the next generation of innovators, regardless of geography or institutional privilege.

The democratization of PAI is already underway and its success depends not on a single component, but on the convergence of engineering ingenuity, community collaboration, and clinical partnership. As affordable systems continue to mature, the prospect of routine functional imaging at the point of care becomes real: evaluating vascular health in outpatient clinics, guiding wound management in rural settings, and enabling early detection of disease where traditional imaging is unavailable.

If the past decade has proven anything, it is that the future of photoacoustics will be defined not only by its performance, but by its reach. By keeping affordability, safety, and global accessibility at the center of innovation, the PAI community can ensure that this remarkable technology fulfills its true purpose—improving lives everywhere.

References

1. M. Xu, L.V. Wang, Photoacoustic imaging in biomedicine. Rev. Sci. Instrum. **77**, 041101 (2006)
2. M. Kuniyil Ajith Singh, W. Xia, Portable and affordable light source-based photoacoustic tomography. Sensors **20**(21), 6173 (2020)
3. B. Eliason, A. Longo, C. Dehner, L. Lilaj, S. Gronenborn, C. Klett, A. Weigl, D. Dörich, U. Malzahn, G. Zahnd, J. Konradl, P. Leisching, Performance characterization of a high-power, compact, low-cost, and high-efficiency VCSEL subsystem for photoacoustic imaging. Proc. SPIE 13319, Photons Plus Ultrasound: Imaging and Sensing 2025, 133190S (2025)
4. J. Konradl, S. Gronenborn, A. Weigl, C. Klett, D. Dörich, U. Malzahn, A. Longo, C. Dehner, B. Eliason, C. Reichl, G. Zahnd, P. Leisching, High-power compact VCSEL subsystem as game-changer for photoacoustic medical imaging and sensing. Proc. SPIE 13384, Vertical-Cavity Surface-Emitting Lasers XXIX, 133840G (2025)
5. X. Gao, X. Chen, H. Hu et al., A photoacoustic patch for three-dimensional imaging of hemoglobin and core temperature. Nat. Commun. **13**, 7757 (2022)
6. S.K. Kalva, M. Pramanik, Photoacoustic Tomography with High-Energy Pulsed Laser Diodes. SPIE Spotlight Series Book (SL56) (2020)
7. Y. Zhu, T. Feng, Q. Cheng, X. Wang, S. Du, N. Sato, J. Yuan, M.K.A. Singh, Towards clinical translation of LED-based photoacoustic imaging: a review. Sensors **20**, 2484 (2020)
8. C. Canal, G. Rousset, S. Passard et al., Portable multiwavelength laser diode source for handheld photoacoustic devices. Proc. SPIE **9887**, 98872B (2016)
9. K. Daoudi, P.J. van den Berg, O. Rabot et al., Handheld probe integrating laser diode and ultrasound transducer array for ultrasound/photoacoustic dual-modality imaging. Opt. Express **22**, 26365–26374 (2014)

10. A. Thomas, G. van Soest, K.J. Francis, Peripheral vascular photoacoustic imaging using a portable laser diode, in *European Conference on Biomedical Optics, Optica Publishing Group*, pp. S1D-4 (2025)
11. R.G. Kolkman, E. Hondebrink, W. Steenbergen, T.G. van Leeuwen, In vivo photoacoustic imaging of blood vessels with a pulsed laser diode. Lasers Med. Sci. **21**, 134–139 (2006)
12. M. Seeger, A. Stylogiannis, L. Prade et al., Overdriven laser diode optoacoustic microscopy. Sci. Rep. **13**, 19542 (2023)
13. S. Khan, S. Mukhopadhyay et al., Development of a cost-effective compact diode-laser-based photoacoustic sensing instrument for breast tissue diagnosis. J. Biomed. Opt. **29**(1), 017002 (2024)
14. X. Li, V.T.C. Tsang, L. Kang et al., High-speed high-resolution laser diode-based photoacoustic microscopy for in vivo microvasculature imaging. Vis. Comput. Ind. Biomed. Art **4**, 1 (2021)
15. M.N. Cherkashin, J. Laufer, T. Kirchner, Pulsed laser diode excitation for transcranial photoacoustics. Biomed. Opt. Express **16**, 3623–3631 (2025)
16. M.K.A. Singh, N. Sato, F. Ichihashi, W. Xia, Y. Sankai, The evolution of LED-based photoacoustic imaging: from labs to clinics, in Xia, W. (eds) Biomedical Photoacoustics (Springer, Cham, 2024)
17. T.J. Allen, P.C. Beard, High power visible light emitting diodes as pulsed excitation sources for biomedical photoacoustics. Biomed. Opt. Express **7**, 1260–1270 (2016)
18. A. Das, M. Pramanik, Performance characterization of light-emitting diode array-based photoacoustic computed tomography (LED-PACT). Biomed. Opt. Express **16**, 3632–3647 (2025)
19. X. Liu, S.K. Kalva, B. Lafci, D. Nozdriukhin, X.L. Deán-Ben, D. Razansky, Full-view LED-based optoacoustic tomography. Photoacoustics **31**, 100521 (2023)
20. A. Stylogiannis et al., Continuous wave laser diodes enable fast optoacoustic imaging. Photoacoustics **9**, 31–38 (2018)
21. A. Das, M. Pramanik, Nanosecond pulsed current source for light emitting diode (LED) driven photoacoustic signal acquisition systems. J. Phys. Photon. **7**(1), 015015 (2025)
22. C. Willert et al., Pulsed operation of high-power light emitting diodes for imaging flow velocimetry. Meas. Sci. Technol. **21**, 075402 (2010)
23. T. Agano, N. Sato, K. Awazu, LED-based photoacoustic imaging system: why it achieves the same signal to noise ratio as solid-state-laser-based system: a review, *Proceeding SPIE 11240, Photons Plus Ultrasound: Imaging and Sensing 2020*, 112401T (2020)
24. F. Liffran, O. Rabot, Device for controlling at least one diode. US Patent US9819146B2 (2017)
25. M. Mitra, A. Haworth, P. Gaddale, F. Badran, N. Lagno, C. Pameijer, F. Aziz, S.R. Kothapalli, Multiwavelength laser diode based portable photoacoustic and ultrasound imaging system for point of care applications. J. Biophotonics **17**(7), e202400058 (2024)
26. C. Linger et al., Volumetric and simultaneous photoacoustic and ultrasound imaging with a conventional linear array in a multiview scanning scheme. IEEE Trans. Ultrason. Ferroelectr. Freq. Control **70**(12), 1607–1620 (2023). https://doi.org/10.1109/TUFFC.2023.3268820
27. G.S. Jeng, M.L. Li, M. Kim et al., Real-time interleaved spectroscopic photoacoustic and ultrasound (PAUS) scanning with simultaneous fluence compensation and motion correction. Nat. Commun. **12**, 716 (2021)
28. Y. Sun, Y. Wang, W. Li, C. Li, Real-time dual-modal photoacoustic and fluorescence small animal imaging. Photoacoustics 36, 100593 (2024). https://doi.org/10.1016/j.pacs.2024.100593
29. R. Lin, Q. Zhang, S. Lv, J. Zhang, X. Wang, D. Shi, X. Gong, K. Lam, J. Biophotonics **16**(4), e202200269 (2023)
30. W. Xia, D. Piras, J.C. van Hespen, S. van Veldhoven, C. Prins, T.G. van Leeuwen, W. Steenbergen, S. Manohar, An optimized ultrasound detector for photoacoustic breast tomography. Med. Phys. **40**(3), 032901 (2013)
31. H. Chen, S. Agrawal, M. Osman, J. Minotto, S. Mirg, J. Liu, A. Dangi, Q. Tran, T. Jackson, S.R. Kothapalli, A transparent ultrasound array for real-time optical, ultrasound, and photoacoustic imaging. BME Front. **8**(2022), 9871098 (2022)

32. S. Cho, M. Kim, J. Ahn et al., An ultrasensitive and broadband transparent ultrasound transducer for ultrasound and photoacoustic imaging in-vivo. Nat. Commun. **15**, 1444 (2024)
33. A. Kurnikov, A. Sanin, X.L.D. Ben, D. Razansky, P. Subochev, Ultrawideband sub-pascal sensitivity piezopolymer detectors. Ultrasonics **141**, 107349 (2024)
34. S. Chandramoorthi, A. Thomas, G. van Soest, K.J. Francis, Ultrasound and photoacoustic imaging of peripheral arteries: a feasibility study. Ultrasound Med. Biol. **51**(10), 1854–1858 (2025)
35. B. Zhai, Y. Shi, H. Deng, H. Liu, C. Xie, N. Zhang, W. Yu, D. Sun, Y. Yu, C. Ma, Handheld multispectral photoacoustic imaging for assessing myocardial metabolism. J. Biomed. Opt. **30**(10), 105001 (2025)
36. D. Jiang, H. Lan, S. Tong, X. Zhang, F. Gao, Signal-domain speed-of-sound correction for ring-array-based photoacoustic tomography. Photoacoustics 44, 100735 (2025)
37. Z. Qin, P. Lai, M. Sun, Photoacoustic thermal-strain measurement towards noninvasive and accurate temperature mapping in photothermal therapy. Photoacoustics. 40, 100651 (2024)
38. R. Nauber et al., Photoacoustics-guided Real-Time Closed-loop Control of Magnetic Microrobots through Deep Learning, bioRxiv 2025.08.13.670011
39. J. Lusk, E. Marschall, C. Miranda, C. Aridi, B. Smith, Photoacoustic detection of genetically encoded fluorophores for neuronal subtype identification. J. Neural Eng. **22**(2) (2025)
40. J. Benavides Lara, R. Prakash, K. Avanaki, Assessment of a single-element scanning system for enhanced photoacoustic imaging of brain hemorrhage. J. Biophotonics **18**(3), e202400153 (2025)
41. M. Zafar, R. Manwar, S.M. Ranjbaran, K. Avanaki, Utility of low-cost multichannel data acquisition system for photoacoustic computed tomography. Photonics **12**(4), 385 (2025)
42. S. Frey, S. Vostrikov, L. Benini, A. Cossettini, WULPUS: a wearable ultra low-power ultrasound probe for multi-day monitoring of carotid artery and muscle activity. *2022 IEEE International Ultrasonics Symposium (IUS)*, Venice, Italy, pp. 1–4 (2022)
43. Integrated Systems Laboratory (IIS) at ETH Zurich, (2025), *WULPUS: Wearable Ultra Low-Power Ultrasound System*. GitHub repository. Available at: https://github.com/pulp-bio/wulpus. Accessed Nov 2025
44. J. Gröhl, L. Hacker, B.T. Cox, K.K. Dreher, S. Morscher, A. Rakotondrainibe, F. Varray, L.C.M. Yip, W.C. Vogt, S.E. Bohndiek, Members of the international photoacoustic standardisation consortium (IPASC). the IPASC data format: a consensus data format for photoacoustic imaging. Photoacoustics. **26**, 100339 (2022)
45. B.E. Treeby, B.T. Cox, k-Wave: MATLAB toolbox for the simulation and reconstruction of photoacoustic wave fields. J. Biomed. Opt. **15**(2), 021314 (2010)
46. NIRFAST Team, NIRFAST and NIRFAST-PA: Open-source software for near-infrared and photoacoustic tomography. GitHub repository, https://github.com/NIRFAST-admin. Acessed Oct 2025
47. D. Hamid et al., Near infrared optical tomography using NIRFAST: algorithm for numerical model and image reconstruction. Commun. Numer. Methods Eng. **25**(6), 711–732 (2008)
48. J. Cao, S. Montero-Hernandez, R.C. Mesquita, A.T. Eggebrecht, H. Dehghani, NIRFASTerFF: an accessible, cross-platform Python package for fast photon modeling. J. Biomed. Opt. 30(11), 115001 (2025)
49. K.B. Chowdhury, J. Prakash, A. Karlas, D. Jüstel, V. Ntziachristos, A synthetic total impulse response characterization method for correction of hand-held optoacoustic images. IEEE Trans. Med. Imaging **39**(10), 3218–3230 (2020)
50. K.B. Chowdhury, M. Bader, C. Dehner, D. Jüstel, V. Ntziachristos, Individual transducer impulse response characterization method to improve image quality of array-based handheld optoacoustic tomography. Opt. Lett. **46**(1), 1–4 (2021)
51. P. Omidi, L.C.M. Yip, E. Rascevska, M. Diop, J.J.L. Carson, PATLAB: A graphical computational software package for photoacoustic computed tomography research. Photoacoustics. 28, 100404 (2022)
52. Else et al., PATATO: a Python photoacoustic tomography analysis toolkit. J. Open Source Softw. **9**(93), 5686 (2024)

53. G. Janek, K.K. Dreher, M. Schellenberg, T. Rix, N. Holzwarth, P. Vieten, L. Ayala, S.E. Bohndiek, A. Seitel, L. Maier-Hein, SIMPA: an open-source toolkit for simulation and image processing for photonics and acoustics. J. Biomed. Opt. **27**(8) (2022)
54. M.T. Graham, M.A.L. Bell, PhocoSpace: an open-source simulation package to implement photoacoustic spatial coherence theory, in 2022 IEEE International Ultrasonics Symposium (IUS) (Venice, Italy, 2022), pp. 1–5
55. M.T. Graham, M.A.L. Bell, Photoacoustic spatial coherence theory and applications to coherence-based image contrast and resolution. IEEE Trans. Ultrason. Ferroelectr. Freq. Control **67**(10), 2069–2084 (2020)
56. Van der Wal, J. Gröhl, LED System Modelling: Simulation framework for LED-based photoacoustic imaging systems. GitHub repository, https://github.com/jgroehl/led_system_modelling. Accessed Oct 2025
57. A. Korobov, Z. Besedovskaia, E. Petrova, A. Kurnikov, A. Glyavina, A. Orlova, S. Nemirova, I. Druzhkova, M. Sirotkina, E. Shirshin, D. Gorin, L. Xi, D. Razansky, P. Subochev, SKYQUANT 3D: quantifying vascular anatomy with an open-source workflow for comprehensive analysis of volumetric optoacoustic angiography data. J. Biophotonics **17**(11), e202400143 (2024)
58. M. Jaeger, S. Schüpbach, A. Gertsch, M. Kitz, M. Frenz, Fourier reconstruction in optoacoustic imaging using truncated regularized inverse k-space interpolation. Inverse Prob. **23**(6), S51 (2007)
59. A. Hauptmann, B. Cox, F. Lucka, N. Huynh, M. Betcke, P. Beard, S. Arridge, Approximate k-space models and deep learning for fast photoacoustic reconstruction, in *International Workshop on Machine Learning for Medical Image Reconstruction* (pp. 103–111). Springer, Cham (2018)
60. K.P. Köstli, M. Frenz, H. Bebie, H.P. Weber, Temporal backward projection of optoacoustic pressure transients using Fourier transform methods. Phys. Med. Biol. **46**(7), 1863 (2001)
61. Y. Xu, D. Feng, L.V. Wang, Exact frequency-domain reconstruction for thermoacoustic tomography. I. Planar geometry. IEEE Trans. Med. Imaging **21**(7), 823–828 (2002)
62. M. Xu, L.V. Wang, Universal back-projection algorithm for photoacoustic computed tomography. Phys. Rev. E **71**(1), 016706 (2005)
63. G. Matrone, A.S. Savoia, G. Caliano, G. Magenes, The delay multiply and sum beamforming algorithm in ultrasound B-mode medical imaging. IEEE Trans. Med. Imaging 34(4), 940–949 (2015)
64. J. Park, S. Choi, F. Knieling et al., Clinical translation of photoacoustic imaging. Nat Rev Bioeng **3**, 193–212 (2025)
65. A. Breger et al., PhotIQA: A photoacoustic image data set with image quality ratings. arXiv 2507.03478 (2025)
66. F. Özdemir, B. Lafci, X.L. Deán-Ben, D. Razansky, F. Pérez-Cruz, OADAT: Experimental and Synthetic Clinical Optoacoustic Data for Standardized Image Processing. arXiv:2206.08612 [eess.IV] (2022)
67. Y. Lou, W. Zhou, T.P. Matthews, C.M. Appleton, M.A. Anastasio, Generation of anatomically realistic numerical phantoms for photoacoustic and ultrasonic breast imaging. J. Biomed. Opt. **22**(4), 041015 (2017). https://doi.org/10.1117/1.JBO.22.4.041015
68. T. Else, J. Groehl, L. Hacker, S. Bohndiek, Dataset for: PATATO: A Python Photoacoustic Analysis Toolkit. Apollo—University of Cambridge Repository (2023)
69. J. Gröhl, T. Else, L. Hacker, E. Bunce, P. Sweeney, S. Bohndiek, Dataset for: Moving beyond simulation: data-driven quantitative photoacoustic imaging using tissue-mimicking phantoms. Apollo—University of Cambridge Repository (2023)
70. J. Meng, J. Yu, C. Liu, Photoacoustic vascular image dataset. Mendeley Data **V2**(2024). https://doi.org/10.17632/dp5jgrkd6k.2
71. DiSpirito, Anthony (2020). Photoacoustic Microscopy Dataset. figshare. Dataset. https://doi.org/10.6084/m9.figshare.12871805.v1
72. O.J.M. Stam, K.J. Francis, N. Awasthi, PA OmniNet: a retraining-free, generalizable deep learning framework for robust photoacoustic image reconstruction. Photoacoustics. 45, 100740 (2025). https://doi.org/10.1016/j.pacs.2025.100740

73. H. Lan, D. Jiang, C. Yang, F. Gao, F. Gao, Y-Net: Hybrid deep learning image reconstruction for photoacoustic tomography in vivo. Photoacoustics. 20, 100197 (2020)
74. N. Awasthi, G. Jain, S.K. Kalva, M. Pramanik, P.K. Yalavarthy, Deep neural network-based sinogram super-resolution and bandwidth enhancement for limited-data photoacoustic tomography. IEEE Trans. Ultrason. Ferroelectr. Freq. Control **67**(12), 2660–2673 (2020)
75. D. Allman, A. Reiter, M.A.L. Bell, Photoacoustic source detection and reflection artifact removal enabled by deep learning. IEEE Trans. Med. Imaging **37**(6), 1464–1477 (2018)
76. S. Agrawal, T. Suresh, A. Garikipati, A. Dangi, S.R. Kothapalli, Modeling combined ultrasound and photoacoustic imaging: simulations aiding device development and artificial intelligence. Photoacoustics. **15**(24), 100304 (2021)
77. International Photoacoustic Standardisation Consortium (IPASC). Official website. https://www.ipasc.science/. Accessed Oct 2025
78. J. Gröhl, PACFISH: Photoacoustic data conversion API for IPASC data format. GitHub repository, https://github.com/IPASC/PACFISH. Accessed Oct 2025
79. S. Bohndiek, Addressing photoacoustics standards. Nat. Photonics **13**, 298 (2019)

MIX
Papier aus verantwortungsvollen Quellen
Paper from responsible sources
FSC® C105338

If you have any concerns about our products,
you can contact us on
ProductSafety@springernature.com

In case Publisher is established outside the EU,
the EU authorized representative is:
Springer Nature Customer Service Center GmbH
Europaplatz 3, 69115 Heidelberg, Germany

Printed by Libri Plureos GmbH
in Hamburg, Germany